The Blueprints to Our Home

A Physical Geology Lab Manual

FIRST EDITION

By Elli Pauli

The George Washington University

cognella
San Diego, CA

Bassim Hamadeh, CEO and Publisher
Michael Simpson, Vice President of Acquisitions
Jamie Giganti, Managing Editor
Jess Busch, Senior Graphic Designer
Brian Fahey, Licensing Associate

First published in the United States of America in 2015 by Cognella, Inc.

Printed in the United States of America

ISBN: 978-1-63189-082-6 (pbk)/ 978-1-63189-083-3 (br)

cognella™

www.cognella.com 800.200.3908

Contents

ACKNOWLEDGMENTS

While I laid down the groundwork for this publication, the final product is the result of several years worth of collaborating, testing, and revising with the help of a multitude of dedicated people. To begin, I offer my sincerest thanks to Frances Junker—without whom this publication would not have been possible -for her countless hours, alternative suggestions, excellent ideas, and the entertainment that kept this project from becoming a frustrating and tedious bore.

In addition, I owe an immense debt of gratitude to Dr. Giuseppina M. Kysar and Dr. Catherine Forster for giving up hours of their personal time to review this publication and contribute invaluable knowledge and insight.

Laboratory Exercise 1

The Basics

Objective

This lab is designed to refresh your memory on some basic mathematical and scientific principles that you will need to successfully complete the laboratory course. Use the information in the conversion charts at the end of this lab exercise to answer the following questions.

Part A: Scientific Notation

Recall that we can express very large or very small numbers in scientific notation; this means that we express numbers in terms of powers of 10.

$10^1 = 10$	and	$a*10^1 = a*10$ (e.g., $3*10^1 = 30$)
$10^2 = 100$		
$10^3 = 1000$		

$10^{-1} = 0.1$	and	$a*10^{-1} = a*0.1$ (e.g., $3*10^{-1} = 0.3$)
$10^{-2} = 0.01$		
$10^{-3} = 0.001$		

An important rule is that $10^0 = 1$.

Problems

1. The radius of the Earth is 6371 km. Express this in scientific notation.

    ____

2. A single salt crystal is on the order of 10^{-8} meters in size. Write this number as a decimal.

    ____

Part B: Dimensional Analysis

It's not necessary to memorize every possible conversion that you'll encounter in the course of this class. You can use a succession of conversions to calculate the one that you need, such as in the following example.

Example: How many meters are in a light year (the distance that light travels in one year), given that light travels at a speed of 3.0×10^8 meters per second?

Answer: Recall that Distance = Rate x Time (D=RT). We know the rate ($R = 3.0 \times 10^8$ m/s) and the time (T=1 year), which leaves distance as our unknown.

Therefore: $D = 3.0 \times 10^8$ m/s x 1 year

But in order to solve this problem, we must convert years to seconds so the time cancels out leaving only distance.

So how do we calculate how many seconds are in a year? We know the following conversions: 60 secs = 1 min; 60 min = 1 hour; 24 hours = 1 day; and 365 days = 1 year.

$$1 \text{ year} \times \frac{365 \text{ days}}{1 \text{ year}} \times \frac{24 \text{ hours}}{1 \text{ day}} \times \frac{60 \text{ mins}}{1 \text{ hour}} \times \frac{60 \text{ secs}}{1 \text{ min}} = 3.1536 \times 10^7 \text{ sec}$$

Now, we can multiply this value by the speed of light to get the distance represented by a light year.

Problems

3. Finish the above example: how many meters are in a light year? ____________m

4. Make the following conversions using the conversion factors provided at the back of this lab exercise.

3,168 feet (ft) = ________________ inches (in) = ________________ miles (mi).

21 meters (m) = ________________ kilometers (km) = ________________ centimeters (cm).

23,400 seconds (s) = ________________ minutes (min) = ________________ hours (hrs).

62,300 kg = ________________ pounds (lbs) = ________________ tons.

220 ft/min = ________________ m/s.

2 centuries = ________________ seconds.

45 seconds = ________________ days.

5. The Earth is 4.6 billion years old. Given that 1 billion = 10^9, how many centuries is this? *Show your work.* ________________ centuries

6. The Earth's tectonic plates move at rates of 1–10 centimeters per year. How many miles per hour is 5 cm/year? *Show your work.* ________________ mph

7. The United States is approximately 3,000 miles across. How many inches is this? *Show your work.* ________________ inches

8. The islands of Hawaii are volcanic islands built up by layers of successive lava flows. When measured from its base on the ocean floor, one of the Hawaiian Islands stands approximately 9,140 meters. Radiometric dating of the volcanic rocks indicates that the island is 0.9 million years old. What is the average rate of build-up of lava flows in centimeters per year? *Show your work.*

________________ cm/yr

9. Historical records show that the volcanoes on the island above do not erupt every year, thus in some years, there is no build-up of lava and the height of the island is actually reduced by erosion. What does this say about the actual rate of build-up of the island as compared to the average rate calculated in the previous question?

____

____

____

10. The Himalaya Mountains were formed by the collision of the Indian subcontinent with Asia. They have the highest rate of uplift in the world (1 cm/yr). If the initial collision between the Indian and Asian subcontinents occurred 40 million years ago, what should the height of the Himalayas be in kilometers (km)? *Show your work.* _______________ km

11. In actuality, the Himalayas are approximately 8.25 km high. Why do you think the actual height of the Himalayas is not the same as the height you calculated in question 10?

____

12. Knowing that the actual height of the Himalayas is approximately 8.25 km, what is the net rate of uplift in cm/yr? Show your work. ___________ cm/yr

PART C: DISTANCE, AREA, VOLUME, DENSITY

In geology, we deal with many spatial relationships. We sometimes need to calculate areas and volumes, and it's important to know how to convert the corresponding units when the calculations are finished.

In the same way that we do dimensional analysis of distances, we can do them for areas or volumes, keeping in mind that the units in the numerators and denominators of our relationships must cancel. For example, we can calculate how many square feet are in one square meter:

$$1\ \text{m}^2 * \frac{(100\ \text{cm})^2}{1\ \text{m}^2} * \frac{(1\ \text{in})^2}{(2.54\ \text{cm})^2} * \frac{(1\ \text{ft})^2}{(12\ \text{in})^2} = 10.7\ \text{ft}^2$$

Note that when we want to square a unit, we have to also square the quantity that goes with the unit. In other words, 1 m2 is (100 cm)2 = 1.0 x 104 cm2; *not* 100 cm2! The same goes for volumes.

Problems

13. Recall that the area of a rectangle is its length x width. If we pretend that a given country is a rectangle with a length of 2000 km and a width of 500 km, what is its area in km^2? *Show your work.*
____________________ km^2

14. What is its area in square miles? *Show your work.* ______________ mi^2

Use the following chart to help you organize your answers to questions 15–21. Any numbers given to you in questions 15–21 have already been filled in for you.

	Radius	Mass	Volume	Density
Earth	6371 km	5.97×10^{24} kg	______ km^3	______ kg/m^3
Core	______ km	______ kg	______ km^3	7.874 kg/m^3

15. Recall that the formula for volume of a sphere is $4/3\pi r^3$. If the radius of the Earth is 6371 km, what is its volume in km^3? *Show your work.* ______________ km^3. Record the volume of the Earth in the chart provided above.

16. The mass of the Earth is 5.97×10^{24} kilograms; if density (ρ) is calculated as ρ = mass/volume, what is the density of the Earth in kg/m^3? *Show your work.* ______________ kg/m^3. Record the density of the Earth in the chart provided above.

17. What is the mean density of the Earth in g/cm^3? *Show your work.* ______________ g/cm^3

18. If the Earth's core has a radius equal to half the radius of the Earth, what fraction of the Earth's volume is occupied by the core? (*Hint*: Calculate the volume of the core and divide it by the volume of the Earth.) Record the radius and volume of the core in the chart provided above. *Show your work.* ______________________________

19. If the density of iron is 7,874 kg/m^3, what is the approximate mass of the Earth's core? *Show your work.* ________________ kg. Record the mass of the core in the chart provided above.

20. What fraction of Earth's mass is represented by the mass of the core? *Show your work.*

____

21. What is the mean density of the rest of the Earth? (*Hint:* Find the mass and volume of the rest of the Earth by subtracting the mass and volume of the core from the mass and volume of the Earth, respectively, and use those values to find the density of the mantle + crust.) *Show your work.* ________________________ kg/m^3

PART D: RATES AND PERCENTAGES

22. Use the Comparative Rates Figure on page 7 to answer the following questions.

a. Fingernails grow at a rate of approximately 15 mm/yr. Which geologic processes are occurring at approximately the same rate? ______________________________
____

b. Assuming that you can run like an Olympic distance runner (let's say at a rate of about 12.5 miles per hour), do you think that you could outrun:

i. debris flow? ____________________

ii. mudflow? ____________________

iii. flowing glacier? ________________

Support your answer using mathematics.

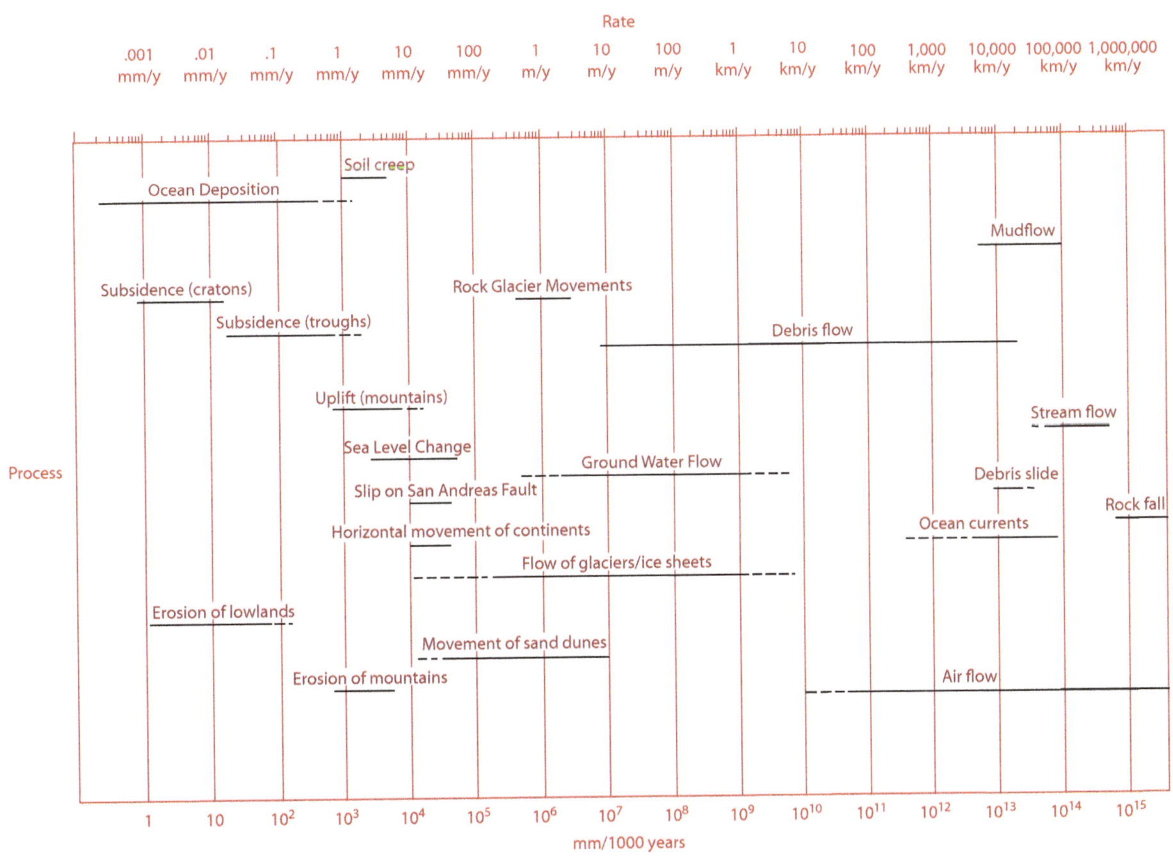

23. The composition of municipal solid waste is approximately 38% paper, 18% yard wastes, 11% metals, 9% plastics, 8% food wastes, 6% glass, 5% wood, rubber, textiles and leather, and 5% miscellaneous inorganic wastes. If 50% of all glass, paper, plastics, and yard wastes were recycled, what percent reduction in waste would be achieved? *Show your work.* ______________________

24. Assuming that recycling values are as follows: paper, \$5/ton; glass, \$8/ton; and plastics, \$10/ton. Using the values from the previous question, if your local landfill receives 600 tons of refuse per year and they wanted to start a recycling program, would it be most advantageous to start with paper, glass, or plastic? *Show your work.* ______________________

Conversion Charts

Numerical Value	Verbal equivalent	Prefix	Symbol	Exponential Expression
0.000 000 000 000 000 001	One-quintillionth	atto	a	10^{-18}
0. 000 000 000 000 001	One-quadrillionth	femto	f	10^{-15}
0. 000 000 000 001	One-trillionth	pico	p	10^{-12}
0. 000 000 001	One-billionth	nano	n	10^{-9}
0. 000 001	One-millionth	micro	µ	10^{-6}
0. 001	One-thousandth	milli	m	10^{-3}
0.01	One-hundredth	centi	c	10^{-2}
0.1	One-tenth	deci	d	10^{-1}
1				10^{0}
10	Ten	deca	da	10^{1}
100	Hundred	hecto	h	10^{2}
1000	Thousand	kilo	k	10^{3}
1 000 000	Million	mega	M	10^{6}
1 000 000 000	Billion	giga	G	10^{9}
1 000 000 000 000	Trillion	tera	T	10^{12}
1 000 000 000 000 000	Quadrillion	peta	P	10^{15}
1 000 000 000 000 000 000	Quintillion	exa	E	10^{18}

Linear Measurements	Area Measurements
1 foot = 12 inches	1 mi^2 = 640 acres
1 mile = 5,280 feet	1 acre = 43,560 ft^2
1 nautical mile = 6,076.115 feet	1 acre = 4,840 yd^2
1 km = 1000m = 10^3m; 1 mm = .001m = 10^{-3}m	1 acre = 4,047 m^2

English/Metric and Metric/English Conversions		
	English à Metric	Metric à English
Inches & Millimeters	1 inch = 25.4 millimeters	1 millimeter = 0.0394 inches
Feet & Meters	1 foot = 0.3048 meters	1 meter = 3.281 feet
Yard & Meters	1 yard = 0.9144 meters	1 meter = 1.094 yards
Miles & Kilometers	1 mile = 1.609 kilometers	1 kilometer = 0.6214 miles
In^2 & cm^2	1 in^2 = 6.4516 cm^2	1 cm^2 = 0.155 in^2
Tt^2 & m^2	1 ft^2 = 0.0929 m^2	1 m^2 = 10.764 ft^2
Yd^2 & m^2	1 yd^2 = 0.836 m^2	1 m^2 = 1.196 yd^2
Mi^2 & km^2	1 mi^2 = 2.59 km^2	1 km^2 = 0.3861 mi^2
In^3 & cm^3	1 in^3 = 16.39 cm^3	1 cm^3 = 0.061 in^3
Ft^3 & m^3	1 ft^3 = 0.0283 m^3	1 m^3 = 35.3 ft^3
Yd^3 & m^3	1 yd^3 = 0.7646 m^3	1 m^3 = 1.308 yd^3
Quarts & liters	1 quart = 0.946 liter	1 liter = 1.057 quarts
Gallons & m^3	1 gallon = 0.003785 m^3	1 m^3 = 264.2 gallons
Ounces & gram	1 ounce = 28.35 grams	1 gram = 0.0353 ounces
Pounds & kilogram	1 pound = 0.4536 kilograms	1 kilogram = 2.205 pounds
Tons & pounds		1 metric ton = 2,205 pounds

Time Conversions
1 year = 52 weeks
1 week = 7 days
1 day = 24 hours
1 hour = 60 minutes
1 minute = 60 seconds

Volume and Cubic Measurements
1 quart = 2 pints = 57.5 in^3
4 quarts = 1 gallon = 231 in^3
1 ft^3 = 1728 in^3
1 yd^3 = 27 ft^3
1 ft^3 = 7.48 gallons
1 in^3 = 0.554 fluid ounces
1 gallon = 128 U.S. fluid ounces
1 liter = .001 m^3 = 1 decimeter
1 liter = 1000 milliliters
1 deciliter = 100 milliliters
1 milliliter = 1 cm^3 (cc)
1 m^3 = 1 000 000 cm^3

Acknowledgment

Many thanks to Dr. Rebecca Ghent and Sarah Lebeau for the production of this laboratory exercise.

Laboratory Exercise II

Mineral Identification

Introduction

Given that the Greek root *geo* means "world" and *ology* means "study of," the word geology translates to "study of **EARTH**." It is therefore the job of a geologist to investigate how nutrients, resources, and energy cycle through Earth's multiple subsystems (lithosphere, hydrosphere, and atmosphere), and ultimately contribute to the sustenance of a thriving biosphere. While there are numerous mechanisms involved in making these necessities available for biological consumption, the most fundamental—and in many ways the most important—is what we call a mineral.

Minerals are essentially the building blocks of the Earth. It is minerals that compose the rocks that make up our planet, serve as the main ingredient in soil and are therefore the supplier of nutrients to the biosphere, and provide countless resource materials we use on a daily basis. As such, whether grown or mined, rest assured its origins were in what we call a mineral.

Minerals differ from other forms of matter in that they must satisfy the following five criteria:

1. Be capable of being produced by nature.
2. Retain a solid form at standard surface temperature and pressures.
3. Be inorganic (lacking a C - H bond).
4. Have a crystalline structure—i.e., an orderly internal arrangement of atoms that repeats at regular intervals.
5. Have a fixed chemical formula

To date, there are over 4000 different minerals, which are categorized into 8 families (see Figures 2.1a and 2.1b), but a working knowledge of 10–20 of them will allow you to identify most rocks by their composition and therefore gain an understanding of how Earth is built.

Objective

The objective of this lab is twofold:

1. Familiarize you with the diagnostic properties of minerals
2. Practice using diagnostic properties to identify and distinguish between our most common rock forming and resource minerals

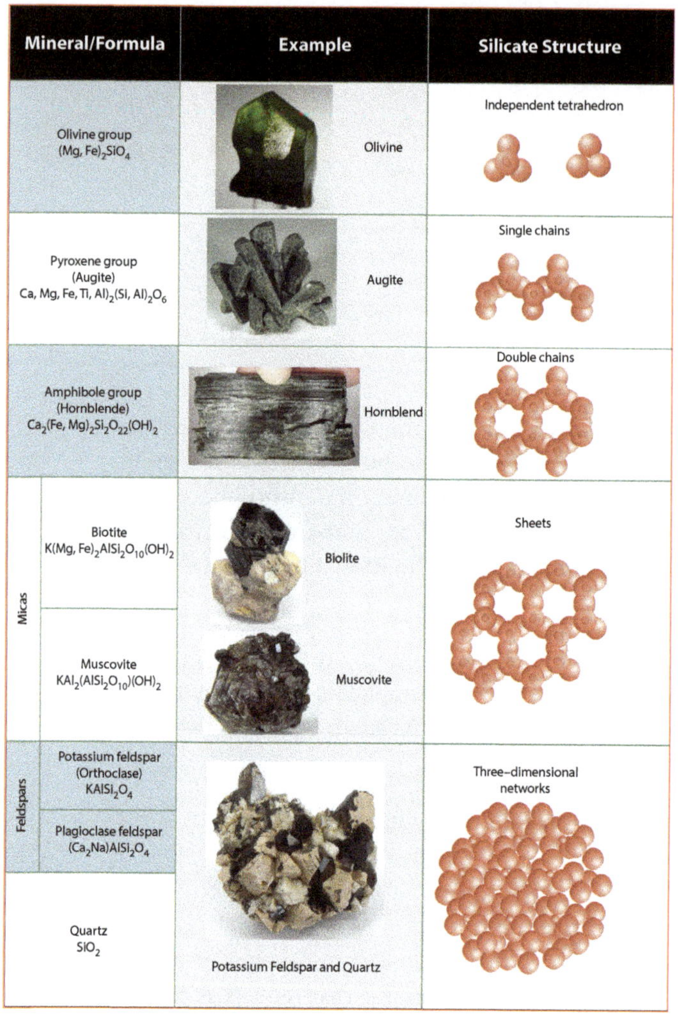

Figure 2.1a – Silicate Family

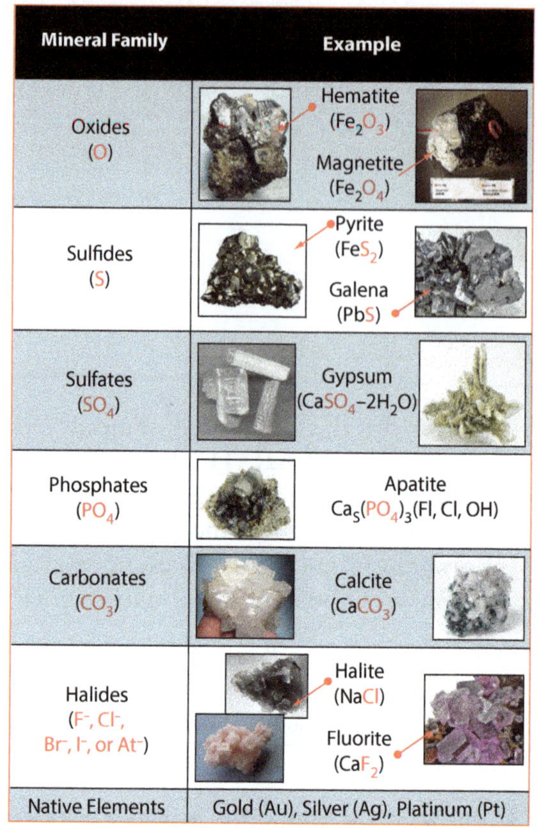

Figure 2.1b – Non-Silicate Families

Part A: Diagnostic Properties of Minerals

How do we tell one mineral from another? Well, just like you have a certain hair color, eye color, height, weight, and body shape that distinguish you from another person, each type of mineral has a unique set of physical properties that distinguishes it from all other types of minerals.

I. Why do different types of minerals exhibit different properties?

A mineral's physical properties (ex: hardness, shape, color, etc.) are determined by A) the mineral's crystal structure—i.e., the arrangement and bonding of the atoms in the mineral; and B) the mineral's composition—i.e., the types of atoms in the crystal structure. If the crystal structure and/or composition differ from one sample to another, then the samples are going to have distinctly different physical properties. These similarities and differences in physical properties, resulting from variance in crystal structure and composition, are how we group like samples together as one type of mineral and differentiate them from other types of minerals.

II. What properties help us determine the identity of a mineral?

Properties Due to Crystal Structure

Crystal Form (Habit) – Refers to the general shape or character of a crystal or cluster of crystals that have grown in nature unimpeded. Crystal forms are classified by such terms as fibrous, needlelike, prismatic, cubic, and platy to name a few. Crystal form is a direct reflection of the arrangement of atoms within the crystal structure, and therefore will vary between minerals. Crystal form can be an efficient diagnostic tool and is often useful when differentiating between minerals that otherwise appear nearly identical. See Figure 2.2 for select examples of crystal habit.

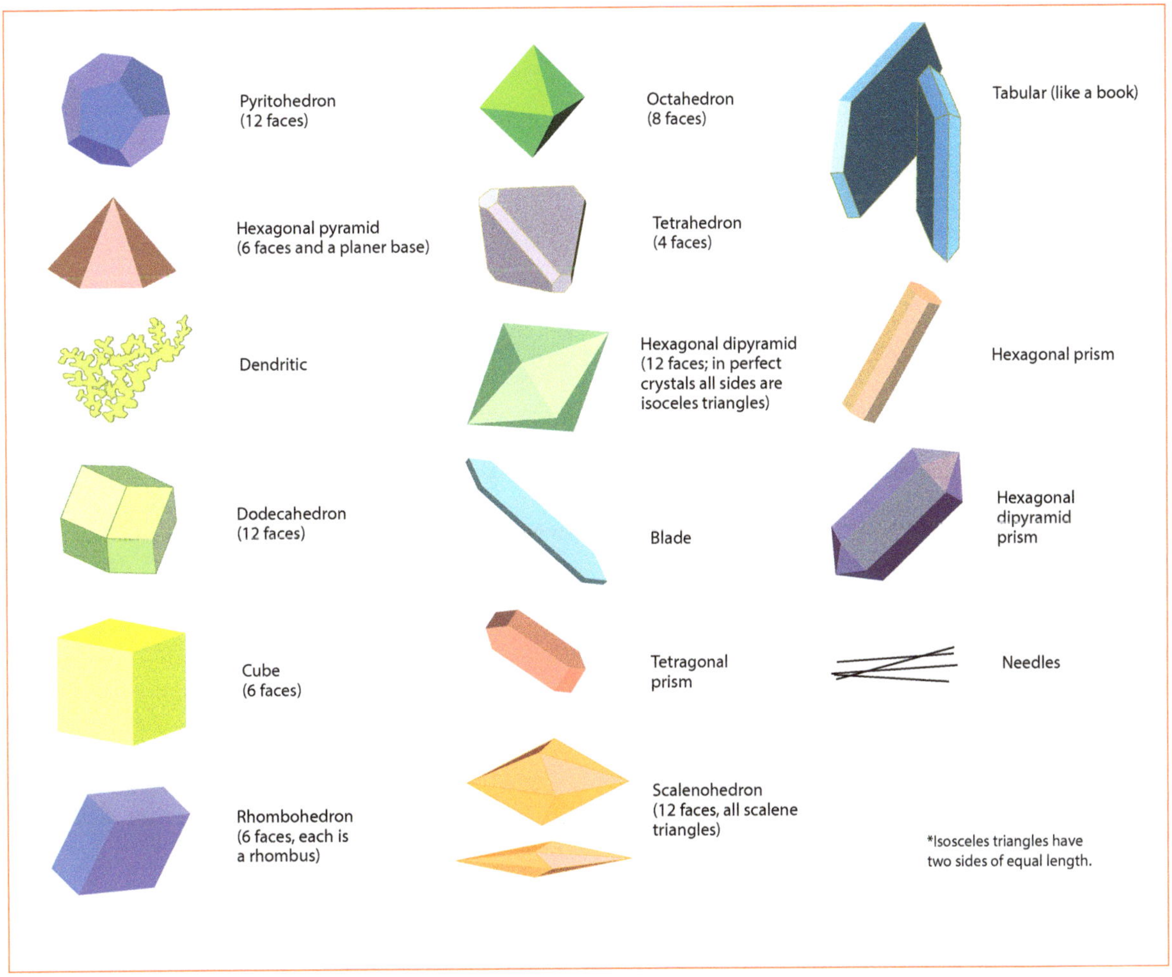

Figure 2.2 – Select Varieties of Crystal Forms

Cleavage and Fracture – Describes the method by which a mineral breaks. Many minerals break in an identifiable, orderly manner along flat surfaces. These surfaces are called "cleavage planes" and result from weak chemical bonding between parallel layers of atoms in the crystal structure. Because different crystal structures have weak bonds located in different places, minerals with dissimilar crystal structures will have diverse cleavage patterns. Some structures have no weak bonds at all, others have only one set of weak bonds, and still others have multiple sets of weak bonds. Each different set of parallel cleavage planes has an orientation relative to the crystal structure and is referred to as a "cleavage direction." See Figure 2.3 for a visual representation of select cleavage examples.

Number of Cleavage Directions	Shape	Sketch	Directions of Cleavage	Sample
1	Flat sheets			 Mica
2 at 90°	Elongated form with rectangle cross section (prism)	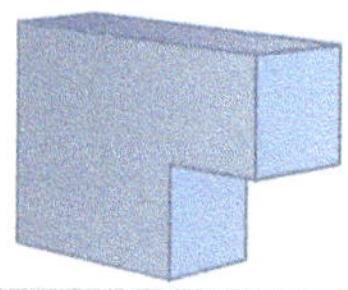	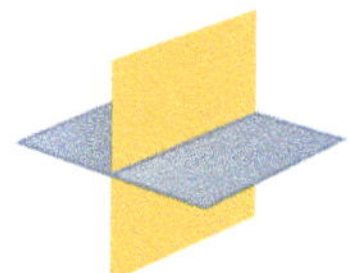	 Feldspar
2 not at 90°	Elongated form with parallelogram cross section (prism)			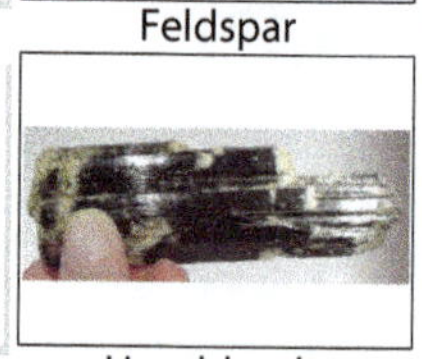 Hornblende
3 at 90°	Cube		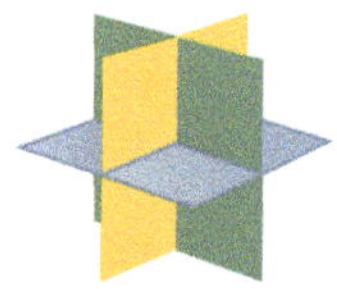	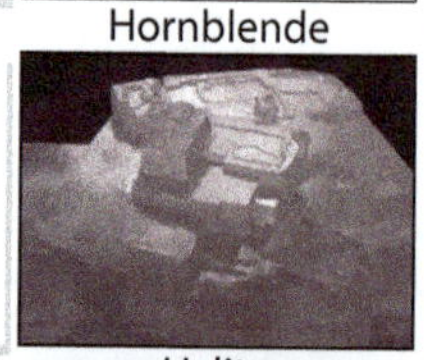 Halite
3 not at 90°	Rhombohedron			 Calcite
4	Octahedron			 Fluorite

Figure 2.3 – Select Varieties of Cleavage

- Cleavage is described as either:
  - Excellent – Easily identifiable; minerals that when broken produce pieces shaped as cubes, rhombohedrons, columns, pyramids, or planes.
  - Good – Minerals that when broken retain a distinctive shape with flat sides, but have distortion around parts of the mineral (ex: rounded edges).
  - Poor/Absent – Distinguishing between minerals with poor cleavage and no cleavage is many times not possible, as poor cleavage surfaces are often inconspicuous.

*Note that when a mineral cleaves, it produces numerous broken pieces of similar or identical shape.

Minerals that have bonds of comparable strength within their crystal structures will fracture, as opposed to cleave, when stressed. This results in the mineral randomly shattering instead of breaking along defined boundaries, so no two broken pieces will be the same shape.

Fracture is described as either conchoidal or non-conhoidal (see Figures 2.4 and 2.5):

Conchoidal **OR** **Non-Conchoidal**

Figure 2.4
Conchoidal Fracture in Volcanic Glass

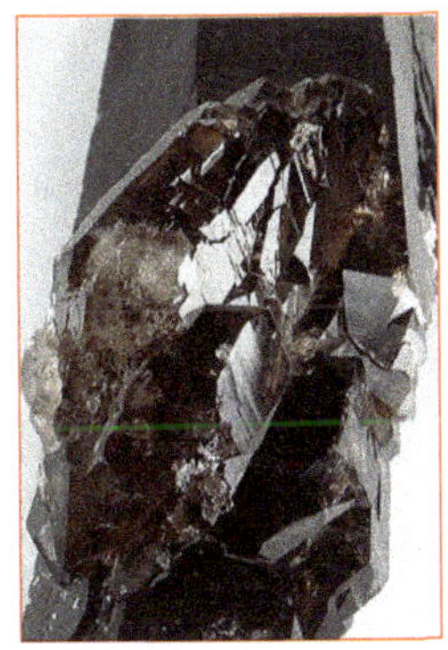

Figure 2.5a **Figure 2.5b**
Non-Conchoidal Fracture in Quartz

*Note that the absence of cleavage can be equally as diagnostic as the presence of cleavage.

Cleavage can often be confused with crystal form, but that can easily be rectified by looking at a broken side. If the mineral has been broken, then excellent/good cleavage (or fracture) will be visible. If however, the mineral is completely intact as nature produced it, then what you are seeing is crystal form (Figure 2.6a). In addition, cleavage planes can be repeated like steps, whereas a crystal face is one solid, smooth surface (Figure 2.6b).

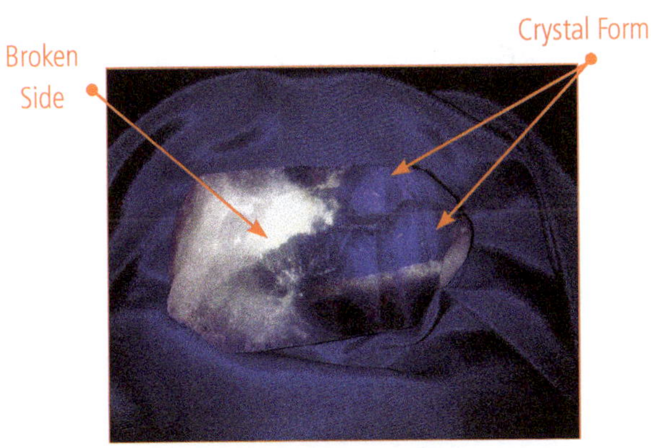

Figure 2.6a –
Cleavage/Fracture vs. Crystal Form

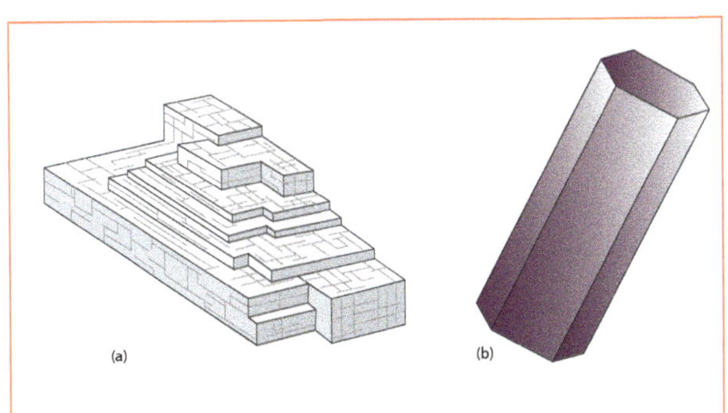

Figure 2.6b –
Cleavage Terracing vs. Smooth Face

Mineral Hardness: A measure of the resistance of a mineral to abrasion or scratching. A mineral's hardness is determined by: A) the strength of the bonds throughout the crystal structure; and B) the configuration of the atoms within the crystal structure. Minerals made up of atoms that are held together primarily by the weak van der Waals force constitute our softest minerals, as it takes very little energy to terminate the attraction between the atoms and break up the structure (ex: talc and graphite). Minerals comprised primarily of strong covalent bonds (ex: quartz and the diamond) are the hardest minerals on Earth, and those made up primarily of ionic bonds (halite and calcite) fall in between, albeit closer to the softer end. In addition, the more densely packed the atoms, the stronger the bonding and the harder the mineral.

Relative mineral hardness is categorized using Mohs Hardness Scale (Fig. 2.7), which assigns each mineral a number between 1 and 10—1 being the softest and 10 being the hardest. Absolute mineral hardness can be read from the Y-axis and divulges exactly how much harder each mineral is than another.

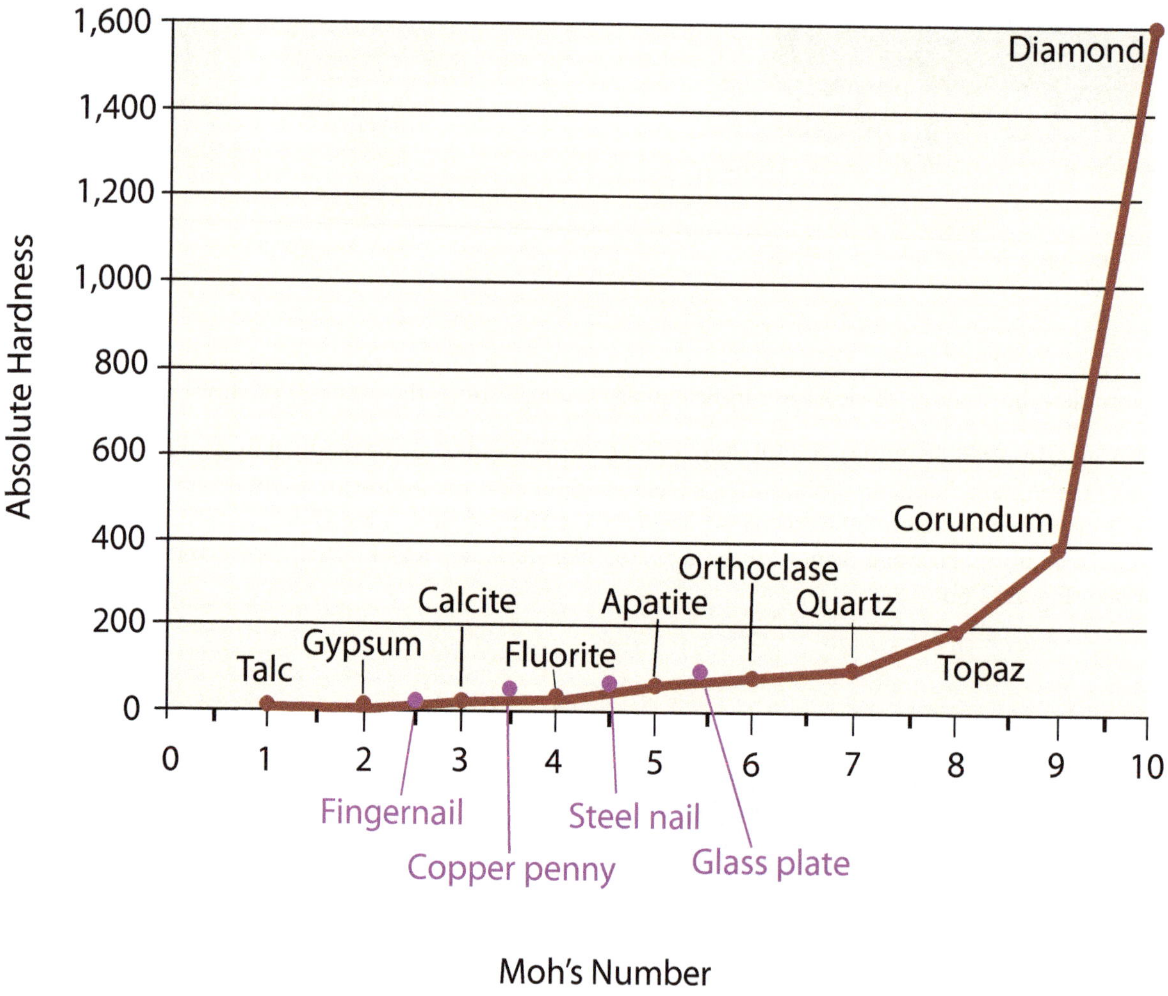

Figure 2.7 – Relative and Absolute Hardness

How significantly does the type of crystal structure and bonding affect the properties of a mineral?

Let's look at an example where two minerals have the same chemical composition, but very different structures—the diamond (C) and graphite (C). See Figure 2.8.

Diamond (C) vs. Graphite (C)

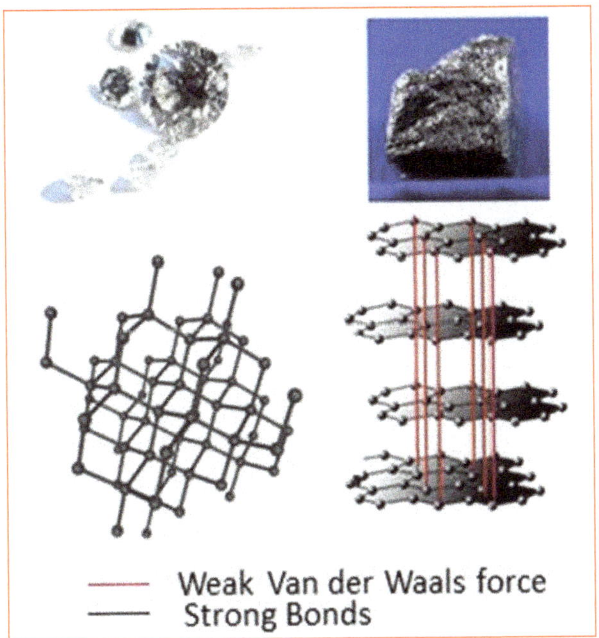

Figure 2.8 – Diamond vs. Graphite

Diamond (C)

- Made exclusively of covalent bonds
- 3D crystal structure
- Dense – Due to compact packing of atoms
- Hardest material on Earth
- Electrical Insulator

Graphite (C)

- Has van der Waals bonds between sheets
- Sheet crystal structure
- Not dense – Due to spacing between the atoms
- One of the softest materials on Earth
- Electrical Conductor

- So … the difference between what you buy in Staples for $3.99 a pack and the "forever" gem that costs upward of $2000 is nothing more than the orientation of the atoms and the strength of the bonds.

Properties Due to Composition

Color – Color results from the way a mineral interacts with light. Sunlight contains the whole spectrum of color and each color has its own unique wavelength. The mineral absorbs most of the colors' wavelengths, and the color you see when looking at it represents the wavelength the mineral did not absorb (i.e., the one it reflected). Certain minerals always have the same color, but many show a range of colors due to minute impurities that slide into the crystal structure. As such, in many cases, color is not a diagnostic property of a mineral, as one mineral may have multiple colors and multiple minerals may have the same color (See Figures 2.9 and 2.10).

Figure 2.9 – Select Color Varieties of Quartz
Select quartz family members including amethyst, citrine, rose, colorless rock crystal, chalcedony, tiger eye, onyx, agate and aventurine.

Figure 2.10 – Three Different Minerals

Figure 2.11 – Sulfur Streak

Streak – Color of the fine powder of a mineral, obtained by rubbing the mineral across a piece of unglazed porcelain (see Figure 2.11). Pulverizing the mineral into powder form will result in all the minerals of the same type having a similar color streak, regardless of the external color. Note: The streak plate has a hardness on Mohs Hardness Scale of ~6.5–7. As such, minerals harder than 6.5 cannot be tested for streak. Given what you now know about mineral hardness, consider why this is the case.

Figure 2.12 – Magnetite Attracts Iron Nails. Photograph from 1921

Magnetism – There is a limited selection of minerals that demonstrate magnetic properties. The presence of iron in the mineral's chemical composition is single-handedly responsible for the unique behavior of any magnetic mineral. The most well-known magnetic mineral is Magnetite, hence the descriptive name (see Figure 2.12).

Reaction to hydrochloric acid – When a member of the carbonate family (group of minerals whose chemical formula includes a CO_3 group) comes in contact with dilute hydrochloric acid, the acidic aqueous solution has enough strength to break the covalent bonds holding the carbonate group together. This reaction produces CO_2 as a byproduct, which manifests itself in the form of erupting gas bubbles—aka "fizzies,"—see Figure 2.13. While all carbonate minerals fizz to some degree upon contact with HCl, the carbonate mineral you will be working with is known as calcite ($CaCO_3$). Note: If one drop of dilute HCl solution does not make the mineral fizz, then 10 drops will not make it fizz … so please do not bathe the minerals in HCl.

Figure 2.13 – Calcite-rich rock reacts with HCl

Specific Gravity – Measures the density (weight per unit volume) of a mineral. This can be a useful diagnostic tool, as samples of two different minerals that are the same size and of similar appearance may have significantly different weights.

Properties Due to BOTH Crystal Structure and Composition

Luster – Describes the way a mineral reflects light. The most basic division of luster is metallic versus non-metallic (see Figures 2.14 and 2.15).

Metallic:
Silvery, gold, or brass sheen displayed by metals

Non-metallic:
Any luster unlike that of a metal.

Examples of Metallic Luster

Figure 2.14 – Metallic Luster

Note: While every mineral has a characteristic luster, minerals of the same type may exhibit different lusters. In addition, there is no scientific method available to determine luster. This being the case, determining the luster of a particular specimen is subjective; two people looking at the same specimen may observe two different lusters. For these reasons, we are going to restrict our analysis of luster to Metallic vs. Non-metallic only.

Examples of Non Metallic Luster	
Nonmetallic: Earthy/Dull • Non-reflective, like concrete or dirt • Develops when minerals are composed of coarse grains that scatter light in all directions.	
Nonmetallic: Greasy • Resembles fat or grease • Results from a significant amount of microscopic impurities included in the structure.	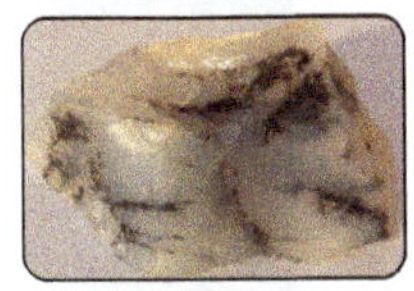
Nonmetallic: Pearly • Thin transparent sheets stacked one top of the other give the mineral a pearl like appearance	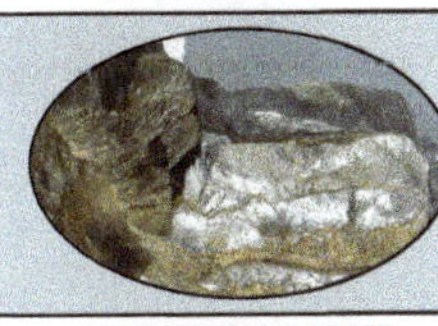
Nonmetallic: Resinous • Has the appearance of resin or smooth plastic	
Nonmetallic: Vitreous • Like glass or a glossy photo • Typically found in transparent and translucent minerals	
Nonmetallic: Silky • Very fine fibers align parallel giving the appearance of silky fabric	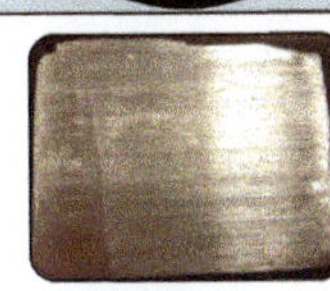
Nonmetallic: Waxy • Has a candly-like texture	

Figure 2.15 – Non-metallic Luster

Diaphaneity – A measure of the amount of light that passes through a mineral (see Figure 2.16). This is determined by the amount of light being absorbed by the mineral—which is, in turn, a function of its composition and crystal structure.

- Diaphaneity is described as either:
  - Transparent – No light is absorbed, permitting all the light to pass through the mineral. Completely transparent minerals do not have color.
  - Translucent – Allows light to pass through, but scatters and diffuses it, creating only a partial transparency.
  - Opaque – All light is either absorbed or reflected, meaning no light passes through the mineral.

Transparent

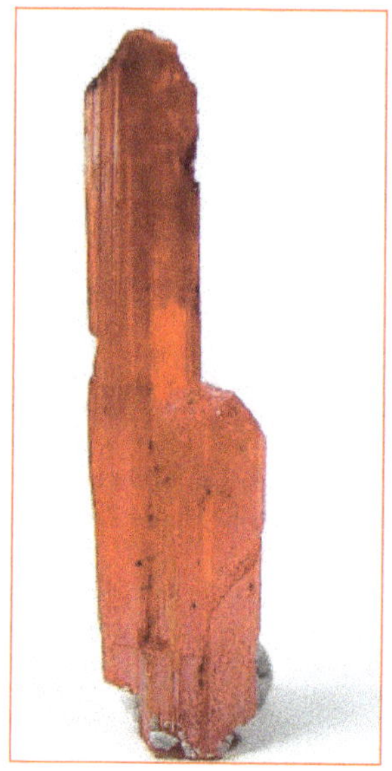

Translucent

Opaque

Figure 2.16 – Comparison of Diaphaneity

Exercise 1: Using Diagnostic Properties

While many different minerals may appear similar or even identical at first glance, a closer examination using the properties described above will demonstrate they are in fact very different materials. On the flip side, minerals that look nothing alike, but exhibit the same properties will be shown to be the same substance. The box on the table in front of you, labeled "mixed minerals," contains 14 mineral samples. The samples in this box are made up of 6 different minerals, meaning there are several samples of each mineral. Your job is to divide up the 14 samples into 6 groups, each comprising the samples made of that mineral only. In the following chart record the letters of the samples that belong to each group and write down what properties made you group those samples together and differentiate them from the others. The number of samples per mineral has been supplied for you. Once you have completed this task, your TA will go over the answers with you, before proceeding to the next section.

# of Samples in Group	Letter of samples that belong to this group	Properties that unite these samples and distinguish them from the others
4		
3		
2		
2		
2		
1		

Exercise 2: Diagnostic Properties and Mineral Identification

I: Rock-Forming Minerals

This exercise is designed to teach you how to identify minerals using the diagnostic properties described in the previous pages. In the container before you are multiple different rock-forming minerals (*Note: If multiple samples are in the same slot, it means they are different varieties of the same mineral*). Your job is to identify them using the following procedure. First, list the requested properties for each mineral sample in the accompanying Mineral Identification Charts provided on the next pages. Next, use the *Mineral Properties Database* entitled "Rock-Forming Minerals" at the end of this exercise to help you identify each mineral. The database is designed to allow you to use the most basic properties, like cleavage, to first rule out impossible samples and then to further narrow down the minerals identity using other less obvious properties, like hardness, streak, specific gravity, etc. Keep in mind that not all properties of every mineral are equally useful. Generally, you will be able to identify a mineral by using only two or three primary properties. These will differ for each mineral. For instance, although galena has a hardness of 2.5, this is not generally a primary property, as many other minerals also have this same hardness. You will likely be able to identify galena by the fact that it has a very high specific gravity and has cubic cleavage (three cleavage planes), thus the density and cleavage (cubic) will be the primary properties for galena. In addition, some minerals depict one specific property (such as magnetism) that can lead you to an immediate identification. Determine which properties are most valuable in helping you identify each mineral and enter these properties in the primary properties column of your Mineral ID chart.

- Note: Samples 16 and 17 appear almost identical to the naked eye. As such, the characteristics of these two minerals have been filled in on your mineral ID charts for you. Save these two samples for last. When you are left with only these two unknowns, look carefully at the description for each of those two minerals in your *Mineral Properties Database* and determine what definitive property, when visible, can distinguish one from the other in hand sample, then put the name of each mineral on the appropriate line.

MINERAL IDENTIFICATION CHART

Specimen Number	Color	Streak	Hardness	Cleavage (good or absent)	Primary Properties	Name
1						
2						
3						
4						
5						
6						
7						
8						
9						
10						

MINERAL IDENTIFICATION CHART

Specimen Number	Color	Streak	Hardness	Cleavage (good or absent)	Primary Properties	Name
11						
12						
13						
14						
15						
16	Black-Greenish Black	Green/Gray	5.5–6	Good 2-directional (almost 90°)		
17	Black-Greenish Black	White/Gray	5.5–6	Good 2-directional (60° and 120°)		
18						
19						
20						

MINERAL DATABASE: ROCK FORMING MINERALS

Family	Mineral Name and Formula	Color	Opacity	Luster	Streak	Hardness	Cleavages	Crystal Habit	Diagnostic Properties
Silicates	AUGITE/PYROXENE $(Ca,Mg,Fe,Ti,Al)_2(Si,Al)_2O_6$	Black-Greenish Black	Opaque	Non-metallic: Vitreous or Resinous	Green/Gray	5.5–6	2-directional (almost 90°)	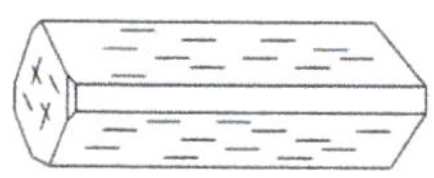 Columnar, often an-or subhedral	Cleavage and columnar habit. Very similar to Hornblende (use cleavage angles to differentiate)
	BIOTITE $K(Mg,Fe)_3(Al,Fe)Si_3O_{10}(OH,F)_2$	Black – Dk. Brown	Transparent to Opaque	Non-metallic: Sub-Metallic or Vitreous	Gray/Brown	2–3	Excellent basal cleavage (unidirectional)	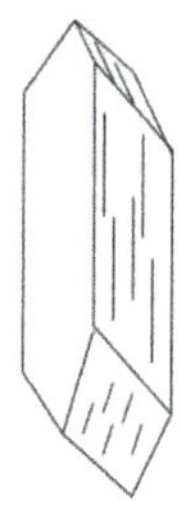 Pseudo-hexagonal, Platy	Sheet-like cleavage (Similar to Muscovite, color is darker)
	GARNET $Mg_3Al_2(SiO_4)_3$	Various, but most commonly dk. Red –Purple	Transparent to Sub-translucent	Non-metallic: Vitreous	N/A	6.5–7.5	None	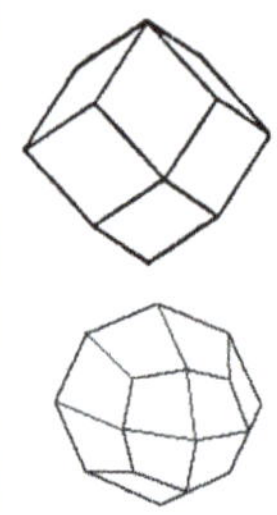 Hexoctahedron or Dodecahedron	"Soccer Ball" shape
	HORNBLENDE/AMPHIBOLE $Ca_2(Mg,Fe^{3+},Al)_5(Al,Si)_8O_{22}(OH)_2$	Black-Greenish Black	Opaque	Non-metallic: Vitreous or Resinous	White/Gray	5.5–6	2-directional (60° and 120°)	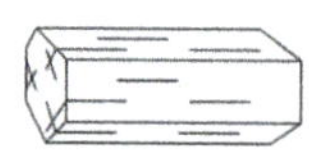 Columnar, sometimes fibrous	Cleavage and columnar habit, Very similar to Augite (use cleavage angles to differentiate)
	KAOLINITE $Al_4(Si_4O_{10})(OH)_8$	White	Opaque	Non-metallic: Earthy	White	1.5–2	Perfect basal cleavage (unidirectional) (not often visible in hand sample)	 Clay-like, often anhedral	Powdery and brittle
	KYANITE Al_2SiO_5	Blue / Green-White	Translucent	Non-metallic: Vitreous or Pearly	White	5.5–7	2-directional but only one is perfect.	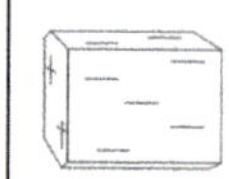 Columnar, often fibrous or in blades	Columnar Shape, Blue Color and differs in hardness between faces (anisotropism)

Mineral Database: Rock Forming Minerals

Family	Mineral Name and Formula	Color	Opacity	Luster	Streak	Hardness	Cleavages	Crystal Habit	Diagnostic Properties
Silicates	MUSCOVITE $KAl_2(AlSi_3O_{10})(F,OH)_2$	Colorless, Lt. Brown/ Green	Transpar-ent to Translucent	Non-metallic: Vitreous, Pearly or Silky	White	2 – 2.5	Perfect basal cleavage (unidirectional)	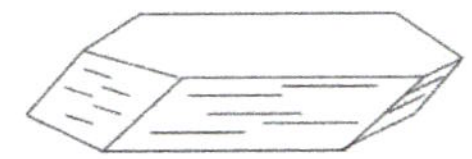 Hexagonal "books," platy sheets	Sheet-like cleavage (similar to Biotite, color is lighter)
	OLIVINE $(Mg,Fe^{2+})SiO_4$	Green	Transpar-ent to Translucent	Non-metallic: Vitreous	N/A	7	Very Poor	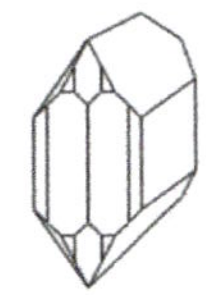 Orthorhombic, Mostly anhedral	Hardness, dense aggregate of rounded sand-like grains
	PLAGIOCLASE FELDSPAR $(Na,Ca)(Si,Al)_4O_8$	White, Gray or Bluish	Opaque	Non-metallic: Vitreous or Pearly	White	6 – 6.5	2-directional (almost 90°)	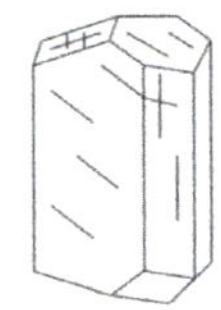 Striated tabular, blocky and sometimes bladed	Good cleavages (often staircase-like), large tabular crystals (similar to Potassium Feldspar, color differentiates)
	POTASSIUM FELDSPAR $KAlSi_3O_8$	Pink–White	Opaque	Non-metallic: Vitreous or Pearly	White	6 – 6.5	2-directional (almost 90°)	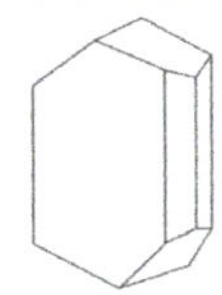 Striated tabular, blocky and sometimes bladed	Good cleavages (often staircase-like), large tabular crystals (similar to Plagioclase, color differentiates)
	QUARTZ SiO_2	Colorless, White, Gray/Black, Yellow/ Brown, Violet or Pink	Transpar-ent to Translucent	Non-metallic: Vitreous, waxy or greasy	N/A	7	None: Conchoidal Fracture	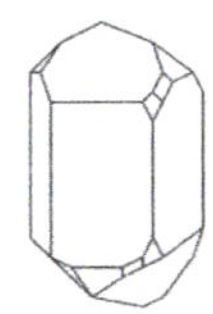 Prismatic crystals with pyramidal terminations, elongate along c-axis	Scratches glass, lack of cleavage: conchoidal fracture (looks like circular ripples)
	TALC $Mg_3Si_4O_{10}(OH)_2$	White–Gray sometimes Greenish	Translucent	Non-metallic: Pearly or Dull	White	1	Perfect basal cleavage (Not often visible in hand sample)	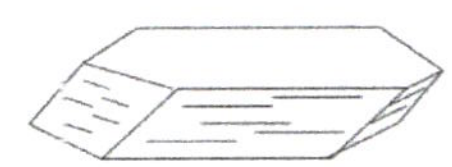 Platy but usually massive and anhedral	Greasy and extremely soft Distinguished from gypsum by lack of visible cleavage

MINERAL DATABASE: ROCK FORMING MINERALS

Family	Mineral Name and Formula	Color	Opacity	Luster	Streak	Hardness	Cleavages	Crystal Habit	Diagnostic Properties
HALIDES	HALITE NaCl	Colorless, White, Yellow, Orange, Reddish, Purple or Blue	Transparent	Non-metallic: Vitreous	White	2.5	Excellent and 3-directional, mirroring each side of the cube.	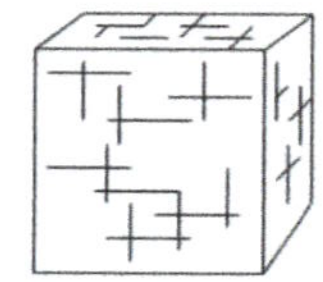 Mostly Cubic and often euhedral, sometimes rounded and granular	Cubic habit and salty!
CARBONATES	CALCITE $CaCO_3$	Colorless, White, Grey, Yellowish or Blue	Transparent to Opaque	Non-metallic: Vitreous or Milky	White	3	Excellent and 3-directional, in directions that mirror it's rhombohedral sides.	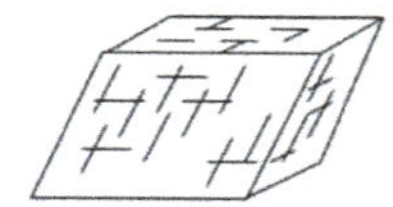 Has many crystal habits but often occurs as rhombohedral, prismatic or scalenohedral.	Fizzes with HCl, hardness and rhombehedral habit
OXIDES	HEMATITE Fe_2O_3	Silvery Gray/ Black, or Brick red	Opaque	Non-metallic or Metallic: Earthy or Metallic	Red -Reddish Brown	1.5–6	None	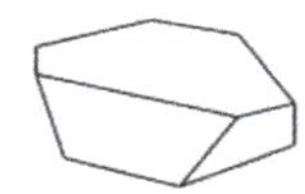 Often anhedral, massive and granular. Can be tabular.	Red streak no matter the mineral color. Occasionally attracted to magnet.
	MAGNETITE $Fe^{3+}_2Fe^{2+}O_4$	Silvery Gray – Black	Opaque	Metallic	Dk. Gray	6	None	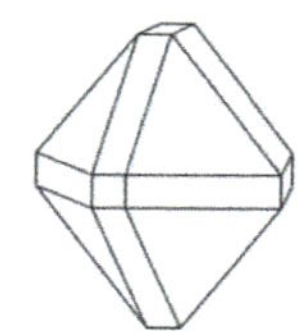 Often anhedral, massive or granular. Can be octahedral.	Magnetic!
SULFATES	GYPSUM $CaSO_4 \cdot 2H_2O$	Colorless, White, Gray or Yellowish	Translucent	Non-metallic: Pearly, milky to subvitreous	White	2	3-directional but only one is perfect (dashed lines in crystal habit figure represent "good" directions)	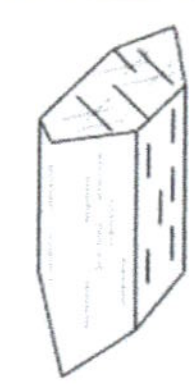 Often massive but can be elongate and fibrous or tabular.	Very soft. Elongate and fibrous habit. Distinguished from Talc by presence of visible cleavage

II - Economic Minerals

Now that you've used diagnostic properties to sort minerals of similar appearance and then used them to identify individual rock forming minerals, let's follow-up with an exercise that unites all the features practiced above and allows you a glimpse at some of our most prominent economic minerals. Some of these minerals you looked at in the previous exercise and others will be new. On the table in front of you is a box of minerals labeled "Economic Minerals." These samples correspond with the questions below. Use the provided Mineral Properties Database entitled "Economic Minerals" at the end of this exercise to identify the mineral where necessary and answer the questions that follow.

Samples 1a and 1b

All the samples you see in Box 1a are varieties of a halide mineral called fluorite (CaF_2). Fluorite is used in the metal-processing industry as a flux for smelting and contributes to the manufacture of glasses and enamels. The purest forms of fluorite can supply the fluoride necessary to produce hydrofluoric acid and transparent fluorite is used in microscope and telescope lenses. Because of its color variety, fluorite also serves a prominent place in the arts and jewelry industry.

a. What two distinctive properties do all the samples have in common that would allow you to distinguish them from the quartz samples in Box 1b

____________________ and ____________________

b. Why do these two minerals display so many different colors?

____

*Note: Quartz is also a widely used resource mineral, used in building materials, glassmaking, and electronics.

Sample 2

Pick this sample up and try to write your name with it! This mineral is the same mineral that serves as your pencil "lead." It is also used in the manufacture of paint, as a lubricant, and as electrodes.

a. Identify this mineral: ____________________

b. This mineral is made of pure C, same as a diamond. What creates the vast difference in the properties of these two minerals?

____________________ and ____________________

Samples 3 and 4

These two samples are heavily used in the construction and electrical industries.

a. What is this group of minerals collectively called? ______________________

b. Why are these minerals capable of being peeled apart like sheets in a stack of paper?

____

Sample 5

This mineral exhibits excellent cubic cleavage and is unusually dense. In addition, it is the main ore of lead and therefore permits the manufacture of batteries, ammunition, radiation shielding, and numerous industrial chemicals, to name a few.

a. Identify this mineral: ____________________

b. Why does this mineral have an usually high density? ______________________________

Samples 6, 7, and 8

At first glance, samples 6 and 7 appear to be the same, but a closer examination of their properties will demonstrate that one of these samples is hematite and the other is magnetite. Use the tools in your kit to determine which is which, then answer the questions that follow.

a. Sample 6: ____________________;

Distinguishing Property ___________________________

b. Sample 7: ____________________;

Distinguishing Property ___________________________

c. What invaluable metal do these two minerals supply to society? __________. This metal is the most common material used in industry and is a major component in steel, machinery, transportation vehicles, office supplies, and many other standard resources.

d. Now observe sample 8. This mineral is called pyrite, aka "fool's gold," and also contributes to supplying the same metal as hematite and magnetite.

What mineral family do hematite and magnetite belong to? ____________________

What mineral family does the pyrite belong to? ___________________________

Samples 9 and 10

These two samples are sulfide minerals, each of which serves as the main ore for an important, but different metal. Identify each mineral, then answer the questions that follow.

a. Sample 9: ______________________;

Distinguishing Property ______________________________

Which metal does this mineral provide? ________________

b. Sample 10: ______________________;

Distinguishing Property ______________________________

Which metal does this mineral provide? ________________

Just for Show

Sample X - This sample is a material called bauxite, which is the world's primary aluminum ore. Notice the "bull's-eye" pattern that makes bauxite so distinctive. Bauxite is a term used to describe a mixture of primarily aluminum oxide and aluminum hydroxide minerals, in combination with smaller amounts of minerals from other families.

This being the case, is bauxite itself a mineral? Yes or No

If no, what is the correct label for bauxite? ________________

Mineral Properties Database: Economics Minerals

Family	Mineral Name and Formula	Color	Opacity	Luster	Streak	Hardness	Cleavages	Crystal Habit	Diagnostic Properties	Economic Uses
SILICATES	BIOTITE $K(Mg,Fe)_3(Al,Fe)$ $Si_3O_{10}(OH,F)_2$	Black – Dk. Brown	Transparent to Opaque	Non-metallic: Sub-metallic or Vitreous	Gray/ Brown	2–3	Excellent basal cleavage (unidirectional)	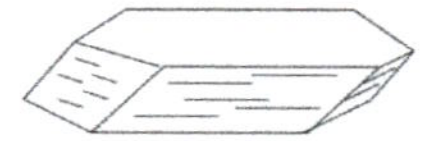 Pseudo-hexagonal, Platy	Sheet-like cleavage (Similar to Muscovite, color is darker)	Insulating in electrical equipment, in lubricants, wall finishes, artificial stone
	MUSCOVITE $KAl_2(AlSi_3O_{10})$ $(F,OH)_2$	Colorless, Lt. Brown / Green	Transpar-ent to Translucent	Non-metallic: Vitreous, Pearly or Silky	White	2–2.5	Perfect basal cleavage (unidirectional)	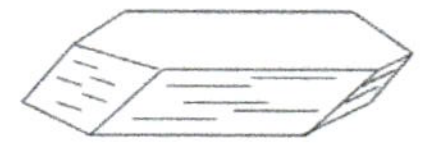 Hexagonal "books", platy sheets	Sheet-Like cleavage (Similar to Biotite, color is lighter)	Insulation in electrical equipment, in lubricants, wall finishes, artificial stone
	PLAGIOCLASE FELDSPAR $(Na,Ca)(Si,Al)_4O_8$	White, Gray or Bluish	Opaque	Non-metallic: Vitreous or Pearly	White	6–6.5	2-directional (almost 90°)	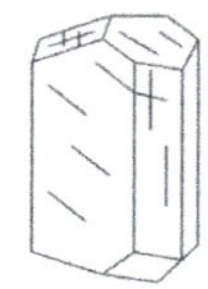 Striated tabular, blocky and sometimes bladed	Good cleavages (often staircase-like), large tabular crystals (Similar to Potassium Feldspar, color differentiates)	Manufacture of porcelain, pottery and glass. Used for glazes in pottery and as mild abrasive.
	POTASSIUM FELDSPAR $KAlSi_3O_8$	Pink – White	Opaque	Non-metallic: Vitreous or pearly	White	6–6.5	2-directional (almost 90°)	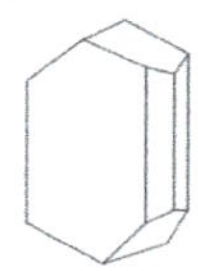 Striated tabular, blocky and sometimes bladed	Good cleavages (often staircase-like), large tabular crystals (Similar to Plagioclase, color differentiates)	Manufacture of porcelain, pottery and glass. Used for glazes in pottery and as mild abrasive.
	QUARTZ SiO_2	Colorless, White, Gray / Black, Yellow / Brown, Violet or Pink	Transpar-ent to Translucent	Non-metallic: Vitreous, waxy or greasy	N/A	7	None: Conchoidal Fracture	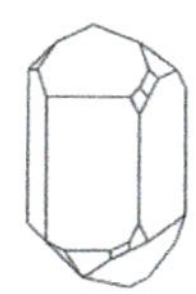 Prismatic crystals with pyramidal terminations, elongate along c-axis	Scratches glass, lack of cleavage: conchoidal fracture (looks like circular ripples)	Building materials, glass making, pottery, ferro-silicon alloys, as abrasive and quartz crystals used in electronics
NATIVE ELEMENT	COPPER Cu	Copper-red, Brown; Often tarnished in a Pistachio Green	Opaque	Metallic	N/A	2.5–3	None	Globular and Massive or Dendritic.	Malleability and Globular shape. Copper color and green tarnish.	Alloyed with various metals to produce bronzes and brasses. Used in electrical industries where high electrical and thermal conductivity is required
	GRAPHITE C	Iron black to steel grey	Opaque	Metallic to Dull	Black	1–2	1 good cleavage direction	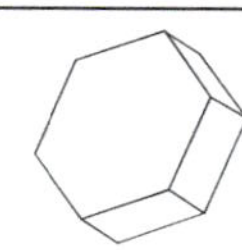 Hexagonal	Greasy, hence you can write with it	Manufacture of crucibles, and foundry molds. Used as a lubricant, electrodes, and in pencils.

MINERAL PROPERTIES DATABASE: ECONOMICS MINERALS

Family	Mineral Name and Formula	Color	Opacity	Luster	Streak	Hardness	Cleavages	Crystal Habit	Diagnostic Properties	Economic Uses
OXIDES	HEMATITE Fe_2O_3	Silvery Gray/ Black, or Brick Red	Opaque	Non-metallic or Metallic: Earthy or Metallic	Red -Reddish Brown	1.5–6	None	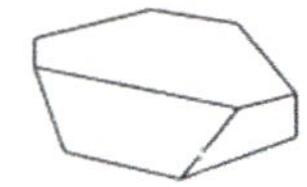 Often anhedral, massive and granular. Can be tabular.	Red streak no matter the mineral color. Occasionally attracted to magnet.	Used as filler for paints, paper, and rubber and in plasters, lubricants, makeup and chalk
OXIDES	MAGNETITE $Fe^{3+}{}_2Fe^{2+}O_4$	Silvery Gray – Black	Opaque	Metallic	Dk. Gray	6	None	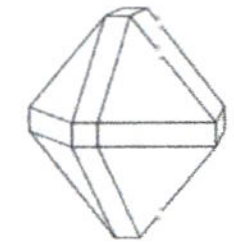 Often anhedral, massive or granular. Can be octahedral.	Magnetic!	Iron and Steel
SULFIDES	CHALCOPY-RITE $CuFeS_2$	Brassy Yellow to Brownish	Opaque	Metallic	Greenish Black	3.5–4.5	Poor	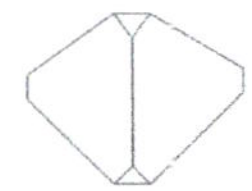 Often anhedral and massive	More "tarnished" look than Pyrite generally. Fairly soft.	Most important ore of copper.
SULFIDES	GALENA PbS	Silvery Gray -Lead Gray	Opaque	Metallic	Dk. Gray	2.5	Perfect 3-directional, mirrors cubic sides.	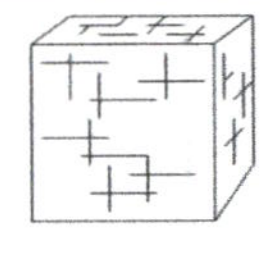 Often cubic or octahedral. Can be massive or granular.	Very good cubic cleavage. Can scratch with fingernail. High density and specific gravity.	Main ore of lead. pipes, alloys, batteries, radiation shielding, pigments
SULFIDES	PYRITE $Fe^{2+}S_2$	Brassy Yellow to Brownish Black	Opaque	Metallic	Greenish Black	6.5	Very Poor	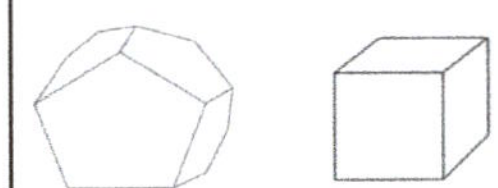 Often cubic or dodecahedral pentagons (pyritohedron).	Also known as "Fool's Gold." Color and distinctive habit.	Iron and Steel
SULFIDES	SPHALERITE ZnS	Yellowish Brown – Red-dish Brown	Opaque	Non-metallic: Resinous	Brown (Smells like eggs!)	3.5–4	Perfect unidirectional	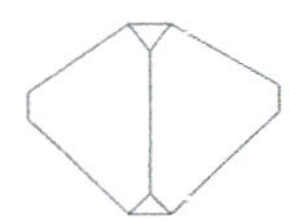	Resinous luster. Distinct sulfur odor when powdered.	Major Zinc Ore.
HALIDES	FLUORITE CaF_2	Colorless, Blue, Yellow, Purple, Pink, Green or White	Transparent to Translucent	Non-metallic: Vitreous	White	4	Excellent 3-directional cleavage (Do not mirror sides of cube)	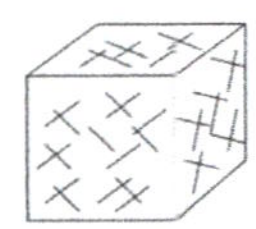 Usually anhedral and massive but can be cubic and prismatic.	Hardness will differentiate fluorite from most. Also, have differing cleavage directions from other cubic minerals.	Used in toothpastes, drinking water, plastics, lubricants, cement, metal alloys, rocket fuel, acid, anesthetics
N/A	BAUXITE Aluminum Hydroxide	Pale to Dark Reddish Brown	Opaque	Non-metallic: Earthy	N/A	N/A	None	Massive and made up of granular spheres with concentric interiors, "bulls' eyes."	Covered in "bulls eyes"	Manufacture of Aluminium Alloys

Part B: Tying it all Together

Now that you've put names to all the minerals based on their properties, let's go back to Exercise 1 and give each group of minerals a name. Transfer your data from the first exercise, then use the mineral databases to name each group of minerals.

# of Samples in Group	Letter of samples that belong to this group	Properties that unite these samples and distinguish them from the others	Mineral Name
4			
3			
2			
2			
2			
1			

Penny for your thoughts?

Olivine, pyroxenes (ex: augite), amphiboles (ex: hornblende), feldspars, micas (biotite and muscovite), and quartz are among the most common rock forming minerals. To which mineral family do these minerals belong?

____

Laboratory Exercise III

Rock Identification

Introduction

Rocks are aggregates of minerals *in which the minerals retain their individual properties*. As such, unlike a cake where the ingredients are blended and mixed producing an end product with no resemblance to its original components, a rock is analogous to a jar of individual beads that have been bound together by physical and/or chemical processes. In this case, each bead retains its own individual color, shape, and characteristics and can be separated out from the others if desired.

Rocks are of critical importance to sustaining a biosphere as they compose the bulk of our planet, serve as an indispensable resource material in countless every day uses and house the minerals biology requires for sustenance. The cyclical movement of rocks—and therefore nutrients, resources, and energy—is accomplished by means of a conveyor-belt system known as the "rock cycle" (see Figure 3.1).

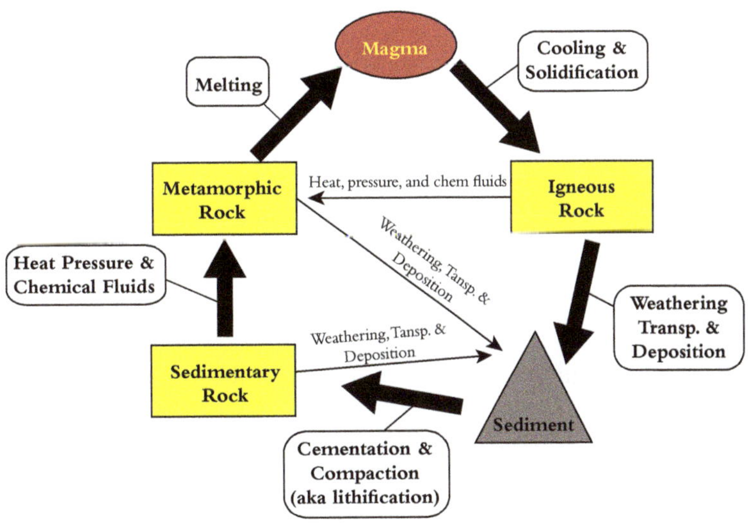

Figure 3.1 – Schematic Representation of Potential Rock Cycle Routes in a Convergent Margin Setting

In addition to supplying us with life-sustaining nutrients and resources, each of these three rock types—igneous, sedimentary, and metamorphic—serve as history books for our planet and tell an important story about the processes that formed them, hazards inherent in their journey and the availability and distribution of these resources.

Objective

The objective of this lab is fourfold:

1. Practice identifying minerals in rocks
2. Learn how to distinguish between igneous, sedimentary, and metamorphic rocks in hand sample.
3. Investigate where each type of rock is found in the landscape.
4. Understand how to use a cross-section to depict rock type and location.

Part A: Identifying Minerals in Rocks

As described above, there are three umbrella categories of rocks:

a. Igneous – Rocks formed from Magma

b. Sedimentary – Formed at Earth's surface from accumulating products of weathering

c. Metamorphic – Previous igneous, sedimentary, or metamorphic rocks that have been altered due to interaction with heat, pressure, and/or chemical fluids

Because rocks are aggregates of minerals in which the minerals retain their individual properties, you should be able to pick out individual minerals when looking at a rock. Samples I–XI are a combination of igneous, sedimentary, and metamorphic rocks. Your assignment is to identify the minerals that comprise each of these samples. All the minerals in the rock samples displayed here you saw in the minerals lab, so none of these minerals should be unfamiliar to you. To aid in this task, individual mineral samples of the minerals contained in each rock have been placed next to most of the rock samples. Complete this exercise by answering the questions below.

Igneous Rocks

The following two samples are igneous rocks. They are undeniably igneous because they are all composed of a homogeneous assortment of randomly oriented crystals that have no space between the grains (meaning they grew in situ and were not cemented together).

1. Examine rock sample I. This rock is composed of several of the most common continental-crust-forming minerals.
    - The gray, crystalline mineral in this rock is the same as the gray, crystalline mineral in the box beside this rock sample. What mineral is it? ______________________
    - The white, opaque mineral in this rock is the same as the white, opaque mineral in the box beside this rock sample. What mineral is it? ______________________
    - The dark, shiny mineral in this rock is the same as the dark, shiny mineral in the box beside this rock sample. What mineral is it?______________________
    - The silver, shiny mineral in this rock is the same as the silver, shiny mineral in the box beside this rock sample. What mineral is it? ______________________

2. Examine rock sample II. This rock is composed of four of the most common continental-crust-forming minerals, all of which were seen in the minerals lab, and several of which were seen in rock sample I. List the four minerals that make up this rock.

______________________________, ______________________________,

______________________________, and ______________________________.

3. While the rock samples in the previous questions are indisputably igneous for the reasons described, rock samples III.a and III.b are not so obvious, but are also igneous for reasons you will understand at the completion of the Igneous Rocks chapter. For now, just take our word for it that they are indeed igneous, and answer the following questions:

Examine rock samples III.a and III.b Notice the dark crystals in each of these samples. Close examination of these dark crystals will demonstrate that one of the samples has puddle-shaped crystals and the other has tabular crystals (almost like darts).

- Which rock sample has the puddle-shaped dark crystals? __________
- Which rock sample has the tabular shaped crystals? ______________
- On the table beside the rocks are two mineral samples. Which of these two samples are the dark crystals in rock III.a? Mineral X or Mineral Y?

  ____

- Which of these two samples are the dark crystals in rock III.b? Mineral X or Mineral Y?

  ____

- Both of these minerals are minerals you looked at in the minerals lab. Can you identify them here?
- Mineral X: ____________________; Mineral Y: ____________________

Sedimentary Rocks

Now let's have a look at some sedimentary rocks. Sedimentary rocks are formed either from physical weathering, followed by lithification, or chemical weathering, followed by precipitation.

4. Examine rock sample IV. This is a sandstone, created from physical weathering of pre-existing rock. Look closely at the particles that make up this rock and you should notice that it is primarily composed of pink sand grains with gray crystals interspersed. What two minerals compose this rock? ______________________________ and ______________________________.

5. Chemical precipitates are formed when water supersaturates with respect to a mineral and that mineral then precipitates out of the water. Examine rock sample V. Subject this rock to the same tests used to identify minerals in the previous lab.

- What mineral is it composed of? ______________________________
- What is the distinguishing property? ______________________________

6. Repeat the same procedure for rock sample VI.

   - What mineral is it composed of? ______________________

   - What are the distinguishing properties? ______________ and ______________

Metamorphic Rocks

The following rocks are termed metamorphic, meaning that they began as igneous or sedimentary rocks, but were altered as a result of exposure to temperature and pressure conditions different from the environment in which they were formed. Evidence of their original rock type and the metamorphic processes that altered them can be seen by examining their composition and texture respectively.

7. One type of metamorphism involves exposure to heat only. Examine rock sample VII. This rock was originally made up of individual grains, but due to exposure to high heat, the grains underwent "recrystallization," in which the atoms broke their bonds then rebonded, forming the original mineral but in an interlocking texture. The mineral this rock is composed of is sitting next to the rock sample. What mineral is this rock composed of? (Hint: You saw it in the mineral identification lab as well as in a sedimentary rock you looked at above.) ______________________ .

8. Another type of metamorphic change involves both heat and pressure. Heat provides the energy needed by the grains/crystals to reposition themselves into a path of least resistance when exposed to elevated pressure. Examine rock sample VIII. This rock began as an igneous granite with randomly oriented crystals. Notice how the grains have been realigned into a similar orientation. What four minerals are visible in these rocks? ________________, ________________, ________________, and ________________.

9. A third type of metamorphic change results in the crystallization of new minerals that are stable only under high temperature and pressure conditions. Examine rock sample IX. This rock contains two minerals that you looked at in last week's lab, one of which is a "metamorphic mineral," as it forms only under high temperature and pressure conditions. The other mineral is found in all three types of rocks. What two minerals are in this rock? ________________ and ________________.

10. Rock sample X has both the grain alignment seen in rock sample VIII, as well as a metamorphic mineral you saw in last week's lab. What is the silver, flaky mineral in this rock? ________________ What is the reddish-purple knob-like mineral in this rock? ______________________________

Part B: Distinguishing between Igneous, Sedimentary, and Metamorphic Rocks in Hand Sample

As discussed above, all rocks are composed of minerals. This being the case, how do you go about distinguishing one rock type from another? Well, just like each type of mineral has a unique set of physical properties that distinguishes it from all other minerals, each of these three rock types (igneous, sedimentary, and metamorphic) depict a unique set of physical properties that distinguishes it from the other two. Generally speaking, the two properties we look at to differentiate one rock type from another (and classify rock subtypes within each group) is *composition*—the mineral assemblage and additional components that make up the rock; and *texture*—the external appearance of the rock due to grain size and orientation.

How Do You Identify an Igneous Rock?

I. Compositional Feature: "Rock Ingredients"

Because igneous rocks are made from the cooling of magma, their composition will primarily include some combination of rock-forming minerals (olivine, pyroxene, amphibole, micas, feldspar, and quartz).

- Igneous rocks can be categorized into the following mineral assemblages:
  - Dominated by dark colored silicate minerals (see Figure 3.2a)
  - ~ 50% dark to 50% light colored silicate minerals (see Figure 3.2b)
  - Dominated by light colored silicate minerals (see Figure 3.2c)

Figure 3.2a – Primarily Dark

Figure 3.2b – Salt and Pepper

Figure 3.2c – Primarliy Light

II. Textural Feature: Grain Size

With the exception of igneous rocks made from volcanic glass (see Figure 3.3c), all igneous rocks will show crystallization, albeit not always visible to the naked eye. Essentially, the size of the crystals is a function of how fast the magma that produced them cooled. Rocks formed in a subsurface environment (intrusive rocks) result from slow cooling, and therefore have large, discernible crystals (coarse grains) that are visible in hand sample (see Figure 3.3a). Rocks formed at Earth's surface (extrusive rocks) develop from much faster cooling, and are therefore comprised of very tiny crystals (fine grained), which can't usually be seen without the aid of a microscope (see Figure 3.3b). *When crystals are visible in hand sample, they will display random orientation throughout the rock (see Figure 3.2a).*

Figure 3.3a – Coarse Grains
Note the "twinkling" as light reflects off the coarse crystal faces

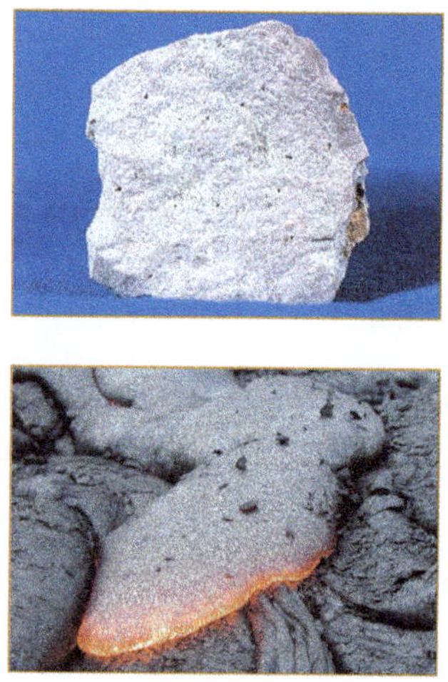

Figure 3.3b – Fine Grains

Figure 3.3c – No Grains

Helpful Hint

You can always confirm the presence of visible crystals by holding the rock up in the light and gently tipping it back and forth. If you can see twinkling, then the rock is composed of coarse crystals.

Note: Remember that not all igneous rocks have crystals that are visible in hand sample and there are several varieties of igneous rocks that have no crystals at all (see Figure 3.3c). In addition, many sedimentary and metamorphic rocks also show crystallization, but these have other textural and mineralogical features that igneous rocks do not display. That being said, the presence or absence of visible crystals cannot by itself be used as a diagnostic tool. However, if visible crystals are present, their orientation can be useful in distinguishing between the three rock types.

Figure 3.4 – Volcanic Glass

III. Textural Feature: Conchoidal Fracture

There is a select group of igneous rocks that are not comprised of crystals. These are rocks made of volcanic glass. Volcanic glass develops under two unique environmental conditions that will be discussed in future laboratory exercises, but the important thing to understand now is that both circumstances prevent crystal formation during magma cooling. Because mineral crystals do not develop, there is no organized arrangement of atoms, and the cooled product is therefore a glass. Identifying properties of volcanic glass include its smooth, shiny surface, sharp edges, dark color, and conchoidal fracture (see Figure 3.4).

Figure 3.5 – Vesicles

IV. Textural Feature: Vesicles

Vesicles are holes in a rock that develop due to exploding gas bubbles in rapidly decompressing lava. The agitated gas bubbles stir the lava up into a foam (similar to the fizz in a coke bottle). Because the foam cools instantaneously upon contact with the air, the holes created by the popping gas bubbles are preserved as vesicles (see Figure 3.5). Since vesicles result from the release of gas in magma, the presence of vesicles is a classic indicator that you are looking at an igneous rock.

How Do You Identify Sedimentary Rocks?

I. Compositional Feature: "Rock Ingredients"

Sedimentary rocks are produced at the surface from the accumulated products of weathering (aka sediment). As discussed earlier, the sediment is formed from the breakdown (weathering) of preexisting rock. There are two main categories of weathering processes, and therefore two types of sediment that are produced.

- Physical weathering processes such as jointing, plant, ice and salt wedging, and thermal shock cause preexisting rock to crumble into small particles/clasts, producing sediment that is a ground-down version of the original (see Figure 3.6a). Once this sediment is transported, deposited, and *lithified* (compacted and cemented), a *detrital sedimentary rock* is formed (see Figure 3.6b).
    - For lack of a better description, detrital sedimentary rocks can be classified as a hodge-podge of "scum on the surface of the Earth." These rocks are made by the gluing together of loose debris on the surface and can contain any or all of the following: mud, sand, rock fragments, shells, fossils, and plant debris.

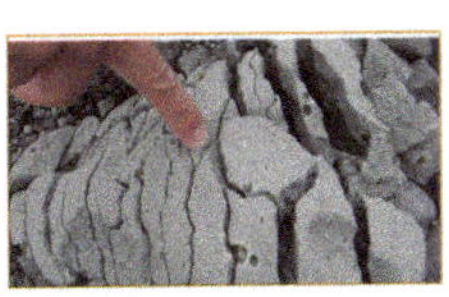

Figure 3.6a – Physical Weathering

Figure 3.6b – Detrital Sedimentary Rocks

- Chemical weathering, which is done through the work of water, destroys the rock on a chemical level by removing atoms from the rock's minerals, thereby dissolving away the rock (see Figure 3.7a). In this case, the atoms serve as sediment and are transported, deposited, and *precipitated* when the water becomes supersaturated, forming a chemical sedimentary rock.

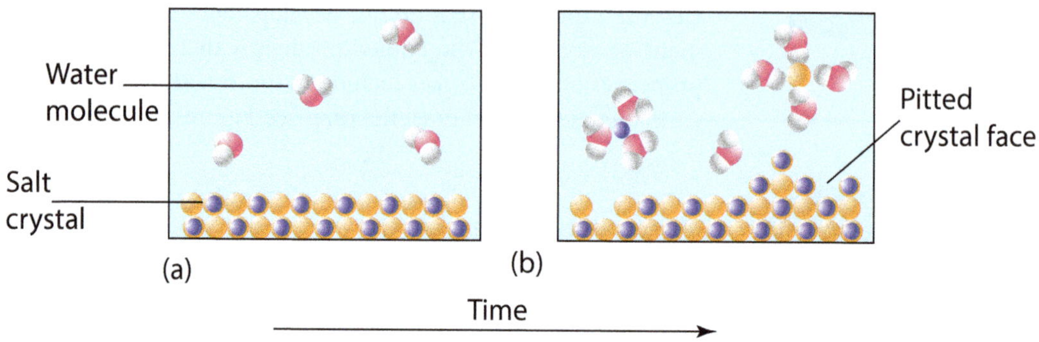

Figure 3.7a – Dissolution

- Rarely are conditions such that water will supersaturate with respect to two minerals at one time. This being the case, chemical sedimentary rocks are deposited one mineral at a time.
- Examples include rocks made out of gypsum (rock gypsum), halite (rock salt), and calcite (limestone). Remember from last week: There is a very quick way to identify all three of these minerals.

Figure 3.7b – Crystalline Limestone

Helpful Hint

Halite, calcite, and gypsum all crystallize but are NOT found in igneous rocks.

II. Textural Features:

- Lithified Rocks: Detrital sedimentary rock is created from the compaction and subsequent cementation of loose surface debris. In many rocks of this type, the "glue" cementing the grains together is visible to the naked eye or can be seen with aid of a hand lens (see Figure 3.8a). This differs significantly from igneous and metamorphic rocks, where there is no space between the crystals (see Figures 3.8b and 3.8c).

Figure 3.8a– Lithfield Rock; Cement between Grains

Figure 3.8b – Igneous Rock

Figure 3.8c – Metamorphic Rock

- Precipitated Rocks: Rocks precipitated from water are generally smoother than grainy detrital rocks and crystalline igneous and metamorphic rocks (see Figure 3.9).

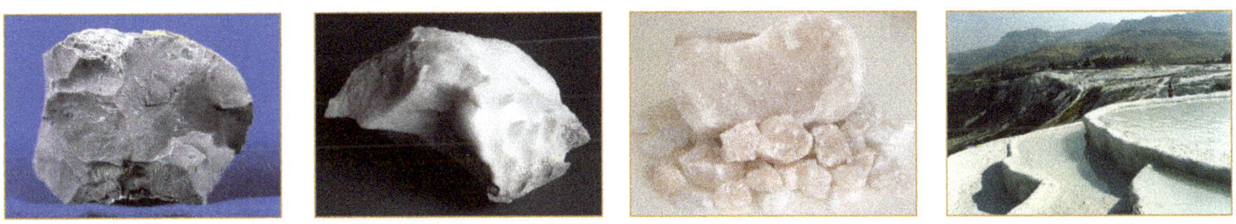

Chert | Rock Gypsum | Rock Salt | Crystalline Limestone

Figure 3.9 – Select Examples of Chemical Precipitates

How Do You Identify Metamorphic Rocks ?

I. Compositional Feature: "Rock Ingredients"

Metamorphic rocks are preexisting rocks that have been altered from exposure to heat and pressure different from that of their formation environment. This being the case, most minerals found in igneous and sedimentary rocks are also found in metamorphic rocks, albeit with a significantly different texture. While metamorphic rocks do contain many minerals found in the other two rock types, they are also home to a unique suite of minerals. These minerals develop as a result of the extra heat speeding up the vibration of the atoms in the existing mineral structure, thereby causing the bonds between the atoms to snap. The freed atoms then immediately seek out and bond with new partners in a stronger and more stable configuration. Select examples of such minerals include kyanite, chlorite, epidote, serpentine and garnet (see Figure 3.10a and 3.10b). The presence of these minerals and other metamorphic minerals not mentioned indicates the rock is clearly metamorphic.

Figure 3.10a and 3.10b – Metamorphic Minerals in metamorphic rocks

II. Textural Feature: Fused Together Grains

Rocks that have been exposed to abnormal amounts of heat, but not enough to induce melting, respond by undergoing a process known as recrystallization, where the atoms in individual grains temporarily break their bonds and then collectively rebond to each other as one huge interlocking mosaic. While the original mineral is still present, the "grains" or crystals are no longer distinct entities (see Figure 3.11a), but instead merge together in a texture referred to as *fused together grains* (see Figure 3.11b).

Figure 3.11a – Before Metamorphism.
Note each grain is individual and separate from the others.

Figure 3.11b – After Metamorphism.
Note how all the grains are now interconnected.

III. Textural Feature: Preferred Mineral Orientation

Although we normally associate rocks with rigidity and brittleness, those characteristics are only present at the surface where standard surface temperatures and pressures prevail. At the depths metamorphic rocks form, high heat permits the rocks to become flexible, more akin to chewing gum or silly putty, and the rocks can therefore bend, kink, curl, distort, and rotate mineral crystals. As such, rocks that were subjected to both high temperatures AND pressures will exhibit stress marks indicating the rock was squeezed while in a high temperature environment (see Figures 3.12–3.14).

Figure 3.12 – Slate

Figure 3.13 – Phyllite

Outside of a Schist

Inside of a Schist

Figures 3.14a and b

One of the most common causes of these unique textures is rotation of elongate and platy minerals in a position of least pressure. Because these minerals are thicker in one direction than in the other, the stress applied to each grain can be substantially reduced if the thinner part of the grain lies perpendicular to the direction of stress. (Imagine you are on a subway in a car packed to capacity. You are standing in the isle, crammed between 2 people, with one person on each side of you pushing into each shoulder. In order to relieve that stress, you will likely turn 90 degrees, so your thinner section is between the stress points and your longer section is unstressed). As such, what began as randomly oriented grains are now all aligned in the same direction, generating a pattern seen only in metamorphic rocks (Figures 3.15a and 3.15b).

Figure 3.15a –
Original igneous rock. *Note the randomly oriented crystals.

Figure 3.15b–
After metamorphism. *Note the preferred orientation of the platy minerals.

Exercise: Practice Identifying Igneous, Sedimentary, and Metamorphic rocks in hand sample

On the tables throughout the room is a mixed assortment of igneous, sedimentary, and metamorphic rock samples. Your job is to determine whether each sample is igneous, sedimentary, or metamorphic, then answer the questions that follow. Note that because we want to familiarize you with the basics of rock identification, we've refrained from putting out any fine-grained igneous rocks, as we want you to use properties that are visible to the naked eye to make your determination. When you approach each sample, use the decision tree on the following page to help guide you through the analysis and lead you to the correct conclusion.

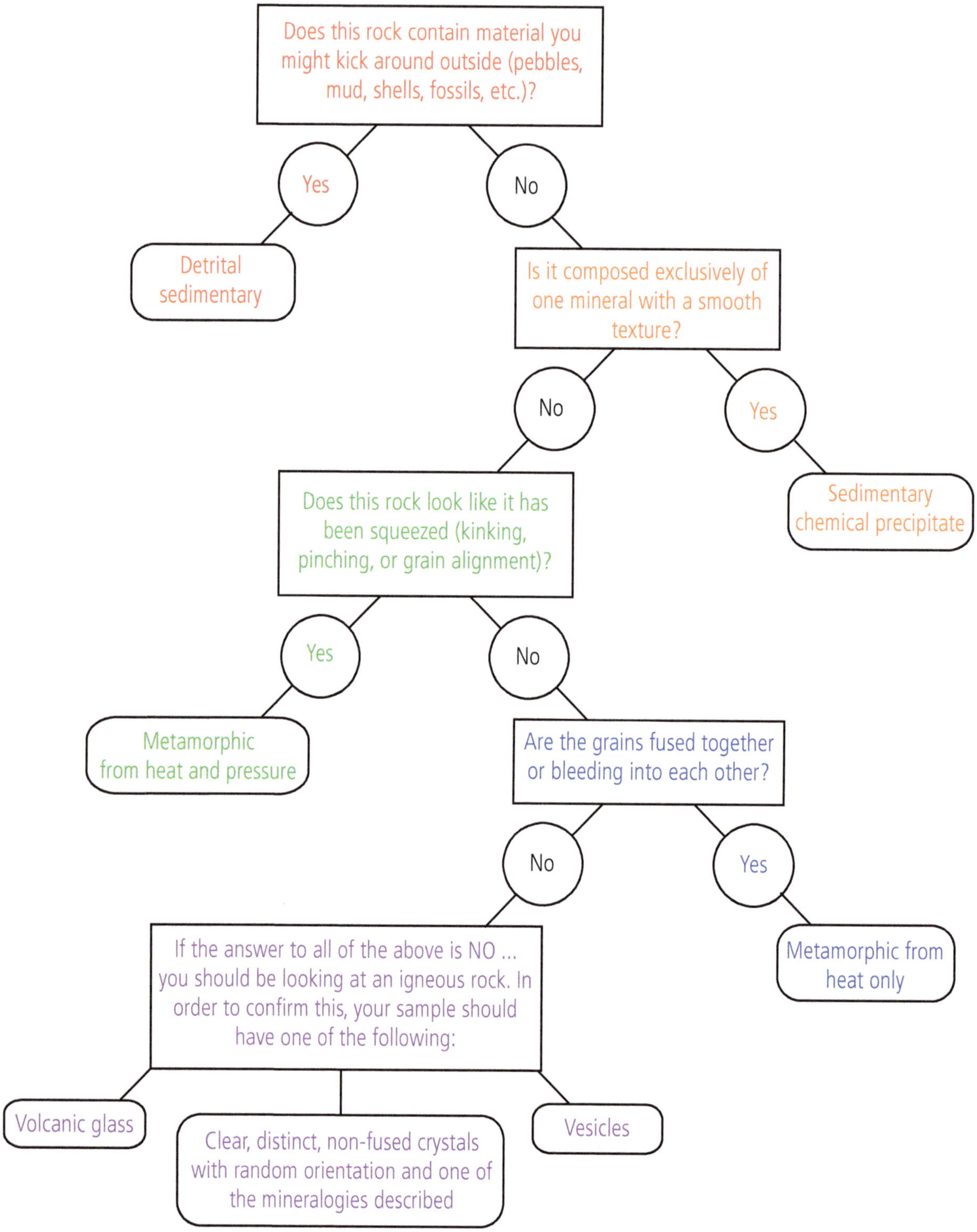
Does this rock contain material you might kick around outside (pebbles, mud, shells, fossils, etc.)?
Yes
No
Detrital sedimentary
Is it composed exclusively of one mineral with a smooth texture?
No
Yes
Sedimentary chemical precipitate
Does this rock look like it has been squeezed (kinking, pinching, or grain alignment)?
Yes
No
Metamorphic from heat and pressure
Are the grains fused together or bleeding into each other?
No
Yes
Metamorphic from heat only
If the answer to all of the above is NO ... you should be looking at an igneous rock. In order to confirm this, your sample should have one of the following:
Volcanic glass
Clear, distinct, non-fused crystals with random orientation and one of the mineralogies described
Vesicles

Sample Number	Rock Classification	How did the rock form?	Rock properties that led you to your conclusion
1		If Igneous: Intrusive or Extrusive	
		If Sedimentary: Detrital or Chemical	
		If Metamorphic: Heat Only or Heat & Pressure	
2		If Igneous: Intrusive or Extrusive	
		If Sedimentary: Detrital or Chemical	
		If Metamorphic: Heat Only or Heat & Pressure	
3		If Igneous: Intrusive or Extrusive	
		If Sedimentary: Detrital or Chemical	
		If Metamorphic: Heat Only or Heat & Pressure	
4		If Igneous: Intrusive or Extrusive	
		If Sedimentary: Detrital or Chemical	
		If Metamorphic: Heat Only or Heat & Pressure	
5		If Igneous: Intrusive or Extrusive	
		If Sedimentary: Detrital or Chemical	
		If Metamorphic: Heat Only or Heat & Pressure	
6		If Igneous: Intrusive or Extrusive	
		If Sedimentary: Detrital or Chemical	
		If Metamorphic: Heat Only or Heat & Pressure	
7		If Igneous: Intrusive or Extrusive	
		If Sedimentary: Detrital or Chemical	
		If Metamorphic: Heat Only or Heat & Pressure	
8		If Igneous: Intrusive or Extrusive	
		If Sedimentary: Detrital or Chemical	
		If Metamorphic: Heat Only or Heat & Pressure	
9		If Igneous: Intrusive or Extrusive	
		If Sedimentary: Detrital or Chemical	
		If Metamorphic: Heat Only or Heat & Pressure	
10		If Igneous: Intrusive or Extrusive	
		If Sedimentary: Detrital or Chemical	
		If Metamorphic: Heat Only or Heat & Pressure	

Sample Number	Rock Classification	How did the rock form?	Rock properties that led you to your conclusion
11		If Igneous: Intrusive or Extrusive If Sedimentary: Detrital or Chemical If Metamorphic: Heat Only or Heat & Pressure	
12		If Igneous: Intrusive or Extrusive If Sedimentary: Detrital or Chemical If Metamorphic: Heat Only or Heat & Pressure	
13		If Igneous: Intrusive or Extrusive If Sedimentary: Detrital or Chemical If Metamorphic: Heat Only or Heat & Pressure	
14		If Igneous: Intrusive or Extrusive If Sedimentary: Detrital or Chemical If Metamorphic: Heat Only or Heat & Pressure	
15		If Igneous: Intrusive or Extrusive If Sedimentary: Detrital or Chemical If Metamorphic: Heat Only or Heat & Pressure	
16		If Igneous: Intrusive or Extrusive If Sedimentary: Detrital or Chemical If Metamorphic: Heat Only or Heat & Pressure	
17		If Igneous: Intrusive or Extrusive If Sedimentary: Detrital or Chemical If Metamorphic: Heat Only or Heat & Pressure	
18		If Igneous: Intrusive or Extrusive If Sedimentary: Detrital or Chemical If Metamorphic: Heat Only or Heat & Pressure	
19		If Igneous: Intrusive or Extrusive If Sedimentary: Detrital or Chemical If Metamorphic: Heat Only or Heat & Pressure	
20		If Igneous: Intrusive or Extrusive If Sedimentary: Detrital or Chemical If Metamorphic: Heat Only or Heat & Pressure	

Part C: Depicting Rock Type and Location in the Landscape

Understanding how the various rock types are distributed throughout the landscape is a critical component in untangling a region's geologic history, forecasting natural hazards, and identifying potential locations for resource extraction. The map geologists use to depict the type and distribution of subsurface rock in a given area is known as a cross section and permits the viewer to see the position of and relationships between subsurface rocks in a given locale.

So what view are you getting? When you look at a cross section, what you are seeing is the subsurface geology by looking at the side of a slice of Earth. Imagine Earth as a layered cake covered in icing. As long as the cake is whole, all you see is the icing (that's the surface), but as soon as you cut a slice, lift it out and look at the side, you can see the cake is actually comprised of several different layers (that's the subsurface) See figure 3.16. As with the layered cake described above, a cross section allows you to visually depict the subsurface layers of Earth, as if looking at the side of a slice.

In addition to depicting the distribution of the three main categories of rock, cross sections also specify the subtype of rock within each category. Figure 3.18 is a legend that depicts select examples of how specific rocks are represented on a cross section, and Figure 3.17, which corresponds to the legend, is an example of a complete cross section (rock type and specific rocks subtype).

Figure 3.16 – Surface View vs. Cross-Section View

Cross sections are a type of geologic map, and just like any map, they too use symbols to mark the locations of depicted features. In the case of cross-sectional maps, the symbols represent the location and distribution of different rock types. Tables 3.A to 3.C introduce you to some of the most common features we see on cross sections and how they are pictorially represented.

Table 3.A – Igneous Features in Cross Section

Igneous Intrusions Note: On a cross section, all intrusions will be outlined with hatch marks as shown in the Metamorphic rocks table		
Massive Intrusion	Sill	Dike
 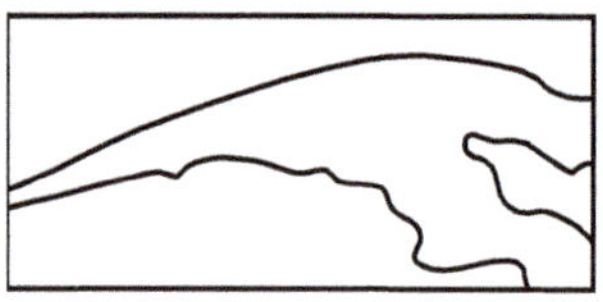	 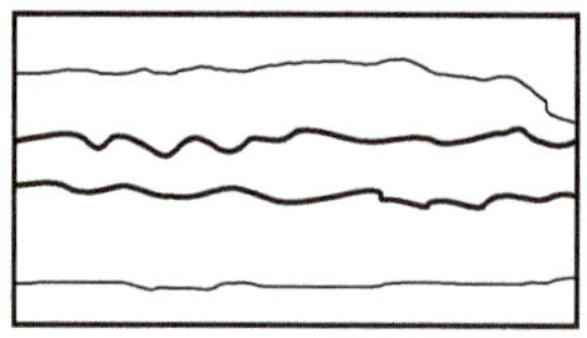	 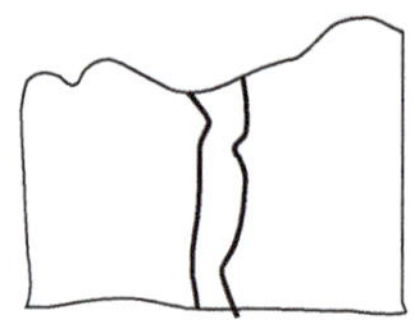
Large, globular plumes of magma	Magma body that squeezes between two parallel layers	Long, narrow intrusions that cut through existing layers at high or nearly vertical angles

Table 3.B – Sedimentary Features in Cross Section

Sedimentary Layers		
Horizontal Strata		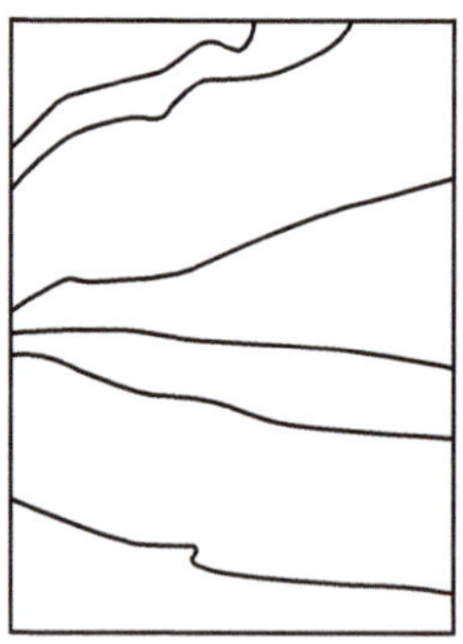
Tilted Strata		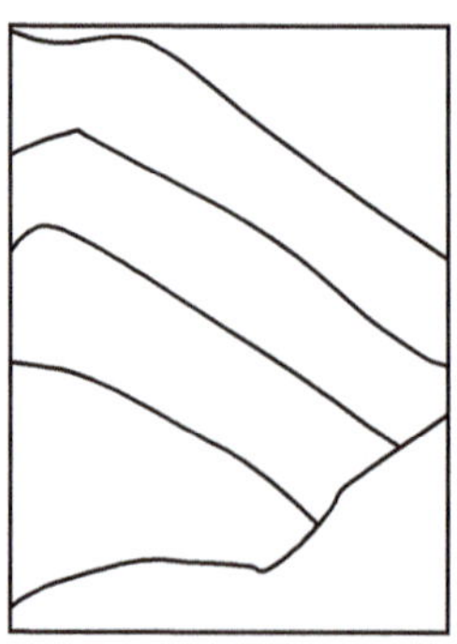
Because sedimentary rocks are deposited under the influence of gravity or precipitated from an even, evaporating water level, sedimentary rocks will be laid down in parallel layers		

Table 3.C – Metamorphic Features in Cross Section

Metamorphic Rocks	
Formed from Heat Only	Formed from Heat and Pressure
 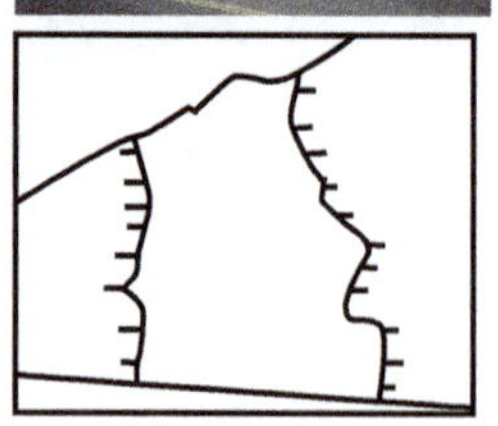	 
Limited area surrounding an igneous intrusion will be "cooked." These areas are delineated by "hatch marks"	Rocks that have distorted due to experiencing significant stress in a high temperature environment are delineated with a small wavy line

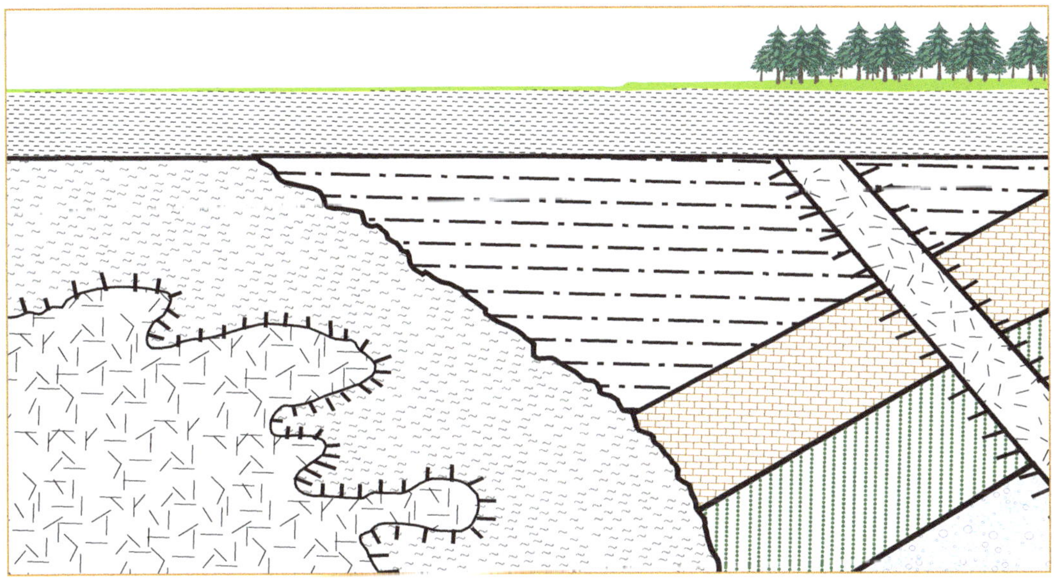

Figure 3.17 – Sample Cross Section

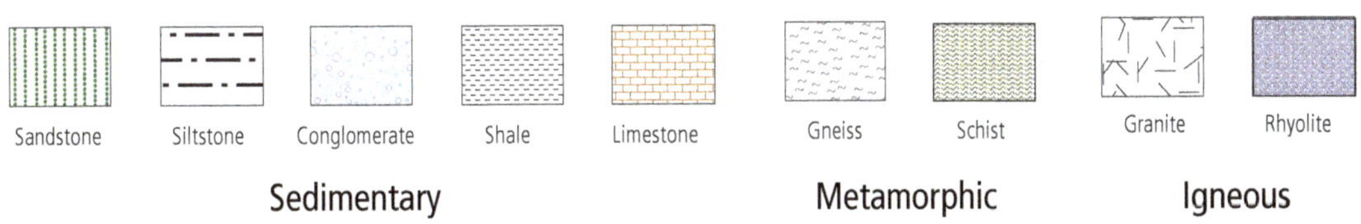

Figure 3.18 – Sample Cross Section Legend

Exercise: Determining rock type and distribution through use of a cross section.

For each of the letters listed below, use the cross section in Figure 3.19 in combination with the "hint" to determine whether each lettered feature is igneous (intrusive or extrusive), sedimentary (detrital or chemical), or metamorphic (heat only or heat and pressure). Next, use the "hint" and the legend associated with Figure 3.20 to help you specify the correct symbol. Finally, fill in the cross section on Figure 3.20 with the appropriate symbol for each feature. Note: Because you are not yet familiar with specific rock names (i.e. subtype), we are restricting this exercise to rock type only (i.e. igneous, sedimentary, or metamorphic).

1. Letters A and C - (Hint: Notice that A and C on the slopes of the volcano correspond with A and C in the air. What texture would you expect these rocks to have?)

   a. Are these rocks:

      - Igneous: Intrusive or Extrusive;
      - Sedimentary: Detrital or Chemical;
      - Metamorphic: Heat Only or Heat and Pressure.

   b. Fill in the cross section on Figure 3.20 with the appropriate symbol.

2. Letter B – (Hint: Unlike A and C, this rock was never in the air and as such, formed on the ground. As such, given the origin of all three, consider how B would differ from and A and C.)

   a. Are these rocks:

      - Igneous: Intrusive or Extrusive;
      - Sedimentary: Detrital or Chemical;
      - Metamorphic: Heat Only or Heat and Pressure.

   b. Fill in the cross section on Figure 3.20 with the appropriate symbol.

3. Letter D – (Hint: This layer was made from debris mass wasting off the eroding mountains.)

   a. Are these rocks:

      - Igneous: Intrusive or Extrusive;
      - Sedimentary: Detrital or Chemical;
      - Metamorphic: Heat Only or Heat and Pressure.

   b. Fill in the cross section on Figure 3.20 with the appropriate symbol.

4. Letter E – (Hint: This layer was precipitated from an evaporating sea.)

   a. Are these rocks

      - Igneous: Intrusive or Extrusive;
      - Sedimentary: Detrital or Chemical;
      - Metamorphic: Heat Only or Heat and Pressure.

   b. Fill in the cross section on Figure 3.20 with the appropriate symbol.

5. Letters F–J – (Hint: These rocks are all contorted for a reason—notice they are under a mountain range.)

   a. Are these rocks

      - Igneous: Intrusive or Extrusive;
      - Sedimentary: Detrital or Chemical;
      - Metamorphic: Heat Only or Heat and Pressure.

   b. Fill in the cross section on Figure 3.20 with the appropriate symbol.

6. Letters K and L – (Hint: Globular-shaped intrusions coming from internal Earth can only be one thing.)

   a. Are these rocks

      - Igneous: Intrusive or Extrusive;
      - Sedimentary: Detrital or Chemical;
      - Metamorphic: Heat Only or Heat and Pressure.

   b. Fill in the cross section on Figure 3.20 with the appropriate symbol.

7. Letters M–R – (Hint: How are these preexisting rocks being affected by K and L?)

   a. Are these rocks

      - Igneous: Intrusive or Extrusive;
      - Sedimentary: Detrital or Chemical;
      - Metamorphic: Heat Only or Heat and Pressure.

   b. Fill in the cross section on Figure 3.20 with the appropriate symbol.

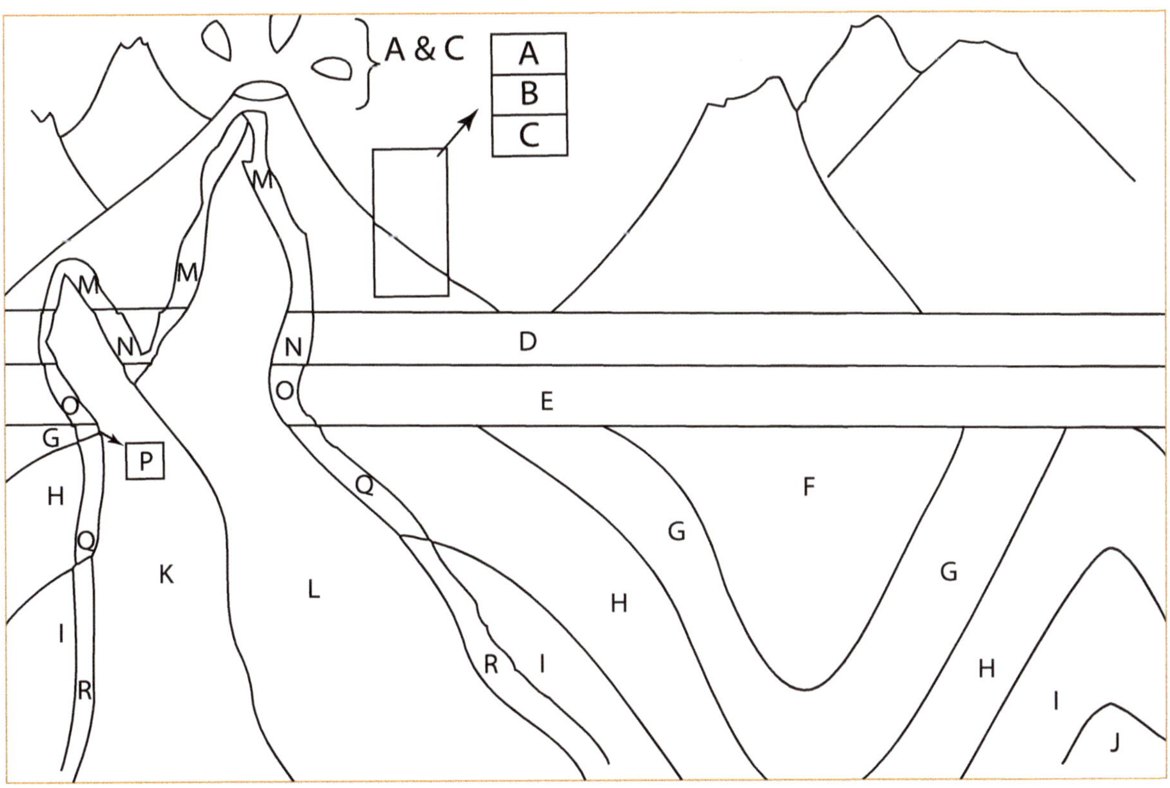

Figure 3.19

Figure 3.20

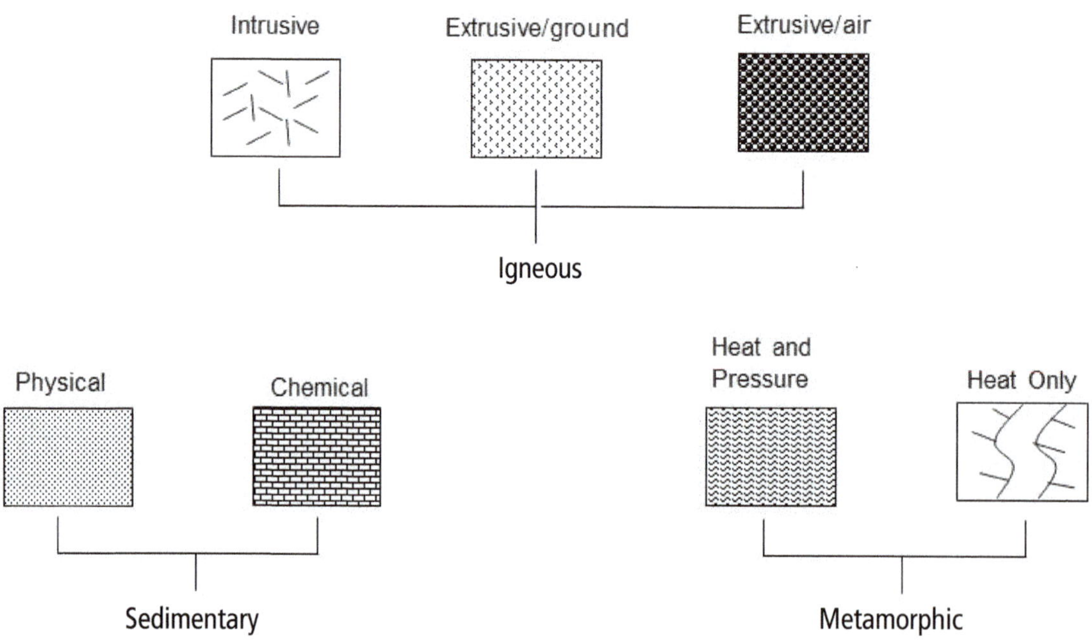

Legend for Figure 3.20

Part D: Tying it all Together

1. How are A and C textually different from B? ______________________________

2. What type of weathering is responsible for producing layer D? ______________ How about layer E? ______________

3. Which rock layers were exposed to both heat and pressure (list the letters)? ______________________________________ .

   Which rocks were exposed to heat only (list the letters)? ______________________

   On what did you base your conclusion? ________________________________

Laboratory Exercise IV

Igneous Rocks and the Volcanoes That Produce Them

Introduction

Igneous rocks are the most abundant rock type on Earth when considered by volume, as they comprise the core, mantle, and the oceanic and continental crusts. Igneous rocks originate as liquid melt (magma), which cools to form solid rock. As such, igneous material is brought to the surface by means of volcanoes. While volcanic activity brings life-giving nutrients, essential resources, and fresh crustal material to the surface—none of which we could live without—they also significantly endanger the local environment through a variety of geologic hazards. There is a common saying in geology: "The key to the present is the past." The igneous rocks in an area surrounding past and present volcanic activity are the history books for the region's volcanic behavior, and when read properly, give us critical insight into the formation of the landscape, the distribution of resources, and potential hazards the surrounding area is likely to face.

Objective

The objective of this lab is threefold:

1. Learn to evaluate igneous rock samples and interpret their history based on the rock's compositional and textural properties.
2. Use igneous rock properties to assess related volcanic hazards.
3. Examine the distribution of igneous rocks in the landscape through use of a cross section.

Part A: Igneous Rock Identification and the Story the Rocks Tell

Igneous, sedimentary, and metamorphic rocks can be identified using two primary properties: composition (chemistry) and texture (fabric). These properties were specifically chosen for identification purposes, as each one tells an important story about the processes that formed the rock. In the case of igneous rocks, composition and texture together give us insight into the region's volcanic origins, behavior, and distribution of volcanic deposits.

Composition

Igneous rock composition reveals the temperature of the source magma, as well as the depth at which the magma was generated. This is crucial information with respect to reconstructing an area's geologic history, as well as to assessing the explosiveness of the volcano.

The composition of an igneous rock is determined by identifying the minerals that make up the rock, which in turn is dictated by the chemical composition of the original melt from which the rock formed. The general composition of an igneous rock can be estimated simply by looking at the ratio of dark and light minerals in the sample (see Figure 4.1). (*Note: "Dark" denotes jet black or forest green. Shades of gray, silver, and brown are considered light.*)

- **Composition/Color Index**
    - **a. Felsic/granitic:** Rock composed mainly of light-colored minerals like quartz and feldspar (dark minerals < 15 %) (e.g., rhyolite, granite)
    - **b. Intermediate/andesitic:** Mixture of light- and dark-colored minerals like quartz, feldspar, amphibole, and biotite (dark minerals ~15%–45%) (e.g., andesite, diorite)
    - **c. Mafic/basaltic:** Dark-colored minerals like olivine, pyroxene, and amphibole account for ~45%–85% of the rock. Plagioclase feldspar makes up most of the rest of the rock (e.g., basalt, gabbro)
    - **d. Ultramafic:** Composed almost exclusively of dark-colored minerals like olivine and pyroxene (most commonly peridotite, which is found in the Earth's mantle)

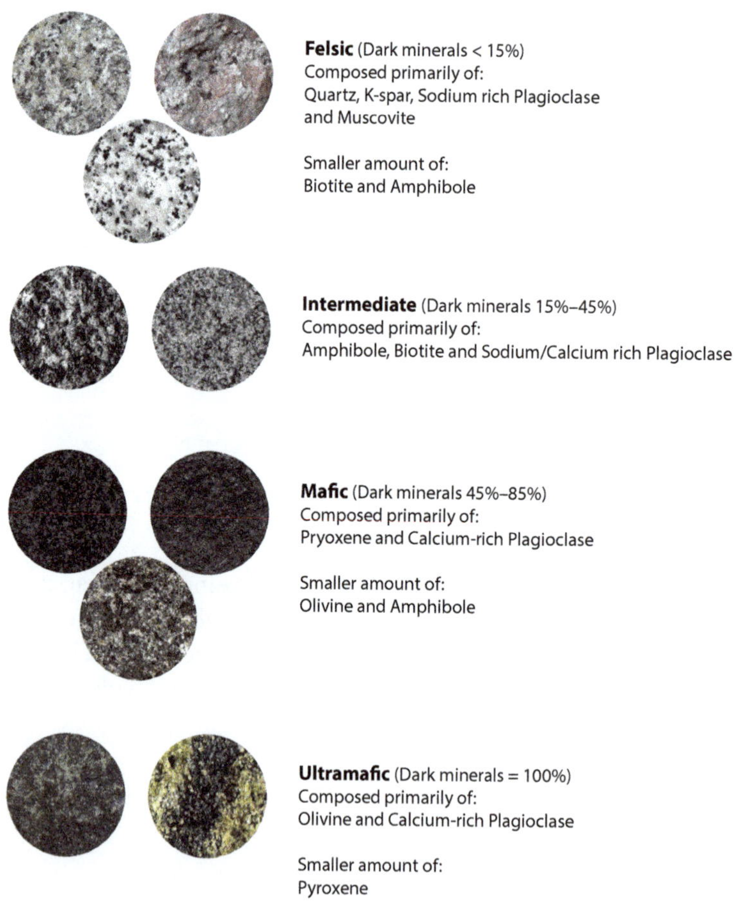

Figure 4.1 – Mineralogical Compositions of Igneous Rocks

What Story does Composition Tell?

The type of eruption that takes place and therefore the various hazards the surrounding area will face, is dictated primarily by the composition of the magma—specifically, the silica (Si) content. Si is the control on the magma's viscosity, i.e., resistance to flow. The more Si the magma has, the higher the magma viscosity and the more difficult it becomes for gas bubbles to escape the magma on its ascent to the surface. Thus, when magma contains a high concentration of Si, the gasses are forced to burst out of the magma, generating explosive volcanic activity. Exacerbating this effect is the fact that high Si magmas inherently contain significantly higher concentrations of dissolved gasses than lower Si ones and are generally lower temperature, which further reduces the fluidity of the magma. These additional factors substantially intensify the explosive nature of intermediate and felsic eruptions. On the flip side, mafic magmas, which contain lower Si concentrations, have little threat of explosion, but the high temperature, fluid lava can travel for miles resulting in significant damage to property. See Figure 4.2 for visual representation of this concept.

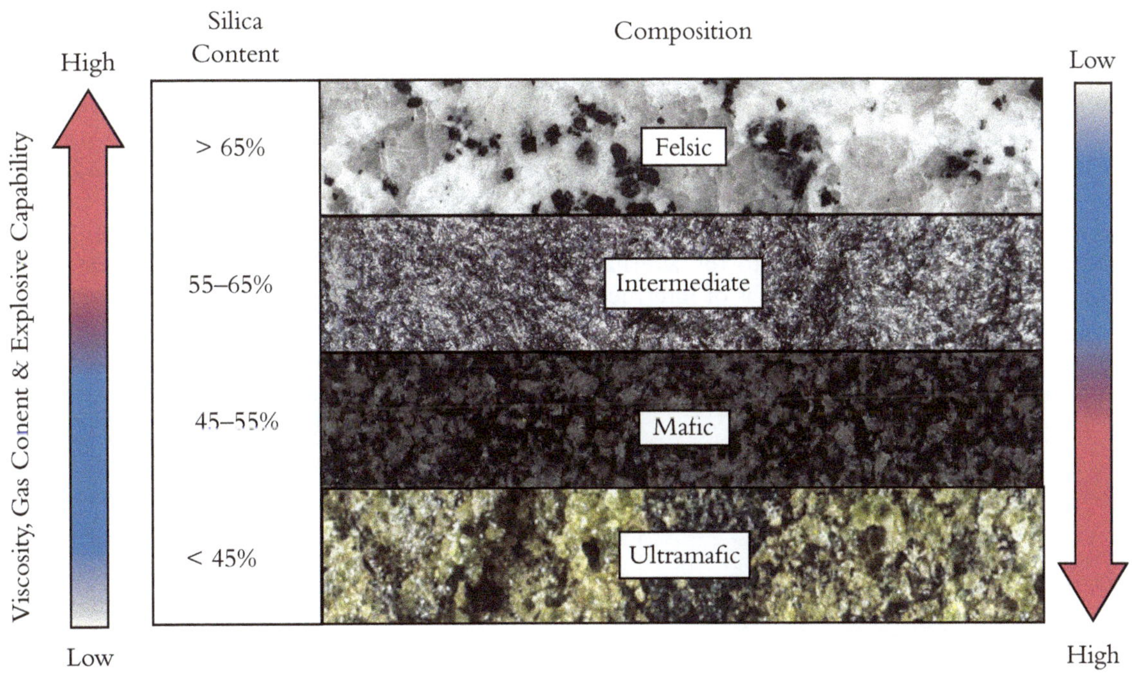

Figure 4.2 – Composition: The Story it Tells

How Does Magma Composition Relate to Eruption Style? (See table 4.A)

Volcanoes are vents in Earth's crust through which subsurface magma ascends and exits at the surface. Over the course of multiple eruptions, volcanic products (lava, ash, or both) accumulate around the vent, generating a volcanic cone that reflects the nature of the magma that produced it.

Mafic magmas, which have been brought up from the mantle and asthenosphere, have a relatively low Si concentration when compared to intermediate and felsic magmas. The lower Si content, in combination with the high temperature of the melt (the deeper the magma formation, the higher the temperature), generates low viscosity magmas that flow with ease and release gas bubbles with very little resistance. Because the primary extrusive product is low viscosity lava, the cone is built up from layer upon layer of thin lava flows, which create a wide, mildly sloping cone that mimics the shape of a warriors shield. Thus, volcanoes that erupt mafic magma are known as "Shield Volcanoes." Shield Volcano eruptions are considered nonexplosive and have little threat of hazards such as ash falls, pyroclastic flows, lateral blasts, or lahars. They do, however, cause significant problems with lava flows, as the fluid melt can travel long distances.

Felsic magmas—the result of shallow crustal melting—contain a high concentrations of Si. The elevated Si, in combination with the low magma temperature, produces a highly viscous magma, which traps gas bubbles and explodes violently upon depressurizing. In many cases, the lava is too viscous to flow any sizeable distance, making ash and other ejecta the primary extruded products. Thus, these volcanic cones are created from steep piles of consolidated ash that tightly hug the vent, creating the shape of a dome. As such, volcanoes that erupt felsic magma are known as "Volcanic Dome Volcanoes." Volcanic Dome eruptions are accompanied by ash falls, pyroclastic flows, lateral blasts, and lahars, yet lava flows are not an issue, as the lava is too viscous to travel any significant distance.

Intermediate magmas, the result of a combined contribution from mantle and crustal melting, produce magmas with a Si concentration and temperature between that of mafic and felsic magmas. As such, both lava and ash are extruded in comparable amounts, generating a steep cone built from alternating layers of viscous lava and steep ash piles. Thus, volcanoes that erupt intermediate magma are called "composite" or 'strato-volcanoes"—in reference to the alternating material that make up their cones. Because the lava erupted at composite volcanoes is of intermediate viscosity, it can and does flow a limited distance from the vent. Therefore, these volcanoes are associated with both lava flow problems, as well as violent, explosive eruptions, which produce all the hazards seen in felsic eruptions but to a lesser degree.

Type	Shape	Si content	Magma Viscosity	Potential Eruption Products
Shield/Mafic	"Warriors Shield" 	~50%	Low (fluid)	Lava flows, (ejecta that are kept extremely close to the vent)
Strato-Volcano/ Intermediate	"Conical" 	~60%	Intermediate	Lava flows, gases, pyroclastic flows, ash falls, debris avalanches, lateral blasts, lahars (late stage highly siliceous lava sometimes produces a small lava dome)
Volcanic Dome/ Felsic	"Volcanic Dome" 	~70%–75%	High (Sticky)	Lava dome, gases, pyroclastic flows, ash falls, debris avalanches, lateral blasts, lahars

Table 4.A – Volcanoes and their Behavior

See Table 4.B for a detailed description of select volcanic hazards associated with the three magma compositions and their respective eruption styles.

Table 4.B – Volcanic Products and the Hazards They Produce

	Products of Volcanic Eruptions	Visual Depiction	Creation process	Affected Area	Hazards Produced
LAVA ACTIVITY	**Lava Flows**	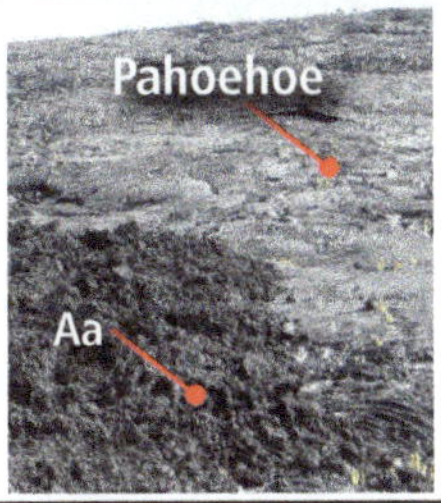	Non-explosive lava eruption from the vent.	Moves from high ground to low ground and can travel up to 10 km from the vent.	Unless on a steep slope, most people can easily outrun them. However, they can have a devastating effect on property, crops, and the landscape.
PYROCLASTIC ACTIVITY	**Ash Fall**		Bits of pulverized rock torn from the vent and glass created from instantaneous cooling of lava drops on contact with air. Particles < 2 millimeters (0.1 in) in diameter.	Covers hundreds and thousands of square km downwind from the source with volcanic ash.	Short-term destruction of vegetation and ecosystem, structural damage from increased load on roofs, respiratory and eye problems, "flame-out" in aircraft engines.
PYROCLASTIC ACTIVITY	**Pyroclastic Flow**		Fast-moving current of hot gas and rock, created from destabilization of large masses of highly viscous lava. Can reach temperatures up to 2,000°F.	Travels at speeds up to 700 km/h (450 mph). downhill for several kilometers and may affect areas as far as 40km downslope.	Moving boulders will flatten trees and buildings and the hot gases will vaporize and set on fire much of the material in their paths.
PYROCLASTIC ACTIVITY	**Volcanic Bomb**		Mid-flight solidification of lava into pieces larger than 65 mm (2.5 inches) in diameter.	Bombs can be thrown many kilometers from an erupting vent.	Volcanic bombs can cause severe injuries and death, as well as significant damage to property on impact.
PYROCLASTIC ACTIVITY	**Lateral Blast**		Eruption that takes place on the flanks of a volcano instead of at the summit. Debris ejected at speeds of several hundred km/hr.	Distribution depends on direction of the blast and distance traveled depends on the strength of the explosion.	Explosive force can scorch and knock over upright structures (trees and buildings) several miles from the blast. Blocks/bombs can be thrown to distances of at least 10 km and pyroclastic flows, which can travel at high speed to distances of more than 30 km, can be produced.

	Products of Volcanic Eruptions	Visual Depiction	Creation process	Affected Area	Hazards Produced
LANDSLIDES	Debris Avalanche		Chaotic movement of rocks, soil, ash, vegetation, water, and/or ice on a volcanic slope. Can occur with *or* without an eruption.	Debris runs down the slope of the volcano and may extend as far as 45 km.	Rapidly moving mass of debris can flatten most obstacles in its path, as well as block roads and dam up rivers.
LANDSLIDES	Lahar (mudflow)		Water, produced during eruption by lava melting an overlying glacier or in the absence of eruption by excess precipitation, mixes with pyroclastic debris generating a rapid flow that can reach depths of >100 feet and move at speeds of 100 km/hr.	Usually travel down valleys and can travel long distances—some have been known to travel hundreds of km from their source.	Can bury and destroy entire cities by covering cars, man-made structures (roads, bridges, and houses), and leaving tons of heavy debris in the trees.
ATMOSPHERIC HAZARDS	Volcanic Gases		Explosive and non-explosive eruptions produce hot, toxic gases such as CO_2, CO, SO_2, H_2S, and HF, among other dangerous compounds.	Carried downwind at speeds of up to 10s of km/hr, depending on the strength of the wind.	Causes death by asphyxiation, damage via acidic corrosion, and water and soil contamination. Produces "Vog"—thick acidic haze made from a combination of volcanic gases, H_20, and oxygen. Has been known to cause breathing problems, headaches, sore throats, watery eyes and flu-like symptoms .
HYDROLOGIC HAZARDS	Floods		Among other triggers, flooding can be caused by sudden introduction of excess water into a region or existing river from melting ice, by deposition of debris in a river that results in a substantial water-level increase, or a natural dam produced by debris blocking the normal river path.	Flooding may extend for 100s of km and is confined to flat areas.	Primary effects (those directly caused by the flood): injury, loss of life, damage caused by swift currents, debris, and sediment to farms, homes, buildings, roads, bridges, and communication systems, and erosion and deposition of sediment in urban and rural landscapes may also result in loss of soil and vegetation.

What Story Does Texture Tell Us?

The texture of an igneous rock conveys where and how the magma/lava cooled and will vary depending on whether the rock formed above ground (**extrusive/volcanic**), and thus rapidly, or below the ground (**intrusive/plutonic**), and thus slowly. More specifically, the texture divulges whether the magma cooled underground, on the surface, in the air, or in the water. Furthermore, the texture is a key indicator as to whether the volcano produces surface hazards, and if yes, what type and to what extent. Since magma from an eruption can remain below the ground to cool, as well as be brought to the surface to cool, any composition of magma can have at least two distinct textures.

Figure 4.3 – Phaneritic Texture

Phaneritic (Coarse-grained)—see Figure 4.3

Description: Visible grains; majority of crystals uniform size and can be seen with a naked eye.

Formation: Slow, subsurface cooling—the result of magma that does not make it to or exit at the surface

Figure 4.4 – Aphanitic Texture

Aphanitic (Fine-grained)—see Figure 4.4

Description: Crystals are very tiny and require a microscope to be seen.

Formation: Rapid, surface cooling—the result of a cooling lava flow.

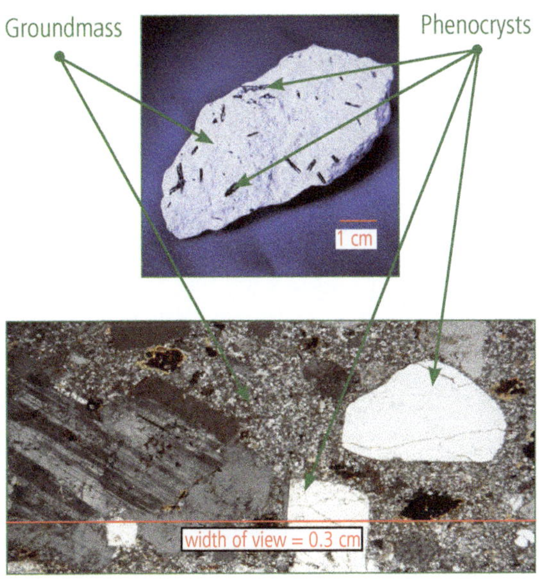

Top: Figure 4.5a – Porphyritic Hand Sample
Bottom: Figure 4.5b – Microscope thin section image

Porphyritic—see Figures 4.5a and 4.5b

Description: Two contrasting grain sizes in the rock (fine- and coarse-grained).

Formation: Slow cooling followed by a period of faster cooling. The result of magma that begins cooling beneath the surface, then is extruded at the surface in the form of a lava flow at a later time.

- Phenocrysts: Coarse crystals
- Groundmass or matrix: Fine crystals

Figure 4.6 – Pegmatitic

Pegmatitic—see Figure 4.6

Description: Abnormally large crystals; grains range in size from several cm up to 10m in length.

Formation: At the end stages of magma crystallization, the remaining melt contains a lot of water, which permits rapid movement and bonding of the atoms, and therefore an accelerated rate of crystal growth.

Note: Pegmatites are extremely unusual rocks, particularly because they are comprised of abnormally large grains that are actually the result of fast cooling. This is the one case where significant crystal growth occurs in a relatively short period of time. In addition, pegmatites are also home to a variety of "odd" elements known as "rare earth metals." Rare earth metals do not comfortably fit into most silicate minerals and therefore get excluded from most silicate mineral structures, which forces them to collect in the remaining melt. As such, pegmatites contain high concentrations of rare earth metals, which are essential to industry, particularly the electronics field.

Figure 4.7 – Glassy

Glassy—see Figure 4.7

Description: No mineral crystals present – smooth, glassy texture.

Formation: Instantaneous cooling in air or water. Result from underwater eruptions or felsic magma of such high viscosity, atoms cannot meet and bond in an orderly manner.

Vesicular—See Figures 4.8a–4.8c

Description: Contains small holes or cavities which form due to gas explosion in the lava as it rises and depressurizes. Typically porous and low density, it may resemble a sponge.

Lapilli Formation: Sprays of foamy lava cool instantaneously on contact with the air. These result from spurts of agitated lava being projected into the atmosphere. Can be mafic (there are a variety of basaltic vesicular lapilli) or felsic (pumice). See Figures 4.8a and 4.8b.

Scoria Formation: Top layer of a cooling basaltic aa lava flow degasses—generating the higher density scoria. See Figure 4.8c.

Figure 4.8a – Felsic and Mafic Lapilli

Figure 4.8b – Pumice

Figure 4.8c – Scoria

Fragmental—See Figures 4.9a–4.9c

Description: Broken, angular fragments of rocky materials welded together.

Formation: Heat during a volcanic eruption welds loose rock fragments together. Heat can be supplied from lava flows, ash falls, and/or pyroclastic flows.

Figure 4.9a – Volcanic Tuff
Comprised primarily of volcanic ash (particles <2 mm in diameter)

Figure 4.9b – Volcanic Breccia
Composed of large fragments embedded in a fine-grained matrix

Figure 4.9c – Volcanic Bomb
Blob of magma ejected into the air that cooled while in midflight.

How Do We Name Igneous Rocks?

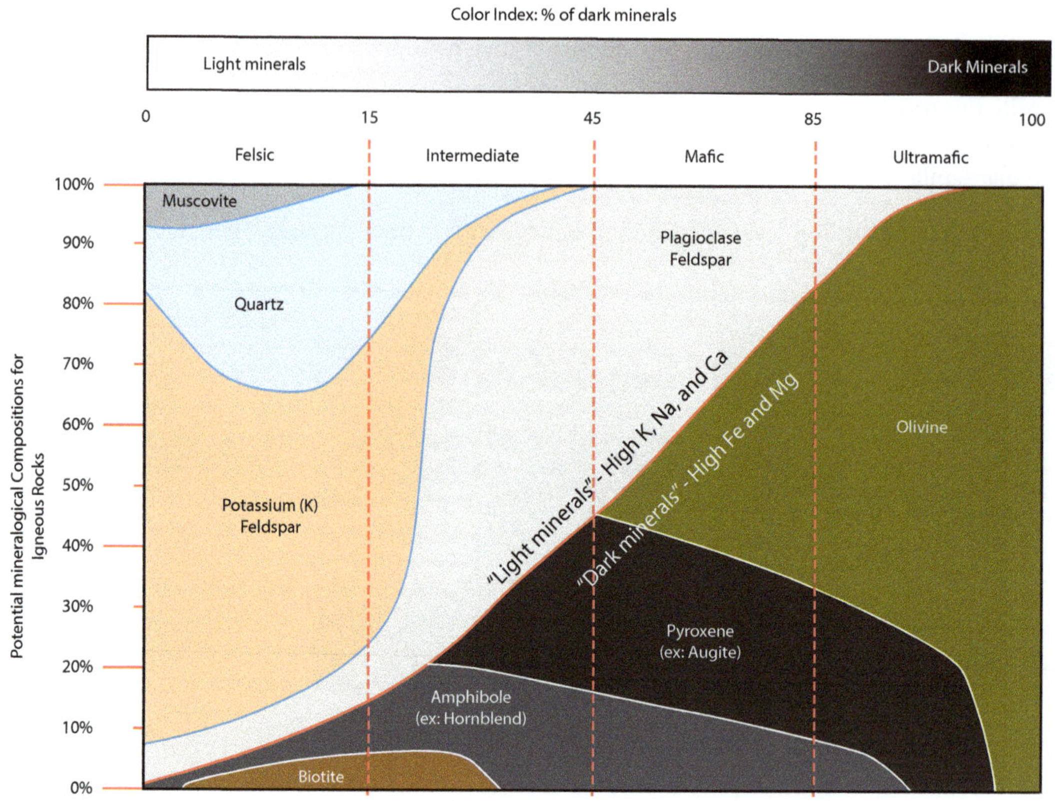

	Texture	Rock Name			
Intrusive (plutonic)	Pegmatitic 	**Pegmatitic Granite**	**Pegmatitic Diorite**	**Pegmatitic Gabbro**	
Intrusive (plutonic)	Phaneritic 	**Granite**	**Diorite**	**Gabbro**	**Peridotite**
Intrusive (plutonic) / Extrusive (Volcanic)	Porphyritic 	**Porphyritic Rhyolite**	**Porphyritic Andesite**	**Porphyritic Basalt**	
Extrusive (Volcanic)	Aphanitic 	**Rhyolite**	**Andesite**	**Basalt**	**Komatiite (rare)**
Extrusive (Volcanic)	Glassy 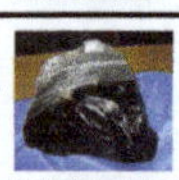	← **Obsidian** →			
Extrusive (Volcanic)	Vesicular 	**Pumice**		**Basaltic Lapilli/Scoria**	
Extrusive (Volcanic)	Fragmental 	**Volcanic Tuff/Breccia/Bomb**			

Figure 4.10 – Texture + Composition = Rock Name

Inherent in the rock name is the answer to five crucial rock formation questions:

FROM COMPOSITIONAL PROPERTIES – *Tells us where the magma was generated*

1. What was the Si content and viscosity of the magma source? (High, medium, or low)
2. From what depth did the magma originate? (Deep, shallow, mixture of both)
3. What was the temperature of the magma source? (High, medium, or low)

FROM TEXTURAL PROPERTIES - *Tells us where the magma cooled*

4. At what speed did the magma cool? (slow, fast, or instantaneous)
5. Where did the magma cool? (Subsurface, surface, air, or water)

Using the information contained in the igneous rocks, we can reconstruct the volcanic history of a region and evaluate how it has evolved over time, identify and map out the distribution of volcanic resources, and forecast eruption type so that appropriate precautionary measures can be taken to safeguard against volcanic hazards.

Tying It All Together

While we've discussed how igneous rocks can be used to understand the volcanic processes in a given area (see Figure 4.11), what we have not addressed is why magma composition—and thereby eruption style and products—differs from one locality to another. In a nutshell, the answer to that question is the Rock Cycle, which is driven by plate tectonics.

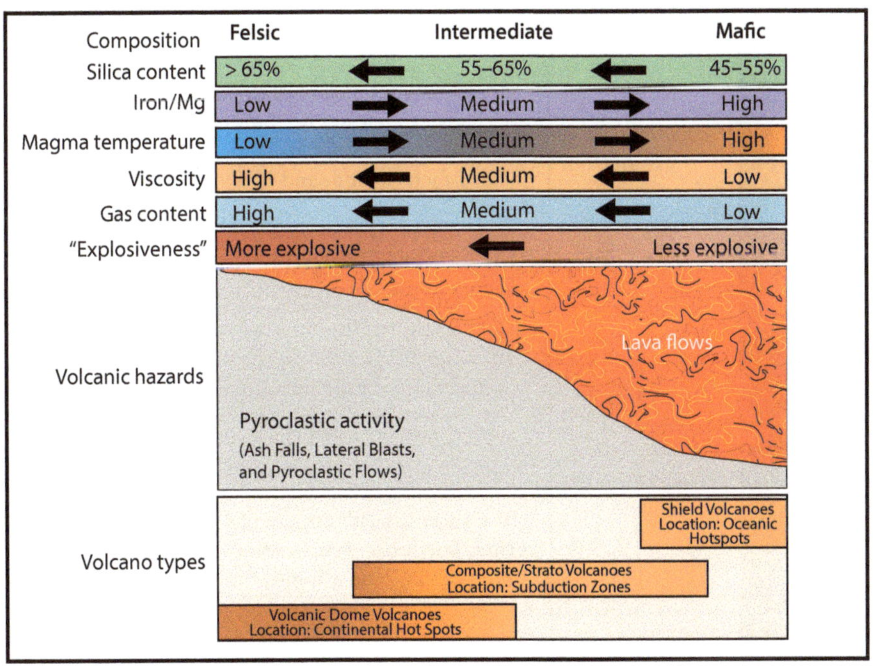

Figure 4.11 – Tying it All Together

Throughout the rock cycle, there are three places where fresh magma rises to the surface: 1) In the middle of the ocean (spreading centers)—where two plates are pulling apart; 2) at subduction zones—deep oceanic trenches where the plates have collided; and 3) hot spots—localized volcanic areas that are not on plate boundaries. Refer to Figure 4.12 and corresponding Table 4.C on the following page to gain a better understanding of the way different magma compositions are produced at these various parts of the rock cycle. Then refer back to Figure 4.11 to get an overall understanding of why volcanoes in these different locations behave the way they do.

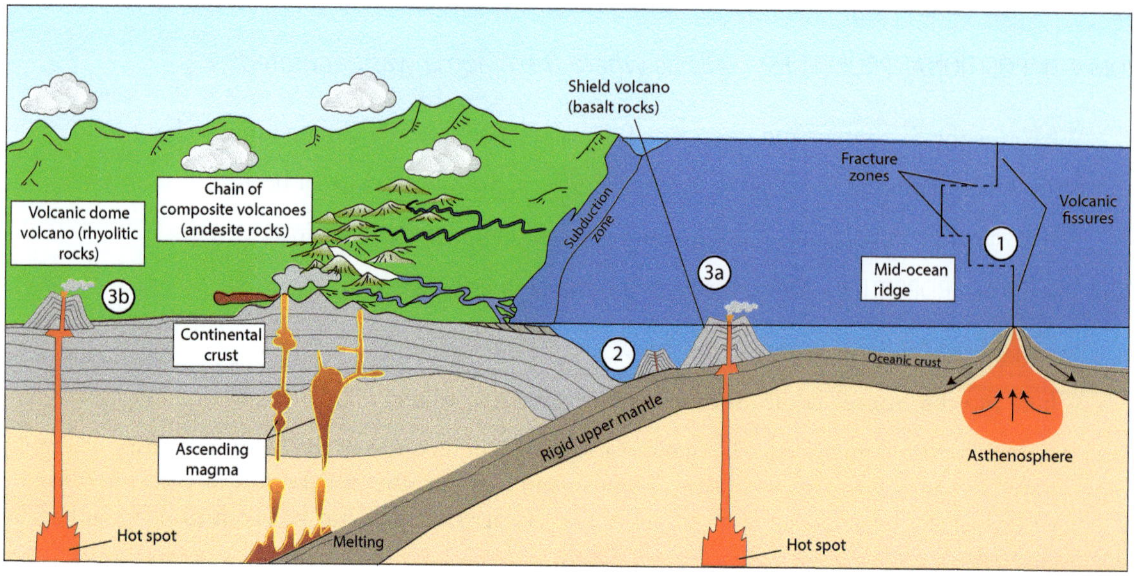

Figure 4.12 – Volcanic Settings and Magma Composition

<table>
<tr><th></th><th>Location of Volcanism</th><th colspan="2">Cause of Magma Generation</th><th>Magma Composition</th><th>Volcano Type</th></tr>
<tr><td>Spreading Centers

#1 in Fig. 4.12</td><td>Middle of Ocean Basins

Where two plates pull apart</td><td colspan="2">Opening of large fissures in the middle of the ocean releases the pressure on deeper, very hot, ultramafic rock, which then expands and rises

Rising rock undergoes DECOMPRESSION MELTING—leads to partial melting of the ultramafic mantle rock</td><td>Mafic</td><td>Fissure eruption</td></tr>
<tr><td>Subduction Zone

#2 in Fig. 4.12</td><td>Over Oceanic Trenches

Where two plates collide</td><td colspan="2">Water is released from the subducting slab, moves up, and lowers the melting temperature of the upper mantle.

Partial WET MELTING of the ultramafic mantle = mafic composition.

Insulation beneath continental crust permits fractional crystallization, magma mixing, and assimilation.</td><td>Range from borderline mafic-intermediate to borderline intermediate-felsic

Constantly evolving</td><td>Composite</td></tr>
<tr><td>Oceanic Hot Spot

#3a in Fig. 4.12</td><td>Mid Ocean above the Hot Spot</td><td rowspan="2">Hot, ultramafic rock rises due to thermal anomaly. Generates partial melting of rising plume.</td><td>Breaks through oceanic crust and forms volcanic island.</td><td>Mafic</td><td>Shield</td></tr>
<tr><td>Continental Hot Spot

#3b in Fig. 4.12</td><td>On Continent above the Hot Spot</td><td>Stall beneath continental crust and partially melts intermediate rock.</td><td>Felsic</td><td>Volcanic Dome</td></tr>
</table>

Table 4.C – Volcanic Settings and Magma Composition

Exercise: Igneous Rock Identification and Analysis

This exercise is designed to help you learn to evaluate igneous rock samples and interpret their history based on the rock's properties. As you approach each sample, use the rock ID chart on page 66 to identify the rock, then answer the questions that follow.

In the interest of time, please answer all the questions that DO NOT have an asterisk (★) next to them first, as you must have the sample to be able to answer these. When you have completed all questions that require samples, go back and answer all those that have an asterisk next to them.

Samples 1a–1e

- Notice these five samples are all coarse-grained and felsic. What is the name of these rocks? ____________________
- List the minerals that are in each sample. All the blanks by each sample number listed below should be filled in.

  1a.____________________, ____________________, ____________________

  1b.____________________, ____________________

  1c.__________, ____________________, ____________________

  1d.__________, __________, __________, __________, __________

  1e. __________, __________, __________, __________, __________, __________

- ★Were these rocks formed intrusively, at the surface, intrusively/surface, or in the air? ____________________
- ★How can the varieties of this rock appear so drastically different, yet still be the same rock? ____
____
____

Samples 2a–2c

- These three rocks all have a felsic composition, but a different texture than the rocks in sample 1. What is the name of these rocks? ____

Sample 2a

- What is the texture of this rock (coarse-grained, fine-grained, porphyritic, glassy, vesicular, or fragmental)? ____________________
- What two minerals are in this rock? ________________ and ________________
- ★Was it formed intrusively, at the surface, intrusively/surface, or in the air? ____________________

Sample 2b

- What is the texture of this rock (coarse-grained, fine-grained, porphyritic, glassy, vesicular, or fragmental)? ____________________
- What mineral is in this rock? ____________________
- ★Was it formed intrusively, at the surface, intrusively/surface, or in the air? ____________________

Sample 2c

- What is the texture of this rock (coarse-grained, fine-grained, porphyritic, glassy, vesicular, or fragmental)? ____________________
- What minerals are in this rock? ________________, ________________, and ________________
- ★Was it formed intrusively, at the surface, intrusively/surface, or in the air? ____________________

Use what you know about samples 1 and 2 to answer the following questions:

- ★Was the magma that produced these rocks considered high temperature, medium temperature, or low temperature? ____________________
- ★Was the magma that produced these rocks generated in the mantle (deep), in the crust (shallow), or a combination of both? ____________________________
- ★What type of volcano could have produced these rocks? Shield, Composite, or Volcanic Dome? ________________________
- ★Which of the following tectonic settings could have produced the magma that created these rocks? (Circle all that apply): Spreading Center, Subduction Zone, Oceanic Hotspot, or Continental Hotspot.
- ★Would the Si content of the magma in this volcano be high, medium, or low? ________________

- ⋆Would the viscosity of the magma from this volcano be high, medium, or low? ____________
- ⋆Would you expect this volcano to be dangerously explosive? Y es or No

Sample 3—These two samples are both called Diorite

- What is their texture (coarse-grained, fine-grained, porphyritic, glassy, vesicular, or fragmental)? ____________________
- What is their composition (felsic, mafic, intermediate, or ultramafic)? ____________________
- What minerals are in these rocks? ________________, ________________, and ____________________
- ⋆Were these rocks formed intrusively, at the surface, intrusively/surface, or in the air? ____________________

Sample 4 —This sample is called Andesite

- What is its texture (coarse-grained, fine-grained, porphyritic, glassy, vesicular, or fragmental)? ____________________
- What is its composition (felsic, mafic, intermediate, or ultramafic)? ________________
- What minerals are in this rock? ________________ and ____________________
- ⋆Were these rocks formed intrusively, at the surface, intrusively/surface, or in the air? ____________________

Use what you know about samples 3 and 4 to answer the following questions:

- ⋆Was the magma that produced these rocks considered high temperature, medium temperature, or low temperature? ____________________
- ⋆Was the magma that produced these rocks generated in the mantle (deep), in the crust (shallow), or a combination of both? ________________________
- ⋆What type of volcano could have produced these rocks? Shield, Composite, or Volcanic Dome? ____________________
- ⋆Which of the following tectonic settings could have produced the magma that created these rocks? (Circle all that apply): Spreading Center, Subduction Zone, Oceanic Hotspot, or Continental Hotspot

- ★Would the Si content of the magma in this volcano be high, medium, or low? ______________
- ★Would the viscosity of the magma from this volcano be high, medium, or low? ______________
- ★Would you expect this volcano to be dangerously explosive? Yes or No ____________________

Samples 5a and 5b

- What is the composition of these two rocks (felsic, mafic, intermediate, or ultramafic)? ___________________________
- Which sample is coarse-grained? _________________
- Which sample is fine-grained? ___________________

Sample 5a

- What is the name of this rock? ___________________
- What minerals are likely in this rock? _____________, _____________, ________________, and a little _________________
- ★Was this rock formed intrusively, at the surface, intrusively/surface, or in the air? ________________

Sample 5b

- What is the name of this rock? ___________________
- What minerals are likely in this rock? _____________, _____________ ________________, and a little _________________
- ★Was this rock formed intrusively, at the surface, intrusively/surface, or in the air? ________________

Use what you know about samples 5a and 5b to answer the following questions:

- ★Was the magma that produced these rocks considered high temperature, medium temperature, or low temperature? ______________________

- ⋆Was the magma that produced these rocks generated in the mantle (deep), the crust (shallow), or a combination of both? ______________________

- ⋆What type of volcano could have produced these rocks? Shield, Composite, or Volcanic Dome? ____________________

- ⋆Which of the following tectonic settings could have produced the magma that created these rocks? (Circle all that apply): Spreading Center, Subduction Zone, Oceanic Hotspot, or Continental Hotspot

- ⋆Would the Si content of the magma in this volcano be high, medium, or low? ____________

- ⋆Would the viscosity of the magma from this volcano be high, medium, or low? ____________

- ⋆Would you expect this volcano to be dangerously explosive? Yes or No

Sample 6 (look only at the crystalline part)

- What is the composition of this rock—felsic, mafic, intermediate, or ultramafic? ____________

- How do you know? ____

- What is the name of this rock? ________________

- ⋆Where is it located in terms of Earth's layered structure? Continental crust, oceanic crust, or the mantle ___________________

- ⋆Why is the fine-grained form of this rock no longer made today? ____________________
____
____
____

Sample 7

- What is this rock called? ________________________

- Does it contain minerals? Yes or No

- Why or why not? ____
____
____

Samples 8a and 8b

- What is the texture of these rocks (coarse-grained, fine-grained, porphyritic, glassy, vesicular, or fragmental)? ____________________
- ⋆What does the texture suggest about the dissolved gas content in the magma that formed these rocks? ____________________
- What is rock sample 8a called? ____________________
- What is rock sample 8b called? ____________________
- ⋆Why is sample 8b so much heavier than sample 8a? ____________________

Sample 9

- What is this rock? ____________________
- ⋆Why is it shaped this way? ____________________

____

____

Sample 10

- What is the texture of these rocks (coarse-grained, fine-grained, porphyritic, glassy, vesicular, or fragmental)? ____________________
- What are these rocks called? ____________________

Sample 11

- What is unusual about the grains in this sample? ____________________
- What is the texture of this rock? ____________________
- What is the composition of this rock? ____________________
- ⋆Why is this type of rock of particular economic importance?

____

____

Part B: Igneous Rock Distribution in the Landscape

Exercise: Determine igneous rock type and distribution through use of a cross section.

A cross section is a diagram depicting the geometry and orientation of materials underground as they would appear on a vertical slice through the Earth. Figure 4.13 is a hypothetical cross section of the most common land-based volcanic setting. Use the cross section and corresponding samples to answer the questions that follow.

1. Just by looking at this x-section, how can you tell the difference between the igneous rocks and the sedimentary rocks? ____

   ____

2. Use a red pen to put hatch marks on the cross section where metamorphic rocks formed from heat only would be located.

3. ⋆What are the structures labeled "J?" ____

4. ⋆What is the structure labeled "K?" ____

5. ⋆What is the structure labeled "L?" ____

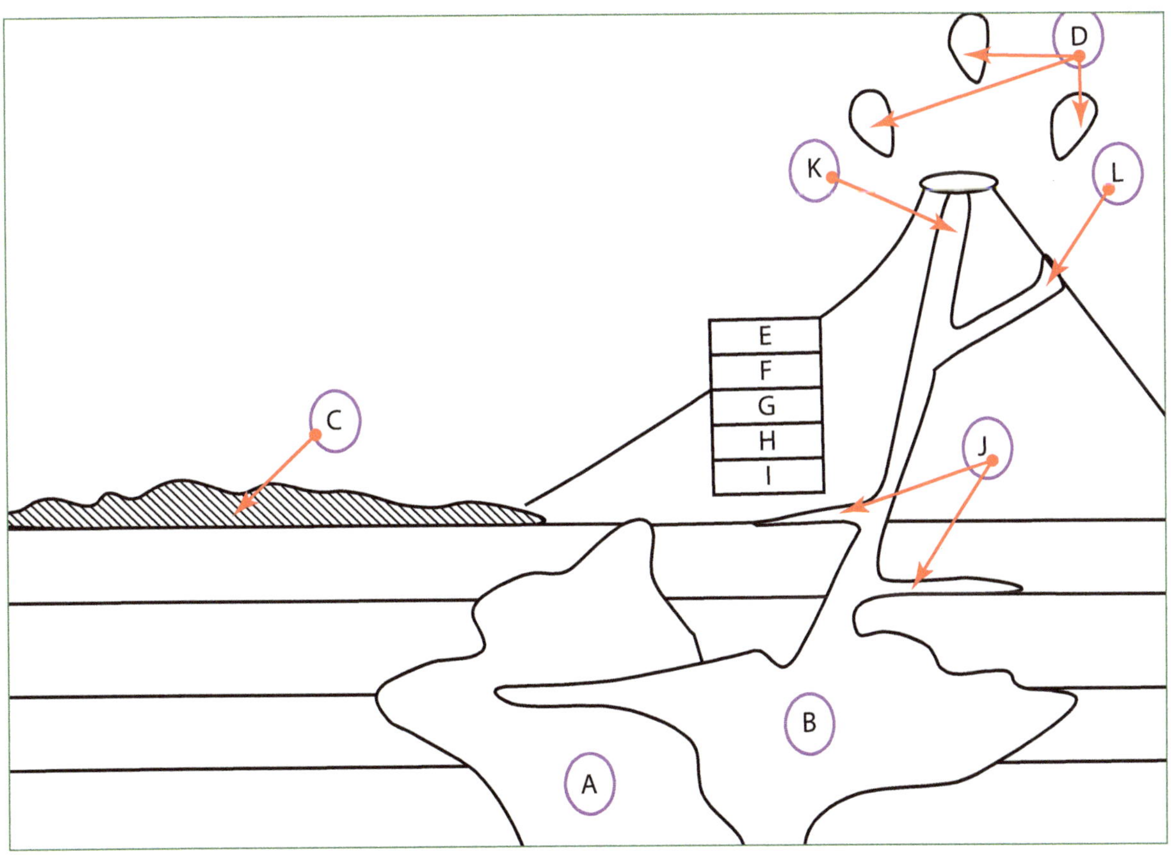

Figure 4.13 – Cross Section

6. On the tables throughout the room are hand samples of all the rocks labeled in the cross section. The letter on the cross section corresponds with the hand sample of the same letter and represents the distribution of that rock in the landscape. Use the hand samples and the cross section to determine the identity of the samples listed below. Remember that rocks underground will be coarse-grained and rocks above ground will be either fine-grained, porphyritic, pyroclastic, or fragmental.

    a. Sample A: ____________________

    b. Sample B: ____________________

    c. Sample C: ____________________

    d. Sample D: ____________________

    e. Sample E: ____________________

    (Hint: Pick up sample E in one hand and the granite behind it in the other. Notice the density difference between these two rocks, which are of approximately equal size. While both rocks were produced by the same magma, the granite is composed of coarse-grained minerals, while sample E is fine-grained pyroclastic material that has been welded together, so it is full of air pockets between the grains. Now can you ID this rock?)

    f. Sample F: ____________________

    g. Sample G: ____________________

    h. Sample H: ____________________

    i. Sample I: ____________________

7. ⋆What extrusive rock is intrusive rock A related to? ____________________________________

8. ⋆What extrusive rock is intrusive rock B related to? ____________________________________

9. ⋆The intrusive rocks have the same ____________________ as their extrusive counterparts, but a different ____________________

10. ⋆List three or four factors that might have caused the composition of the magma to evolve between the time magma A erupted and the time magma B erupted.

    - ____
    - ____
    - ____
    - ____

11. ★What type of volcano is this? Shield, composite, or volcanic dome?

12. ★List three lines of evidence to support your answer to #11.

- ______________________________
- ______________________________
- ______________________________

13. ★In what tectonic setting would you find the volcano is this cross section? Spreading Center, Subduction Zone, Oceanic Hotspot, or Continental Hotspot (Circle all that apply)

LABORATORY EXERCISE V

SEDIMENTARY ROCKS AND THE ENVIRONMENTS THAT PRODUCE THEM

INTRODUCTION

Sedimentary rocks are the most abundant type of rock on the surface of the Earth. They are composed of sediment—products of weathering consisting of loose rock fragments (detritus), mineral grains broken down from once-intact rock, mineral crystals that precipitate directly out of water, as well as shells and skeletons of organisms. Sedimentary rocks have either accumulated or precipitated at the surface of the Earth and hold the key to understanding much of Earth's surface history, including the history of life on this planet.

OBJECTIVE

The objective of this lab is twofold:

1) Learn how to identify sedimentary rocks using compositional and textural properties; and

2) Practice using sedimentary rocks in conjunction with sedimentary structures to interpret paleoenvironments.

PART A: SEDIMENTARY ROCK IDENTIFICATION

Sedimentary rocks can be divided into four categories: **Clastic/detrital**, **Chemical**, **Biochemical**, and **Organic**. The primary differences between these types of sedimentary rocks concern the way the original rock was **weathered**, the type of **sediment** created, and the processes which ultimately unite the sediment to form sedimentary rock.

Types of Weathering and the Rocks They Produce

Physical weathering is the equivalent of nature acting as a sledgehammer by breaking intact rocks into smaller grains or chunks. Nature accomplishes this by a variety of methods, several of which are listed below:

Figure 5.1 – Glacier Striations

Glaciers

Pluck and gouge out pieces of rock in the underlying substrate, then grind, roll, and break large rocks into smaller pieces during transport.

*Note the "striations" which formed where embedded rocks in an over-riding glacier abraded large grooves in the surface rock. See Figure 5.1.

Figure 5.2a

Figure 5.2b

Figure 5.2c

Figure 5.2d – Below
The surface pattern seen here is called "honeycomb weathering"—a product of salt crystallization

Figure 5.2e – Salt weathering

Ice and Salt Wedging

Ice: Water settles into the cracks of rocks, then expands upon freezing, thereby wedging the rock apart. (Figures 5.2a–5.2c)

Salt: Water saturated with NaCl (salt) will infiltrate the cracks of rocks and precipitate salt upon evaporation. Salt crystals will build in the openings, enlarging them as the crystals grow. (Figures 5.2d and 5.2e)

Figure 5.3a – Above
If a road were on top of this, it would be crumbled to pieces

Figure 5.3b

Vegetation

From Beneath: Large roots buckle upward and crack overlying rock into pieces. (Figures 5.3a and 5.3b)

From Between: Shrubs and trees will germinate in the cracks of a rock. As they grow, they wedge the rock apart. (Figures 5.3c–5.3e)

Figure 5.3c

Figure 5.3d

Figure 5.3e

Wind

When full of sand and sediment, wind behaves like sandpaper and shaves down rocks exposed at the surface.

Figure 5.4a–5.4e – Features due to wind weathering

Figure 5.4a

Figure 5.4b

Figure 5.4c

Figure 5.4d

Figure 5.4e

Water

Undercuts fractured rock, which results in the collapse of the cliffs and the breakup of tons of rock. (Figures 5.5a–5.5c)

Tears rock away from cliffs and river channels. (Figure 5.5d)

Wears the rocks into smaller pieces by smashing them together during transport. (Figure 5.5e)

Figure 5.5a

Figure 5.5b

Figure 5.5c

Figure 5.5d

Figure 5.5e

Thermal Expansion and Contraction: Cause rocks to swell during the heat of the day and shrink when it cools off at night. The thermal shock induced by the temperature difference results in cracking.

The smaller **clasts** that result from physical weathering are known as **sediment** and can be transported and deposited a sizeable distance from their source. Eventually, through a number of physical and chemical processes, the sediment will be compacted and cemented via a process called **lithification** to become what is known as a **detrital/clastic** sedimentary rock (Figures 5.6a and 5.6b).

Figure 5.6a – Sediment

Figure 5.6b – Detrital Sedimentary Rock

Chemical Weathering is a process in which the original rock is corroded by water or air and taken apart through chemical reactions atom by atom. Below are some of the methods by which chemical weathering occurs:

- *Chemical dissolution and precipitation* is the most basic chemical weathering process. In this case, water dissolves various minerals, which results in the individual atoms that make up those minerals loosely floating around in the water. This usually occurs as a result of rivers running over rocks or groundwater winding its way through an aquifer. The water, now full of atoms or dissolved **sediment** from chemically weathered rocks, will make its way to a lake or ocean and deposit the dissolved atoms. If the water level decreases due to evaporation, the atoms are able to re-bond, and once supersaturation is reached, the newly formed compound will precipitate out of the water as the *original mineral*. Minerals that form this way, such as halite, gypsum, and calcite, are called *inorganic precipitates*. It's important to note that in this process, water acts only as a transport mechanism and mixing bowl. It takes a mineral apart, moves the pieces from one point to another, then evaporates so the pieces can put themselves back together. It does NOT, however, take part in the actual chemical reaction (see Figure 5.7).

Ex: $Na^+ + Cl^- = NaCl$ (Halite)

Figure 5.7 – Rock Salt made from the mineral Halite

- *Hydrolysis* is an example of a chemical reaction where water breaks down the minerals via chemical dissolution, (*-lysis* means "loosen" in Greek), but then combines with some of the dissolved elements to create a *new mineral*. This process occurs faster if the water is slightly acidic. The formula below shows what happens when slightly acidic water, due to the presence of carbonic acid in rainwater, comes in contact with potassium (K) feldspar. Some atoms making up the feldspar become ions that remain dissolved in water, but some recombine and incorporate water to form a completely different mineral, *kaolinite*, a type of clay (notice the OH group, i.e., the water, in the kaolinite chemical formula). Therefore, chemical weathering of feldspars produces clay, which hardens into claystone (see Figure 5.8). Note that in this case, water actively participates in the chemical reaction.

$2KAlSi_3O_8$	+	$2H^+ = 2HCO_3^-$		+		H_2O
K-spar		Carbonic acid				water
		$\rightarrow$				
$Al_2Si_2O_5(OH)_4$	+	$2K^+$	+	$2HCO_3^-$	+	$4SiO_2$
Kaolinite (Clay)		K ions		bicarbonite ions		diss. Si

Figure 5.8 – Colored Claystone

- *Oxidation* is a common chemical reaction that takes place when elements combine with oxygen. Oxidation reactions in rocks transform Fe-bearing minerals (e.g., biotite and pyrite) into a rusty-brown mixture of various iron-oxide and iron-hydroxide minerals, like hematite, limonite, and goethite (pronounced "ger-tight"), ultimately producing a rock called an ironstone (see Figure 5.9).

Ex: Rusting of Fe

$4Fe^{2+}$	+	$3O_2$	$\rightarrow$	$2((Fe^{3+})_2O_3)$
Ferric Fe		Oxygen		Hematite
$4Fe^{2+}$	+	O_2	$\rightarrow$	$4(Fe^{3+})O(OH)$
Ferric Fe		Oxygen		Goethite

Figure 5.9 – Ironstone

Each of the three chemical weathering processes described above results in the formation of a chemical sedimentary rock.

Types of Sedimentary Rocks

Clastic/Detrital Sedimentary Rocks

All clastic rocks are composed of cemented particles (clasts) that have been subjected to moderate to intense physical **and** chemical weathering. Physical weathering breaks and chips larger pieces into clasts, while chemical weathering dissolves away the outer layers of the rock's minerals, allowing the rock to crumble apart. See Figure 5.10.

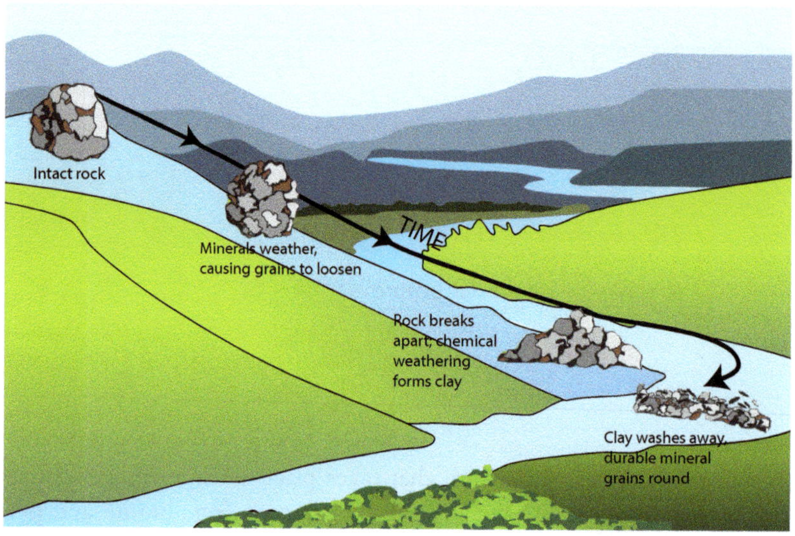

Figure 5.10 – Physical and Chemical Weathering Work Together

The clasts are transported (either by wind, water, or ice) some distance from their source, and are then deposited and **lithified**. Lithification involves a compaction component, where the grains are packed tightly together, as well as a cementation component, where mineral precipitates from chemical weathering, such as calcite, hematite, quartz, and clay cement the particles together. As such, chemical weathering not only helps to break the original rock apart, but provides the dissolved minerals to cement clasts back together. See Figure 5.11.

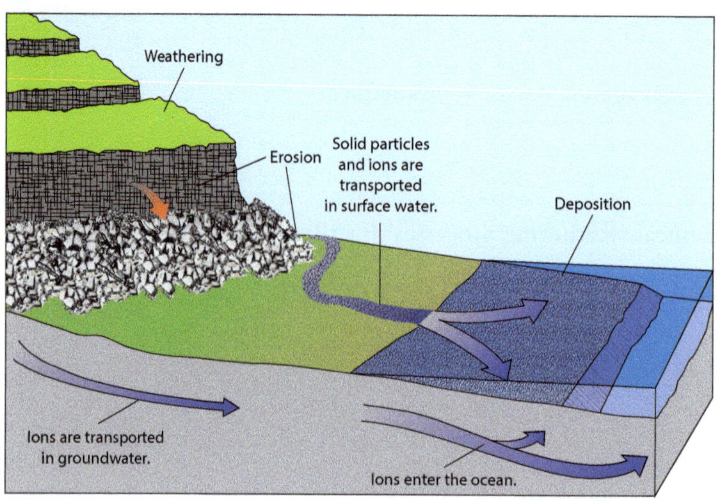

Figure 5.11 – Physical Weathering, Erosion of Sediments (Clasts), Deposition, and Lithification

For clastic rocks, the character of the particles comprising the rock (clasts) will help to determine the exact classification of the rock. The three most important particle properties are 1) particle size; 2) particle sorting; and 3) particle shape.

Particle Size

Sediment particles are specifically classified by the diameter of the particles, but can be generally classified into ranges from coarse to very fine. By and large, the size of a particle is controlled by the energy of the medium moving it and the distance the particle has been transported from its original source. A high-energy environment can move and deposit very coarse particles, whereas a low-energy environment will move and deposit only finer particles. Similarly, the closer you are to the source of the material, the larger the particles that will be deposited, as there will have been less time for the larger pieces to be broken into smaller ones. See Figures 5.12 and 5.13a–5.13c.

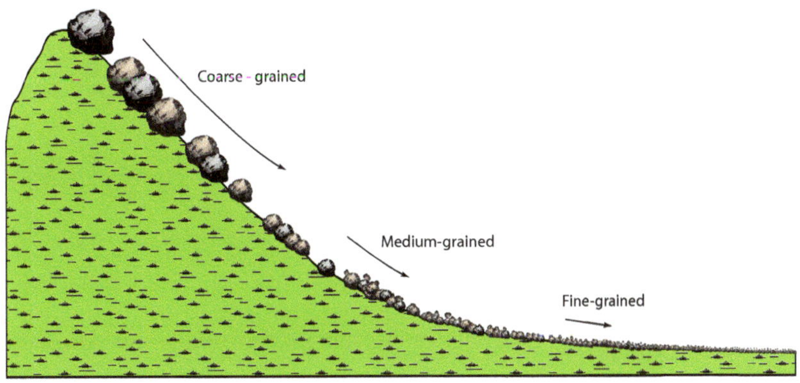

Figure 5.12 – Particle Size Distribution

Figure 5.13a – Mountain Stream

Coarse Grains

Figure 5.13b – Mid-Slope Meandering Stream

Finer on inner bend where water moves slower

Coarser on outer bend where water moves faster (but less coarse than in the mountain stream)

Figure 5.13c – Floodplain

Fine grains

Particle Sorting

The sorting of particles in a sediment refers to the range of particle sizes that are present. A **well-sorted** sediment will contain particles of only one size. A **poorly sorted** sediment will contain particles of varying sizes. The sorting of a sediment is primarily controlled by the method of transport or erosion. Wind will create the best sorted sediments, as wind can only carry sand-sized grains. Thus, sand dunes will be composed of well-sorted sediment, usually in the fine sand to medium sand-size range. Movement of sediment by water will generally produce a well-sorted sediment, for the faster it moves, the larger the particles it can transport. As the water moves out of the highlands, it begins to slow down, dropping the largest particles first, then the medium sized as it slows down more, and ultimately, the finest silt and mud/clay on the floodplain where the water comes to a standstill (refer back to Figures 5.12 and 5.13a–5.13c). However, water does not sort quite as well as wind (see Figures 5.14a–5.14c). Transport by ice, such as in glaciers, commonly produces the most poorly sorted sediments, with particle sizes ranging from mud to boulders, all deposited simultaneously. See Figure 5.15.

Figure 5.14a–5.14c – Well-Sorted Sediment

Note how the sand-sized particles are being moved, and the larger particles are remaining in place

Figure 5.15 – Unsorted Sediment

Particle Shape

The shape of a particle can also be linked to the time and distance the sediment spends in transport. Roundness, the extent to which the corners of individual grains have been rounded off, is an important characteristic of the texture. The roundness scale runs from **angular** through **subangular** to **subrounded** and **well-rounded**. The more time sediment spends in transport, the more rounded its constituent clasts.

Sediment containing very angular grains has not been transported very far or for very long, and is thus an immature sediment. Both a conglomerate and a breccia are unsorted rocks made up in part by coarse particles, but the particles in a conglomerate will be rounded (smooth) and the particles in a breccia will be very angular. Ice wedging in mountainous terrain will causes large pieces of cliff to break off from the cliff face and smash apart when the chunk hits the rocks below. This will result in very angular particles, and thus a breccia. Sometimes breccia can be created by explosive eruptions, such as the volcanic breccia that you looked at in the previous lab. Typical weathering (physical and chemical) and transport will produce well-rounded particles and thus a conglomerate.

Keep in mind that the softer the mineral, the more prone it is to being rounded. As such, a clast of quartz may have traveled significantly further than a clast of calcite, yet still remain more angular.

Grain Sizes	Grain Shapes	Grain Arrangements
Grains visible; larger than sand  Gravel-sized	 Angular Quartz Grains	 Poorly Sorted
Grains visible; like in a sandbox  Sand-sized		
Grains barely visible; feels gritty  Silt-sized	 Rounded Quartz Grains	 Well Sorted
Grains not visible; feels smooth; dull luster on freshly broken surface  Clay-sized Pieces		

*Photographs compliments of Dr. Catherine Forster

Table 5.A – Textural Features of Sedimentary Rocks

Particle Composition

Although the size, sorting, and shape of the particles are the primary diagnostic properties, the composition can also be used to further distinguish similar clastic rocks. All sedimentary rocks are created from weathering of the continental crust, which has a fairly homogeneous composition. Thus, sedimentary rocks are primarily composed of only a few different types of minerals, namely *quartz, feldspars, micas, clay, and calcite*. However, the quantities of these minerals will vary greatly from sample to sample. For instance, both quartz sandstone and arkose sandstone are composed of sand-sized particles, but arkose contains a large percentage of feldspars, whereas most sandstones are composed of mostly quartz grains. In this case, arkose is a type of sandstone, but provides one more level of description, thus identifying it as a feldspathic sandstone. The composition of the sediment particles will be primarily controlled by the composition of the rocks from which it is derived as well as how much weathering took place prior to lithification.

Sediment Maturity

The movement of particles during transport will preferentially weather and erode the softer minerals, thereby progressively enriching the sediment in the stronger and more durable minerals. As such, the longer a sediment sample has been subjected to weathering, the greater the percentage of durable minerals composing the sediment. This is referred to as the "maturity" of the sediment. A more mature sediment will have a greater percentage of quartz, since quartz is one of the most durable common minerals. A very immature sediment will have a greater percentage of minerals that are significantly less resistant to weathering, such as mica and feldspar, and therefore less durable than quartz. Furthermore, mature sediments are characterized by good sorting and grain rounding, all by-products of extensive transport.

Naming Clastic/Detrital Sedimentary Rocks

Clastic sedimentary rocks are named primarily on grain size, grain sorting, and grain shape, as these are the properties that are most useful in reading the rock's story. They are then further differentiated using the rock composition.

Sediment	Grain size	Rock name
Gravel (pebble, cobble, boulder)	> 2 mm	Breccia (if fragments are angular) Conglomerate (if fragments are rounded) Breccia  Conglomerate 
Sand	$\frac{1}{16}$ to 2 mm	Sandstone **Quartz sandstone:** > 95% quartz grains **Arkosic Sandstone:** > 25% feldspar grains with quartz **Lithic sandstone:** > 5% rock (lithic) fragments with quartz 
Silt Mud	$\frac{1}{16}$ to $\frac{1}{256}$ mm $< \frac{1}{256}$ mm	Mudstone Siltstone (if blocky) Shale (if splits into sheets) Mudstone  Shale 

Table 5.B – Naming Clastic Sedimentary Rocks

Chemical Sedimentary Rocks:

Chemical rocks are all created through some form of dissolution and precipitation, the result of chemical weathering of surface rocks and minerals (see Figure 5.16). Because their crystals were all created via chemical reactions, the primary diagnostic property is going to be the chemistry of the rock (as opposed to the particle size used to classify classic rock). In addition, rarely are conditions such that two minerals can coprecipitate out of a solution simultaneously, so unlike clastic rocks, chemical rocks are composed of only one mineral.

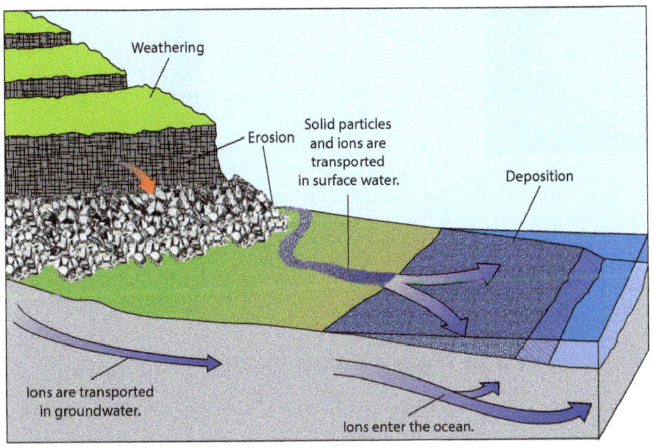

Figure 5.16 – Chemical Weathering, Erosion of Sediments (Ions), Deposition, and Precipitation

Chemical sedimentary rocks are subdivided into five groups:

- Inorganic Precipitates
- Products of Hydrolysis
- Products of Oxidation
- Biochemical Precipitates
- Organic Rocks

Inorganic precipitates—Formed from chemical dissolution and precipitation

- Rock gypsum (alabaster) ($CaSO_4$): Identifiable because it's softer than your fingernail – See Figure 5.17
- Rock salt (NaCl): Identifiable by the salty taste – See Figure 5.18
- Chert (SiO_2): Identifiable because of its smooth texture and ability to scratch glass – See Figure 5.19
- Crystalline limestone ($CaCO_3$): Identifiable because it reacts with HCl – See Figure 5.20

Figure 5.17– Gypsum

Figure 5.18 – Salt

Figure 5.19 – Chert

Figure 5.20 – Crystalline limestone

Formed from hydrolysis

- Claystone: Identifiable by its powdery, brittle texture (see Figure 5.21)

Figure 5.21 – Claystone

Formed from oxidation

- Ironstone: Identifiable by its rusty look and high density (see Figure 5.22)

Figure 5.22 – Ironstone

Biochemical precipitates

Some chemical precipitates are produced from the hard skeletal tissue of once living organisms and we refer to these as **biochemical** rocks. Rocks that are formed in this way require a combination of chemical precipitation and physical packing of the skeletal grains to create the rock.

Ex. 1: Biochemical Limestones

In the case of the biochemical limestones, Ca and CO_3 ions, which were picked up via chemical weathering of the continental crust and delivered by the rivers and groundwater to a surface water reservoir, will become incorporated into the calcitic shells and skeletons of marine organisms via a chemical precipitation process forced by the organisms. Once the organisms die, the calcitic skeletons and/or shells will sink to the bottom of the water body, where they will be physically weathered by other living organisms, ocean currents, and waves. Eventually, the broken-up pieces will become "lithified" just as in a clastic sedimentary rock forming a rock of skeletal debris clasts that is classified as *biochemical*. Varieties of biochemical limestones and how they are classified can be seen in Table 5.C. Remember that these shell and skeletal fragments are made up of calcite. As such, these rocks are all forms of limestone and therefore will react with HCl.

Composition	Particle Constituents	Rock name	Visual Depiction
Calcite (Fizzes with HCl)	Cemented mass of loose, coarse shell fragments	Coquina	Figure 5.23a
	Mix of course and fine-grained fossil debris	Fossiliferous limestone (Fossils + micrite)	Figure 5.23b
	Mostly very fine-grained to microcrystalline detrital mass of calcite and microfossils	Micrite	Figure 5.23d
	Mostly fine-grained, earthy, chalky, light-colored mass of coccolithophore skeletons	Chalk	Figure 5.23c

Table 5.C – Biochemical Sedimentary Rocks (Limestones)

Figure 5.23a – Coquina

Figure 5.23b – Fossiliferous Limestone

Figure 5.23c – Chalk

Figure 5.23d – Micrite

Ex. 2: Biochemical Chert

In the case of chert, different marine organisms will precipitate SiO_2 shells and skeletons, as opposed to $CaCO_3$ shells and skeletons. When the organisms die, their silica-rich remains sink to the ocean floor, where they dissolve into silica-rich ooze. Following burial, the ooze fully dissolves and then reprecipitates inorganically in the form of chert.

Notice that both limestone and chert can be precipitated by inorganic (i.e., chemical) or organic (i.e., biochemical) means.

Organic Rocks

There are also several cases where dead organisms with high quantities of soft tissues are buried prior to decay, generating a fifth group of chemical rocks called **organic rocks**. One such example is coal.

Coal forms when plant material is rapidly buried, such that decay of the soft tissue is prevented. Historically, this has taken place in massive swamps with huge quantities of vegetations on which thick layers of mud are deposited, which seals the dead organic detritus from Earth's corrosive, oxygen-rich atmosphere. Over time, the pressure of overlying, later deposited sediments force the volatiles out of the vegetation, leaving behind a heap of carbon. Coal is identifiable because of its iridescent sheen, jet black color, dusty surface, and low density. See Figures 5.24a and 5.24b.

Figure 5.24a – Coal

Figure 5.24b – Coal Seams

Naming Chemical Sedimentary Rocks

Chemical sedimentary rocks are named primarily on composition, then further differentiated using the rock's texture. See Table 5.D.

Classification	Rock Name	Texture	Composition
Chemical Precipitates	Crystalline Limestone	Fine to Coarse (Crystalline)	Calcite – $CaCO_3$
	Chert	Very Fine (Crystalline) often with Concoidal Fracture	Quartz – SiO_2
	Rock Salt	Fine to Coarse (Crystalline)	Halite – NaCl
	Rock Gypsum (Alabaster)	Fine to Coarse (Crystalline)	Gypsum – $CaSO_4 \cdot 2H_2O$
Product of Hydrolysis	Claystone	Powdery and Brittle	Clay Minerals
Product of Oxidation	Ironstone	Rusty and Dense	Hematite, Limonite and Goethite
Biochemical Rocks	Coquina	Cemented mass of loose, coarse shell fragments	Calcite – $CaCO_3$
	Fossiliferous Limestone	Mix of course and fine grained fossil debris	
	Micrite	Mostly very fine-grained to microcrystalline	
	Chalk	Mostly fine-grained, earthy, chalky and light	
	Chert	Very Fine (Crystalline) often with Concoidal Fracture	Quartz – SiO_2
Organic Rocks	Coal	Black, dusty and low density	Compressed Plant Material

Table 5.D – Classification of Chemical Sedimentary Rock

Just for Show:

Oolitic limestone is a rock that forms when sand grains settle into a calcitic "ooze" at the bottom of the ocean. The calcite hardens around the sand grains in layers, creating "ooids," or calcite-coated spheres. When the ooids become lithified together, they form an oolitic limestone, which is easily recognizable by the "bull's-eyes." See Figure 5.25a.

And when the ooids are > 2mm in diameter, they are called "Pisoids" and form Pisolitic Limestone. See Figure 5.25b.

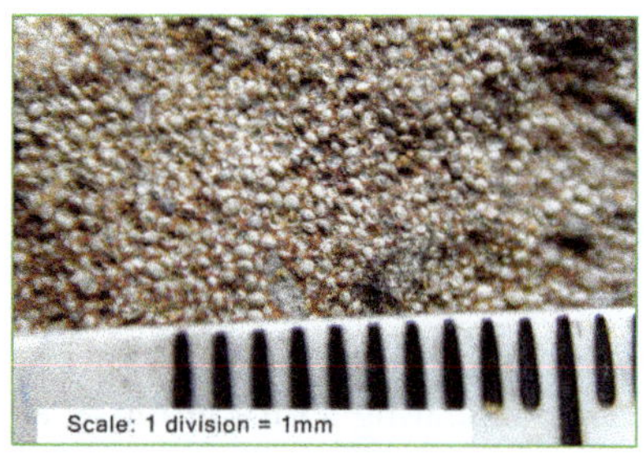

Figure 5.25a – Oolitic Limestone

Figure 5.25b – Pisolitic Limestone

Exercise: Sedimentary Rock Identification and Analysis

On the tables throughout the room are examples of the sedimentary rocks discussed above. Using what you observe concerning the texture and composition of these rocks, first determine whether your rock sample is **clastic** or **chemical**, then narrow down through the use of particle size, if it's a clastic rock, or composition, if it's a chemical rock, eventually identifying each rock by its unique name. As you approach each sample, answer the corresponding questions that follow. Below is a flow chart that will help you determine what type of rock you are examining. In the interest of time, please answer all questions that DO NOT have an asterisk (★ symbol) next to them, as questions marked in this way do not require a sample and can be answered outside of lab.

Note: If samples are unavailable, refer to the appendix at the end of this lab manual for photographic representation.

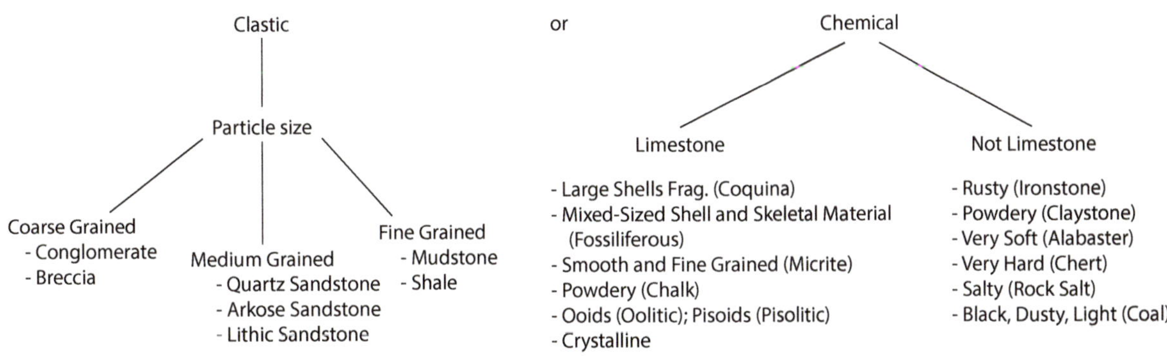

Samples 1a and 1b

- Were these rocks created primarily from physical weathering or chemical weathering? ______________________
- What is the main grain size in these rocks (coarse, medium, fine, all of the above)? ______________________
- What rock is Sample 1a? ______________________
- What rock is Sample 1b? ______________________
- ★Why do these rocks have two different names? ______________________
- ★What does that difference imply about the distance the grains in each rock traveled from their source? ______________________

Samples 2a–2c

- Are these rocks clastic or chemical? ____________________
- Are these samples poorly sorted or well sorted? ____________________
- What is the grain size of the clasts in these rocks (coarse, medium, fine, all of the above)?____________________
- What are these rocks collectively called? ____________________
- Adding the additional layer of description from composition, what is the more specific name of each of these three rocks?

  2a. ____________________

  2b. ____________________

  2c. ____________________
- *Which mode(s) of transportation could have delivered the sediment that formed these rocks? Wind, Rivers, and/or Glaciers (circle all that apply).

Samples 3a and 3b

- These two rocks are both clastic. Of what particles are these rocks composed (gravel, sand, or mud)? ____________________
- What is the textural difference between these two rocks? ____________________ ____________________
- What is the name of each of these rocks?

  3a. ____________________

  3b. ____________________
- *Water is a transport medium that moves this size sediment. Knowing this, what environmental condition must be present in order for the water to be able to deposit this material? ____________________

Samples 4a–4d

- *These rocks are all inorganic chemical precipitates. What chemical weathering process contributed to the formation of these rocks? ______________________ (dissolution, hydrolysis, or oxidation)
- Because they are chemical rocks, they are classified based on the one mineral they are made of. Identify each of the following samples and name the property that distinguishes each one from the others.

  Sample 4a: ______________________

  Distinguishing Property ______________________________

  Sample 4b: ______________________

  Distinguishing Property ______________________________

  Sample 4c: ______________________

  Distinguishing Property ______________________________

  Sample 4d: ______________________

  Distinguishing Property ______________________________

- *Which of the above rocks can also be created by means of biological precipitation? ________________ and ________________

Samples 5a–5d

- What rock classification category do these rocks belong to (clastic, chemical, biochemical, or organic)? ______________________
- Of what mineral are they composed? ________________
- *What chemical weathering process contributed to the formation of these rocks (dissolution, hydrolysis, or oxidation)? ______________________
- Why do these rocks have different names? ____
  ____
- Identify each of the following samples:

  Sample 5a: ______________________

  Sample 5b: ______________________

  Sample 5c: ______________________

  Sample 5d: ______________________

Samples 6a and 6b

While these rocks may appear to be the same at first glance, closer examination will show them to be two very different substances.

- Identify sample 6a. ______________________ (*Hint*: You've already seen this rock.)
- What distinguishing property does Rock 6a exhibit? ______________________
- What mineral is rock 6a composed of? ______________________
- Now examine Rock 6b. Identify this rock. ______________________
- ★What weathering processes produced rock 6b (physical weathering, dissolution, hydrolysis, or oxidation)? ______________________
- ★Of what mineral is rock 6b made? ______________________

Sample 7

- Identify this rock. ______________________
- ★What weathering process produced the minerals that compose this rock (physical weathering, dissolution, hydrolysis, or oxidation)? ______________________
- ★What two elements must be present in order for this weathering process to take place? __________ and __________

Sample 8

- ★Is this rock clastic or chemical? ______________________
- Identify this rock. ______________________
- ★What was buried and compressed to form this rock? ______________________
- ★Why is this rock grouped in a separate category from the biochemicals, even though it was made from organic debris? ______________________
____

Sample 9

- Identify sample 9a. ______________________
- Identify sample 9b. ______________________
- What mineral is comprising the spheres? ______________________

PART B: SEDIMENTARY ROCKS— WHAT STORY DO THEY TELL?

Since sedimentary rocks make up over 75% of the rocks exposed on the surface, they are also responsible for driving, participating in, and recording nearly all the surface processes, from the high mountain glacier environments to the deep sea abyssal plains. Each environment leaves characteristic clues in the rocks it formed; which, when read properly, divulge the story of that area's past sedimentary processes and how the environments have evolved over time. Sedimentary environments are divided up into three main categories: Terrestrial, Transitional, and Marine (see Figure 5.26). Below is a general guide to help you learn how to read the history written in these rocks and ultimately determine the processes that formed them and the ancient sedimentary environment from which they came.

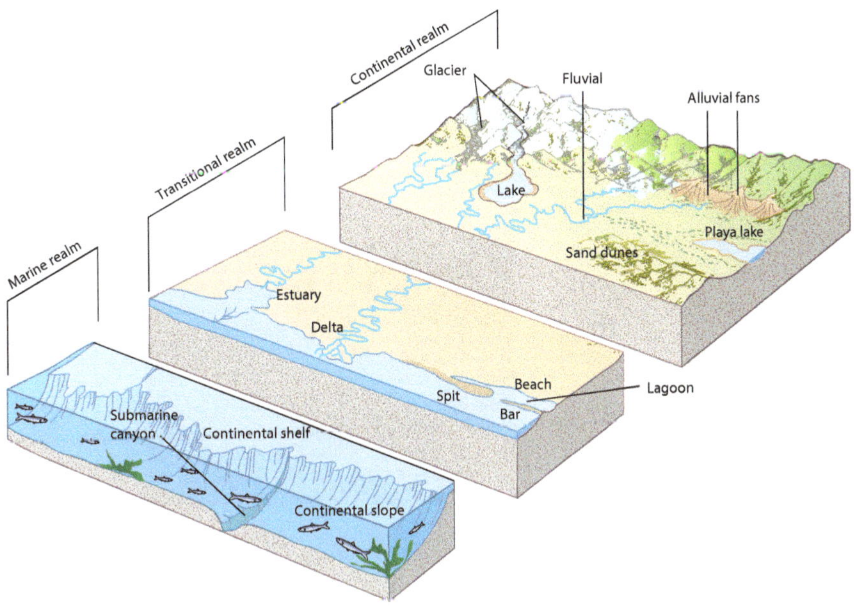

Figure 5.26 – Sedimentary Environments

Terrestrial Sedimentary Environments

1. Glacial Environments: Because solid ice is capable of moving sediment of any size, sedimentary deposits left over from glacier transport, i.e., glacial till, *tends to be very unsorted and unstratified, with grain sizes ranging from boulders to silt and mud.* See Figure 5.27.

Figure 5.27 – Glacial Till

2. Mountain Stream Environments: Water moving off the steep slopes of the mountaintop flows extremely fast and has enough power to carry particles ranging in size from very large to very small. As the water slows, the larger particles settle into the stream bed, while the finer-grained particles are carried away. *Sedimentary deposits from these environments are primarily coarse-grained breccias and conglomerates.* See Figure 5.28.

Figure 5.28 – Mountain Stream Environments

3. River Environments: Rivers carry the sediments from the mountain fronts to the sea—primarily transporting pebbles, sand, silt, and mud. As the river flows down its course, its water velocity decreases in two directions. 1) As the river moves toward the ocean, the land surface becomes less steep, resulting in a general decrease in water velocity from the river's head to its mouth, where upon arrival at the ocean, it will come to a complete standstill. 2) When the river floods, as soon as the water leaves the channel, it will experience a decrease in velocity on both sides of the channel until the water again comes to a complete standstill on the floodplain. As such, rivers leave deposits both in the direction the river is flowing (channel deposits) and outward from the channel on both sides (flood deposits). The sediment in the channel gets progressively smaller from the mountains to the sea. Since the river sediment is deposited in a current, *the sediment surface develops current ripple marks.* Flooding events are recorded in the sedimentary record as lens-shaped sediments, with coarser particles in the middle surrounded by layers of fine-grained floodplain deposits. *On the floodplain, mud layers dry out between floods, leading to the formation of mud cracks.* Minerals in the river have traveled a long distance, and have therefore undergone extensive physical and chemical weathering. As such, *river sediments tend to include very little feldspar (most of it changed into clay) and typically lithify to form sandstone, siltstone, and shale.* See Figures 5.29a–5.29e.

Figure 5.29a – This sediment sequence represents oscillating periods of gentle trickle (fine-grained layers) separated by periods of raging torrent (large clast layers).

Figure 5.29b – Alternating Layers of Sand and Silt

Figure 5.29c – Tepees in Petrified Forest National Park in northeastern Arizona, United States. The colorful bands of mudstone and sandstone were laid down during the Triassic, when the area was part of a huge tropical floodplain.

Figure 5.29d – Mud cracks

Figure 5.29e – Current ripples

Current ripples are sedimentary structures that form in environments where a fluid medium, such wind and/or water, flows in one direction over an unconsolidated substrate. Current ripples have an asymmetrical cross-section, with a gentle up-current slope (stoss) and a steeper down-current slope (lee). They form as a result of irregularities in the surface, which causes grains to pile up at the boundary of an irregularity. Eventually, when the angle of repose is reached, the accumulation of grains destabilizes and then avalanches down the lee side of the ripple. As such, over time, the ripples will migrate downstream. See Figure 5.29f.

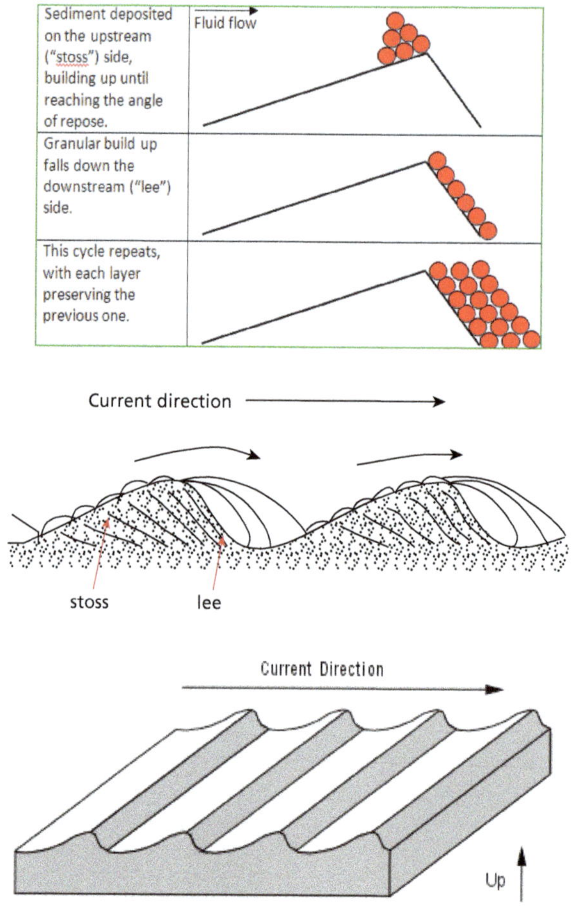

Figure 5.29f – How current ripples form

4. Alluvial Fan Environments: In arid regions where water is insufficient for the stream to flow continuously, when a stream forms, it picks up the mountain sediments and deposits its load right at the mountain front, creating a large, wedge-shaped apron called an alluvial fan. As the water from the mountains splashes and spreads on the plains below, friction with the ground causes the water to slow down to a crawl. Slow-moving water does not have the power to move coarse sediment. As such, the coarse sediment falls out on contact with ground, but the sand is carried to the outer reaches of the alluvial fan. Because the sediment is still close to the mountains—i.e. the source of the sediment—only a small amount of physical weathering and a smaller amount of chemical weathering has taken place. As such, *when the sediments are later buried and transformed into sedimentary rock, it becomes arkose sandstone (as the potassium feldspar has not been weathered to clay), as well as conglomerates and breccias.* See Figures 5.30a and 5.30b.

Figure 5.30a and 5.30b – Alluvial Fan

5. Sand Dune Environments: These are environments that develop in arid regions. Because relatively few plants grow, the ground lies exposed to wind. Since even the strongest winds can only transport sand-sized clasts in large quantities, *sand dune environments produce thick layers of well-sorted sandstone, which includes large (meters high) cross-beds* (see Figure 5.31a). In addition, because of the aridity, *rock gypsum and rock salt deposits are commonly found in these areas* (see Figures 5.32 and 5.33).

Figure 5.31a – Cross Bedding

Cross-bedding is formed by the downstream migration of ripples or dunes (note: dunes are current ripples that are taller than 10cm). Ripples and dunes form in a unidirectional flowing current as a result of sediment accumulation in front of substrate irregularities, which eventually destabilize and avalanche down in the downstream direction. Because the sediment deposited on the down-current side of ripples and dunes is deposited at an angle, if preserved, repetitive avalanches will eventually form the sedimentary structure known as cross-bedding, with the layer dipping in the direction of the paleo-current. See Figure 5.31b.

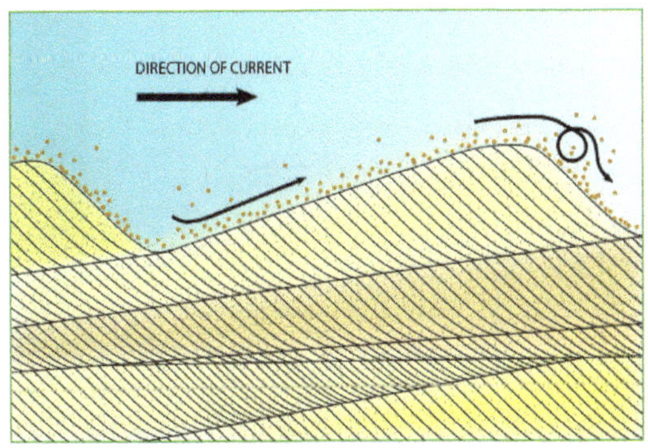

Figure 5.31b – How Cross-Beds Form

Figure 5.32 – Halite, Anhydrite, and Limestone interspersed

Figure 5.33 – Rock Gypsum

6. Lake Environments: Lakes contain relatively quiet water, so only the finest sediment makes it to the interior of the lake, where the sediment eventually settles in the lake bed (See Figure 5.34). *Thus, lake sediments, aka "lacustrine sediments," typically consist of finely laminated mudstone.*

Figure 5.34 – Finely Laminated Mudstone

Transitional Sedimentary Environments

7. Delta Deposits: Where the river empties into the sea (shaped like the Greek letter *delta*—see Figure 5.35). Very complex, as the processes of both the land and the sea are interacting here. Can be recognized by thick sequences in which offshore marine sediments grade progressively into fluvial sediments in a shoreward direction

Figure 5.35 – This photograph highlights two river deltas formed along Lake Ayakum's south-western shoreline near the northern boundary of the Tibetan Plateau. Gray to tan surfaces indicate prior positions of the river channels; the uniform coloration and smooth texture suggest that they are relatively old and now inactive. The younger and currently active delta surfaces have reddish-brown sediment and visible river channels. Lateral channel migration is particularly evident in the approximately 8-kilometer wide active delta at image upper left. The reddish color of the fresh sediment may indicate a change from the sources that formed the older parts of the deltas.

8. Coastal Beach: Ocean currents transport sand along the coastline, resulting in sand washing back and forth in the surf so it becomes well sorted and well rounded (waves winnow out mud and silt). *This back-and-forth motion produces "wave ripples" and creates well-sorted, medium-grained sandstone, often with wave ripple marks.* See Figures 5.36a.

Figure 5.36a – Wave Ripples

Unlike asymmetrical current ripples and dunes, wave ripples are symmetrical and form from the circular movement of water molecules in oscillating waves. When wave energy moves through water, water molecules within the wave do not travel with the wave but instead move in small circular orbits between the wave crest and wave trough. This circular motion at the surface extends downward into the water column, with each downward orbit decreasing in strength. As wave energy approaches the shore, the column of rotating water molecules is compressed by the shallow depth, resulting in the lower most orbits being flattened into ellipses. This elliptical motion then moves the sediment steadily back and forth. It is this back and forth motion that moves the sediment into symmetrical ripples, which retain their symmetry due to the repetitive upward wash and subsequent backwash of wave motion. See Figure 5.36b.

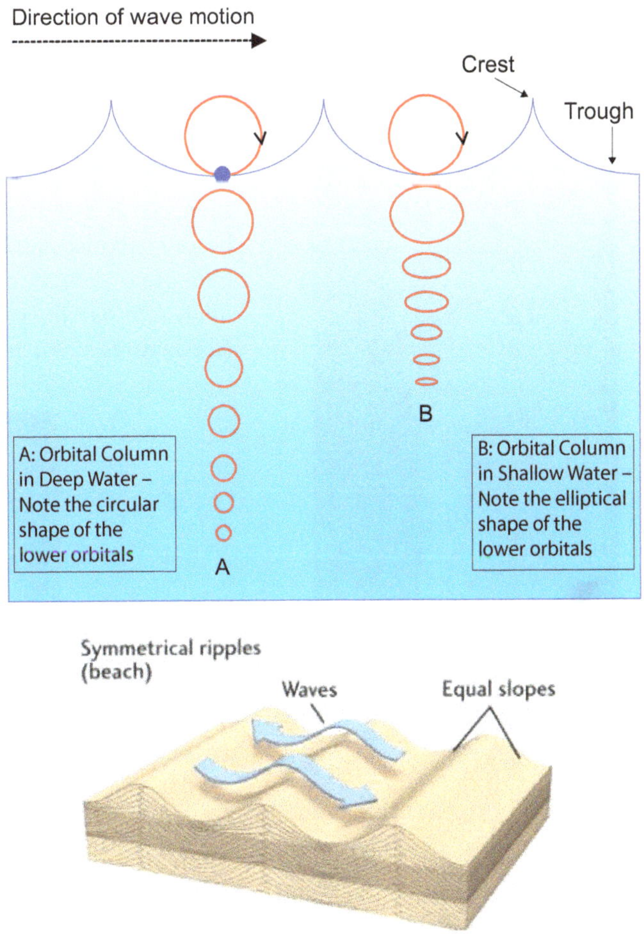

Figure 5.36b – How Wave Ripples Form

9. Tidal Basin: Tidal basins are areas that are exposed daily to cycles of water submergence followed by rapid drying. Tidal basins are located in protected coastal environments, and as such, coastal waves, as well as the tide, are slow and gentle, leaving behind huge quantities of mud when they retreat back to sea. This being the case, *tidal basins are frequently composed of mudstone that contain mud cracks, raindrop imprints, and wave ripples.* See Figures 5.37a–5.37c.

Figure 5.37a – Raindrop Imprints

Figure 5.37b – Wave ripples

Figure 5.37c – Mud Cracks

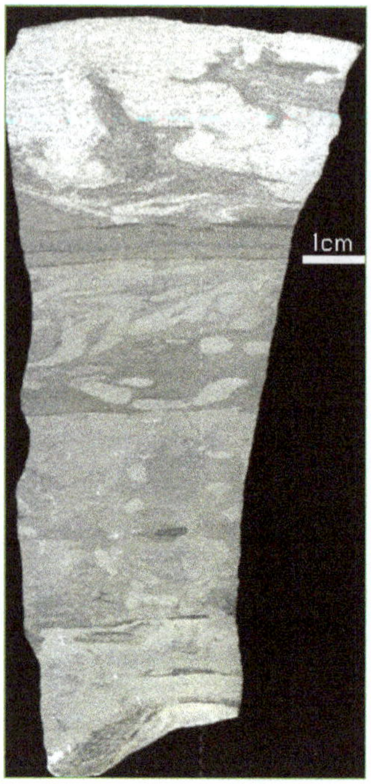

Figure 5.38a
Note the "bioturbation"—the disturbance of sediment primarily through burrowing and boring of living organisms.

Marine Sedimentary Environments

10. Shallow Marine Clastic Deposits: Wave action transports coarser sediments (sand) shoreward, while finer sediments get washed out to sea. In addition the wave energy does not stir the sea floor. As such, clastic sediments that accumulate in this environment tend to be fine-grained, well-sorted, well-rounded silt that is also inhabited by a great variety of organisms like mollusks and worms. *Sediments lithify into smooth beds of siltstone and mudstone containing marine fossils* (see Figures 5.38a and 5.38b).

Figure 5.38b – Marine fossils in mudstone

11. Shallow Water Carbonate Environment: Located far from the mouth of the river, where relatively little clastic material enters these waters. The warm, clear, nutrient-rich water is host to a tremendous amount of shelled organisms, which consist primarily of carbonate material. When these organisms die, their shells make up most of the sediment (i.e., carbonate environment). The nature of the sediment depends on the processes taking place in the various parts of the carbonate environment. Beaches in the shallow water carbonate environment collect sand composed of shell fragments (see Figures 5.39a and 5.39b). Lagoons (quiet water) are areas where calcitic mud accumulates. Reefs and the sloping layer behind them consist of coral and coral debris. Shallow water carbonate environments transform into sequences of coquinas, fossiliferous limestone, and micrite.

Figure 5.39a – Coral Reefs and Debris

Figure 5.39b – Coral Debris

12. Deep Marine Deposits: The corridor between the coastal regions and the abyssal plain is a steeply sloping region that frequently endures underwater landslides called turbidity currents. These avalanches typically flow down submarine canyons and end in a submarine fan of graded beds. Further offshore, in the deep abyssal plain, only fine mud and plankton provide a source for sediment. Mud settles out on the sea floor forming deposits of finely laminated mudstones, and plankton shells settle to form chalk and micrite (if calcitic) or chert (if siliceous). Therefore, deposits of mudstone, chalk, micrite and chert indicate a deep marine deposit. See Figures 5.40a and 5.40b. Note: At 3500m the carbonate compensation depth (CCD) is reached. Below the CCD, there is no longer carbonate deposition and only siliceous material is present.

Figure 5.40a – Limestone Interbedded with Chert

Figure 5.40b – Chalk and Chert

Evolving Environment: It is also important to note that environments are constantly evolving; as such, when looking at strata that cover large stretches of geologic time, you will likely be able to identify multiple environments that occupied the same space at different times (see Figure 5.41).

Figure 5.41 depicts carboniferous limestone. Pure beds are characteristically light in color, with darker beds having a higher content of mud and organic material.
There are at least three depositional cycles visible in the middle part of the outcrop. Each cycle begins with a dark mudstone representing deepwater conditions that are unfavorable for the carbonate-secreting organisms, and as the water shallows progressively, the amount of carbonate increases until clear, lagoonal, and reef deposits dominate. This is a record of sea level changes over hundreds to thousands of years.

Table 5.E – Sum Up

	Environment	Structures	Rocks
Terrestrial	Glacial	------	Unsorted and Unstratified Till
	Mountain Stream	------	Breccias and Conglomerates
	River	• Current Ripples • Mud Cracks	Sandstone Shale
	Alluvial Fan	------	Breccias and Conglomerates Arkose Sandstone
	Sand Dune	• Crossbeds	Rock Salt Rock Gypsum Sandstone
	Lake	------	Finely Laminated Mudstone

	Environment	Structures	Rocks
Transitional	Delta	------	Offshore, marine sediments grade into continental, fluvial sediments
	Coastal Beach	• Wave Ripples	Sandstone
	Tidal Basin	• Mud Cracks • Raindrop Imprints • Wave Ripples	Mudstone

	Environment	Structures	Rocks
Marine	Shallow Marine Clastic	------	Siltstone and mudstone with Marine Fossils
	Shallow Marine Carbonate	------	Coquina Fossiliferous Limestone Micrite
	Deep Marine	• Graded bed	Mudstone Chert Chalk Micrite

Exercise: Interpreting Paleo-Sedimentary Environments

On the other side of the room from the rock identification samples are nine groups of sedimentary samples. Each group contains both rocks and sedimentary structures that record the processes in the environment from which they came. Your job is to identify the sedimentary structures and rocks in each group, then determine what environment those samples are representing from the options in the box below. There are nine groups of samples and nine environments to choose from, so all the options will be used once and there will be no repeats. Complete this exercise by answering the questions that follow.

Environments "Word Bank"

- Terrestrial Environments: Alluvial Fan, Desert (sand dune), River Environment, Lake Environment
- Transition Environment: Tidal Basin, Beach Environment
- Marine Environment: Shallow Marine Clastic, Shallow Marine Carbonate, Deep Marine

Group A

- What are the three sedimentary structures seen here? ________________, ________________ and ________________
- What sedimentary rock is in this group? ________________
- *What sedimentary environment is represented? ________________

Group B

- What sedimentary structure is seen here? ________________
- What sedimentary rocks are in this group? ________________, ________________ and ________________
- *What sedimentary environment is represented? ________________

Group C

- What three sedimentary rocks are in this group? ________________, ________________, and ________________
- *What sedimentary environment is represented? ________________

Group D

- This group has no sedimentary structures, as the water in this environment is extremely quiet and still.
- The rocks in this group are all of one kind and are "laminated," meaning they were created out of very thin layers of similar—(but not identical)—sized particles.
- ⋆What sedimentary environment is represented? ______________________________

Group E

- What sedimentary structure is seen here? ______________________________
- What sedimentary rock is this? ______________________________
- ⋆What sedimentary environment is represented? ______________________________

Group F

- What three sedimentary rocks are in this group? ____________________, ____________ and ____________
- ⋆What sedimentary environment is represented? __________________

Group G

- What two sedimentary structures are seen here? ______________ and ______________.
- What two sedimentary rocks are in this group? __________ and __________
- ⋆What sedimentary environment is represented? ______________________________

Group H

- What two sedimentary rocks make up the coarse-grained rocks in this group?
  ______________________ and ____________________
- What sedimentary rock makes up the medium-grained rock in this group? (Use the additional description and specify the type based on its composition.) ______________________________
- What mineral gives this medium-grained rock its color? ______________________________
- *What sedimentary environment is represented? ______________________________________

Group I

- What sedimentary rock are all these fossils contained in? (Hint: Use the acid test.) _____________
- *What sedimentary environment is represented? ______________________________________

LABORATORY EXERCISE VI

ROCKS IN MOTION AND THE HAZARDS THEY PRODUCE

INTRODUCTION

Plate tectonics is a relatively new theory, developed in the 1960s, to explain why the continents appear to be "drifting" or changing positions over geologic time. Essentially it was discovered that Earth's lithosphere is broken up into several pieces called "lithospheric plates," each of which is constantly in motion on the slushy asthenosphere. Any processes associated with the creation, movement, and/or destruction of these plates is collectively called "plate tectonics."

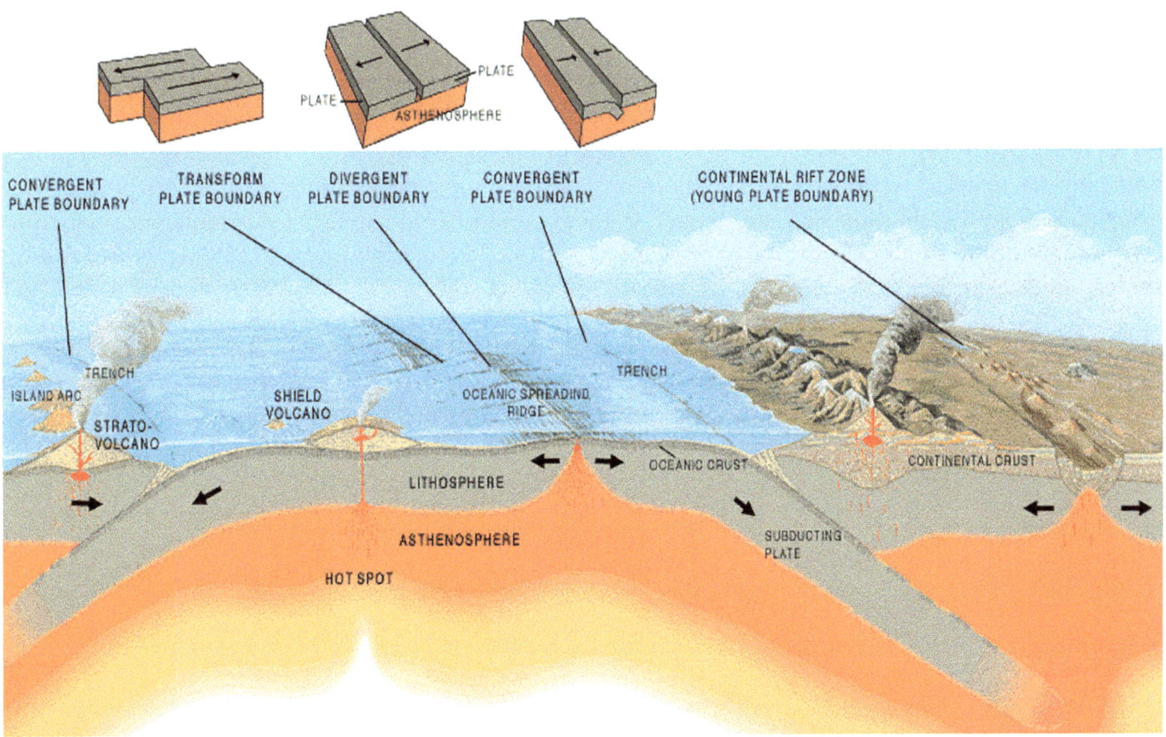

Figure 6.1 – USGS - Plate tectonics overview

Introduction: Plate Motion—The Rock Cycle and the Cycle of Life

According to the developing plate tectonics model (Figure 6.1), the continents are part of rigid lithospheric plates that move about relative to one another. Plates are generated along divergent boundaries, such as mid-ocean ridges, where two plates move away from each other. Magma then rises up between the two plates, fills in the gap, and cools to create new oceanic crust (basalt) on the edges of both plates. Plates are destroyed along convergent boundaries, where the edge of an oceanic plate subducts beneath an oceanic or continental plate, causing the subducting plate to be pulled back into internal Earth. When an oceanic plate subducts beneath another plate, a series of heat-induced reactions on the slab take place that result in the production of magma, which floats up and cools some of it forming new continental crust (grano-diorite). Over time, wind, water, and other surface elements will erode the continental crust and transport the sediment across the continent, where it can be used for nutrients and resources. Those same processes will ultimately bring the spent material to the deep ocean, where it too will be subducted during an oceanic plate collision or an oceanic continental plate collision. If there is a continent at the boundary of *both* plates over a subduction zone, in that case the two continents will collide and form a mountain range due to the crumpling and suturing of the two landmasses, as continental crust is too buoyant to be subducted. Plates may also slide past one another along transform fault boundaries, where plates are neither formed nor destroyed. Earth's size remains constant, because there are processes that simultaneously form and destroy lithospheric crust, thereby keeping production and destruction in equilibrium.

The tectonic cycle is in many ways the backbone of Earth's ability to support biology. It is through the cycles described above that fresh material is regularly brought to Earth's surface for biological consumption, and old material is returned to Earth's interior for a fresh supply of nutrients. If this process ever stopped, biology would soon become extinct, as there would be no supply of nutrients with which to grow crops and no supply of material to provide us with shelter and the myriad of items we depend on for day-to-day survival.

However, as natural and necessary as this process is, it too comes at a price. Earthquakes, volcanic eruptions, landslides, and tsunamis are just a few of the natural hazards plate tectonics brings to local environments. Understanding why and how these hazards come about is a pivotal part in being able to adapt to the changes they produce.

Objective

The objective of this lab is threefold:

1) To acquaint you with the fundamental characteristics of plate tectonics;

2) Use metamorphic rock type to evaluate potential tectonic-related hazards; and

3) Understand the relationships between tectonic processes and rock type/distribution in the landscape through use of a cross section.

Part A: Faulting

As discussed above, plates can interact with each other in one of three ways. They can collide, pull apart, or grind past each other. As such, there are three kinds of directed force (stress) that can be applied to the solid mass of rock (Figures 6.2 and 6.3 below). Compression compacts a block of rock and squeezes it into less space. Tension pulls a block of rock apart and increases its length. Shear smears a block of rock from side to side and will eventually tear it apart into two blocks of rock that slide past each other.

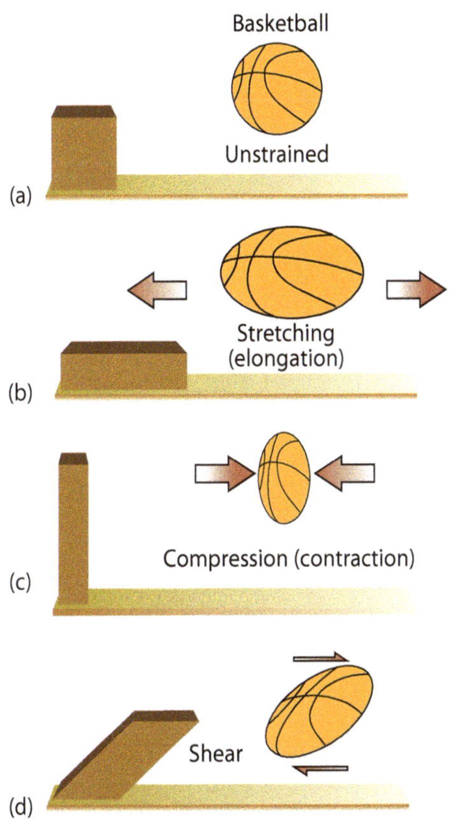

Figure 6.2

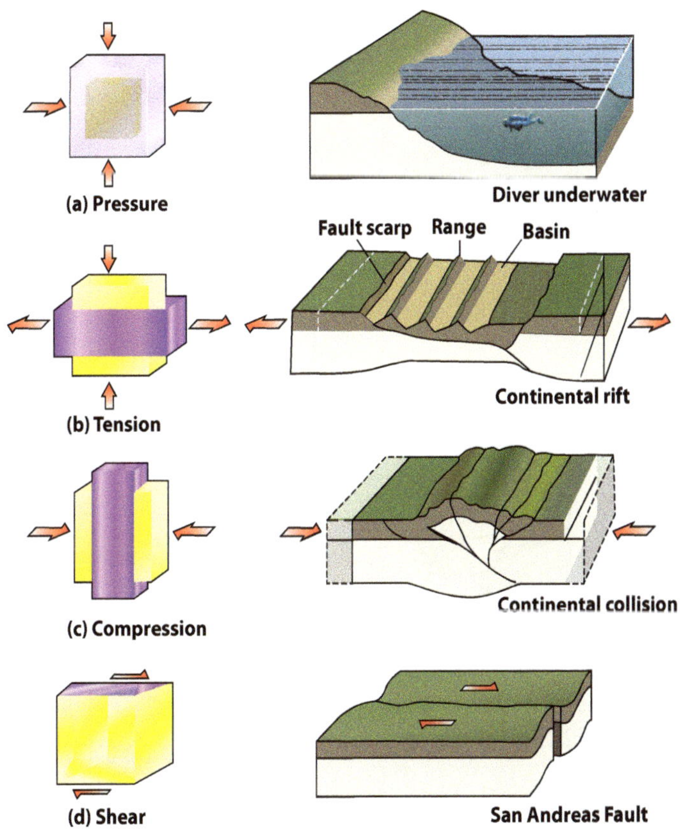

Figure 6.3

1. Analyze the three block diagrams in Table 6.A. Complete the three blank columns in the diagram to infer what kind of stress and strain is associated with each kind of faulting and plate boundary.

Fault Type	Shape Change? (Shortened, Elongated, or Neither)	Type of Stress? (Tension, Shear, or Compression)	Plate Boundary Type? (Convergent, Transform, or Divergent)
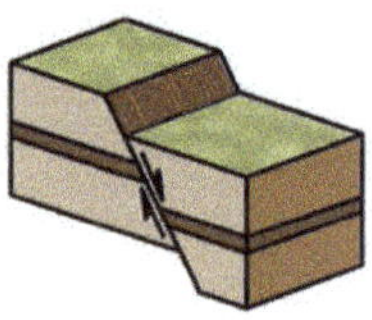			

Table 6.A: Fault types and motion.

Part B: Rocks Subjected to Heat and Pressure

As discussed in the previous exercise, major rock masses will collide, pull apart, or grind past each other as a consequence of tectonic movement—generating enormous stress on the rocks. If rocks on the cold, rigid surface were exposed to pressures of this magnitude, they would crack apart in response to the stress. However, rocks that encounter these stresses deeper in the Earth's interior behave very differently, as these rocks are also heated and are therefore able to deform without rupture. Metamorphism is a process that induces changes in a rock's texture and composition without shattering or melting the original material. Rocks that undergo textural and mineralogical changes due to experiencing temperature and pressure conditions different from those in their environment of formation are termed metamorphic rocks. Metamorphic rocks are the history books for the stressed and deformed rock zones associated with tectonic environments and contain many layers of information from what they were originally to the process that took place, which resulted in their alteration.

Metamorphic rocks are generated from the combined effects of heat and pressure—both of which increase with depth. While temperature increases due to processes taking place far inside the Earth, pressure increases with depth due to the increasing thickness of overlying rock. All rocks that are brought from shallow to greater depths will experience "confining pressure"—a unique type of pressure in which the rock is subjected to the same degree of force from all directions. Thus, when confining pressure is applied, because the rock is being "squeezed" equally in all directions, the rock will become denser and decrease in volume, but distortion to the shape and/or orientation does not result.

In contrast to the confining pressure that all metamorphic rocks are exposed to, isolated circumstances may expose certain rocks to differential stress—a situation in which the forces are not equally applied from all directions—consequentially stressing the rock more from one direction then from another. The rock will respond by stretching, compressing, or shearing - thus resulting in distortion to the shape of the rock and/or the minerals composing it.

Since all metamorphic rocks are exposed to elevated temperatures and confining pressure, it is the presence or absence of evidence for differential stress that provides the first tier of classification for metamorphic rocks—foliated or nonfoliated.

Foliated Rocks

Foliation (see Fig. 6.4) refers to a metamorphic texture in which flat or elongate minerals are preferentially aligned in response to exposure to high temperatures and differential stress in a high pressure environment. Rocks that display foliation are said to have undergone "regional metamorphism"- as the most common place for foliated rocks to form is in transform or reverse fault zones (both of which extend hundreds and sometimes thousands of miles)—where the shearing in the former, or compression in the latter, generate tremendous differential stress.

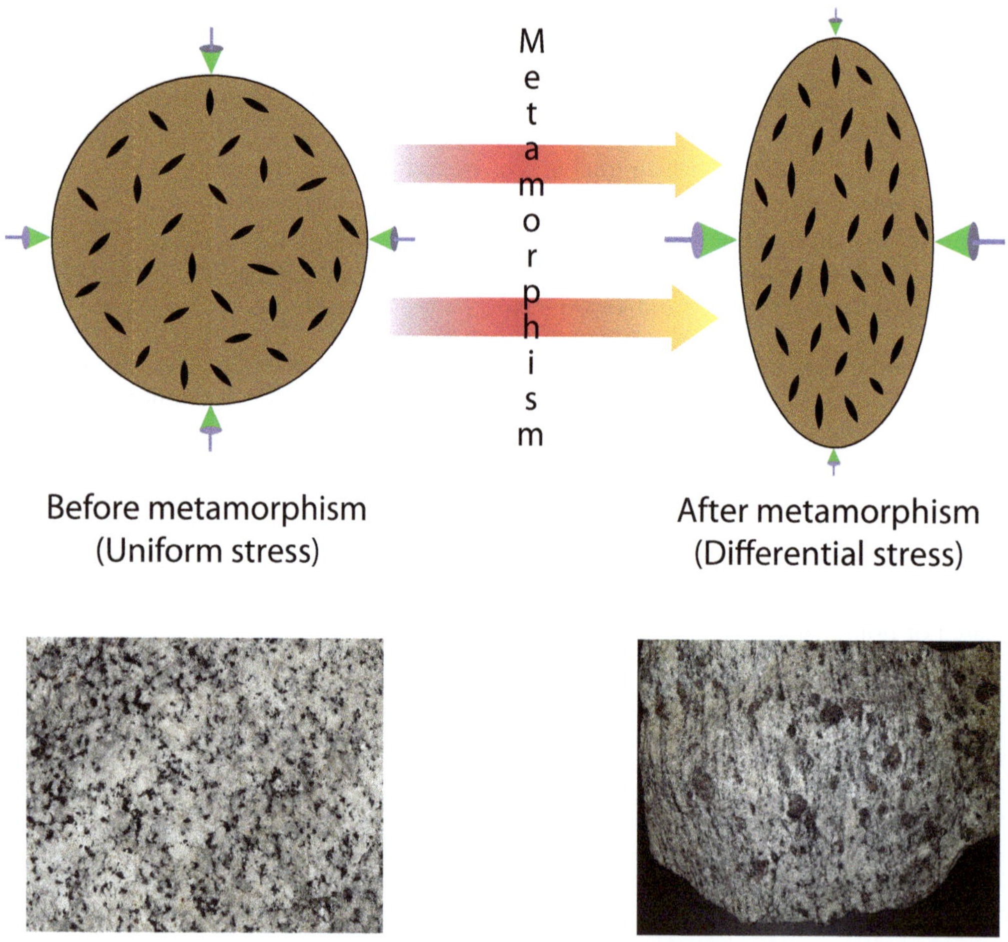

Figure 6.4 – Foliated Metamorphic Rocks

Non-Foliated Rocks

If the rock has been exposed to high temperatures and pressures—but no differential stress—the rock will not show any foliation. Instead, the textural identifier for these rocks is fused-together grains, which are created by a process called "recrystallization," in which the atoms of the original, individual grains, break their bonds and then re-bond as one entity, forming a mosaic-type texture (see Figure 6.5).

These rocks are usually formed when a magma intrusion makes it way up to the surface and "cooks" the surrounding rock. Metamorphic rocks that do not display foliation are said to have undergone "contact metamorphism"—as only the rocks that came in direct contact with the heat from the rising plume are metamorphosed. As such, unlike regional metamorphism, contact metamorphism affects only a localized area.

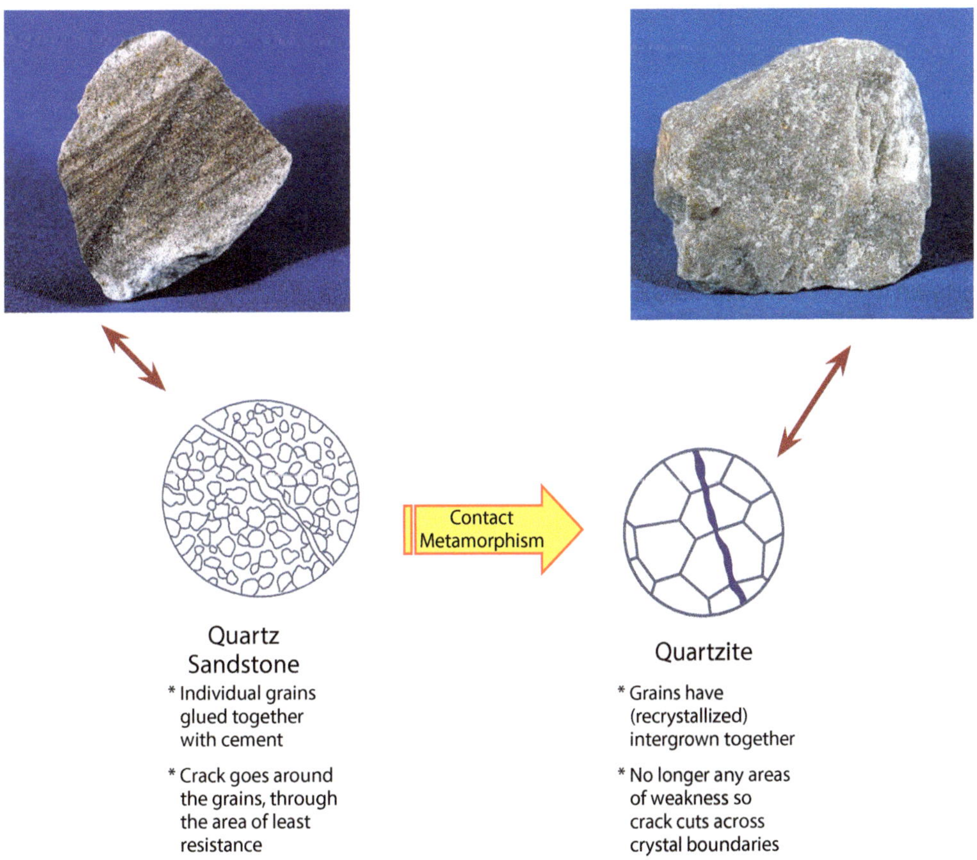

Figure 6.5 – Non-foliated Metamorphic Rocks

Table 6.B – Naming Metamorphic Rocks

	Rock	Hand Sample	Type of Foliation	Degree of Metamorphism	Description	Protolith (Parent Rock)
Foliated	Slate		Slaty cleavage	very low	Alignment of clay particles; fine grained, very slight buckling and pinching	Shale
Foliated	Phyllite		Phyllitic luster	low	Alignment of small micas; more coarse than slate, but still fine grained; clearly visible buckling and pinching	Shale
Foliated	Schist		Schistosity	moderate	Alignment of large micas; medium grained; Buckling, swirling, and pinching	Shale
Foliated	Gneiss	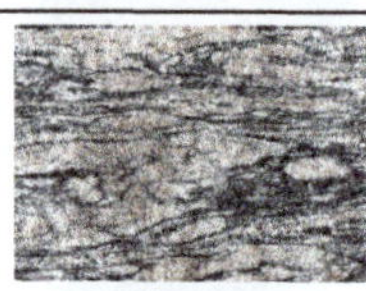	Gneissic	high	Separation of mafic and felsic minerals (i.e. gneissic banding). Coarse grained	Granite, diorite, gabbro
Foliated	Amphibolite		--	moderate	Shiny, black, bladed amphibole	Basalt, gabbro
Foliated or Non-foliated	Marble		--	--	Coarse, interlocking, fused together grains. Fizzes with HCL	Limestone
Foliated or Non-foliated	Quartzite		--	--	Fine, interlocking, fused together grains. Scratches glass.	Sandstone
Foliated or Non-foliated	Serpentinite		--	--	Contains asbestos	Basalt
Foliated or Non-foliated	Hornfels		--	--	Cooked fine grained rock	Shale, Siltstone, Basalt
Foliated or Non-foliated	Metaconglomerate/ Metabreccia	 Metaconglomerate	--	--	Large clasts that have been fused together	Conglomerate, Breccia

PART C: TYING IT ALL TOGETHER—TECTONIC ENVIRONMENTS, ASSOCIATED STRESSES AND HAZARDS, AND THE METAMORPHIC ROCKS EACH PRODUCES

Exercise 1: Spreading Centers

Seafloor spreading takes place at an underwater network of fractures called "midocean ridges" (i.e., spreading centers—see Figure 6.6) that are generated at the interface of the cool, rigid crust and hot mantle. As cracks open up in the oceanic lithosphere, the asthenosphere decompresses and melts. The magma then rises up to fill in the crack—where it cools to create new oceanic crust. This process is then repeated indefinitely.

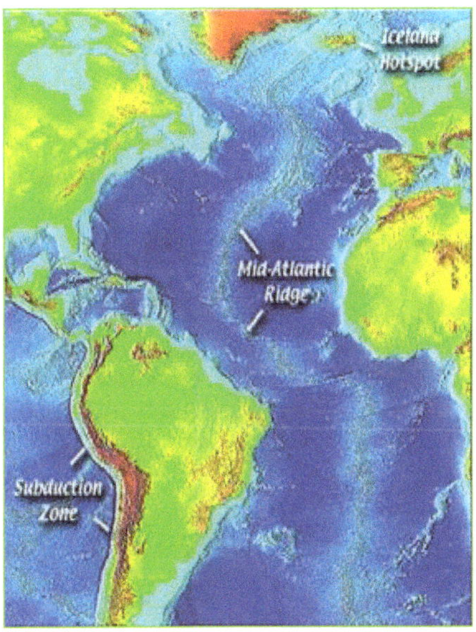

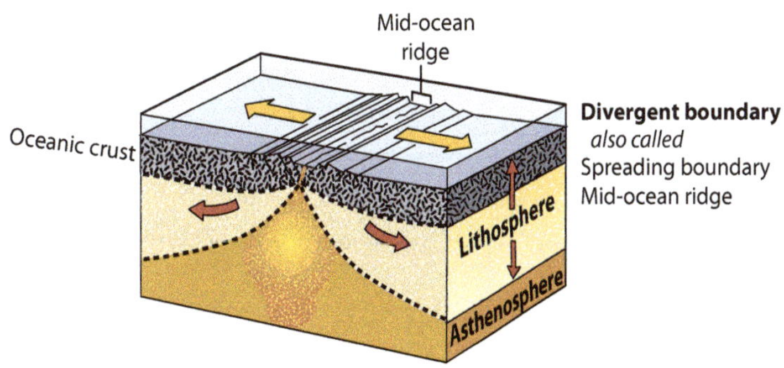

Figure 6.6 – Divergent margins

Left: The Mid-Atlantic Ridge, which splits nearly the entire Atlantic Ocean north to south, is probably the best-known and most-studied example of a divergent-plate boundary.

2. What kind of motion is taking place here? **Divergent**, **Convergent**, or **Strike-slip.**

3. What kind of stress is this fault undergoing? **Tension**, **Compression**, or **Shear.**

4. What primary type of fault would you expect at this location? **Normal**, **Reverse**, or **Transform.**

5. Look again at Figure 6.1, located at the beginning of the lab. Observe that the fissures in a mid-ocean ridge (spreading center) do not form a straight line, but are actually composed of segments that are offset from each other. What additional type of fault does this setup create in spreading center regions? ____________________________. What kind of stress would you expect to find there? ____

6. What hazards are spreading centers responsible for producing? **Volcanism, Earthquakes**, or **both**?

7. Are any of the hazards generated by a spreading center a serious threat to humans? **Yes** or **No?** Explain. ____
____
____

Exercise 2: Subduction Zones

Plates are destroyed along convergent boundaries, where the edge of one oceanic plate subducts beneath another plate, resulting in the subducting slab being pulled back into Earth's interior. As the oceanic plate descends into the hot mantle, water released from the subducting slab floats up and lowers the melting temperature of the upper mantle, resulting in the production of magma, which in turn migrates upward and produces volcanoes.

When an oceanic plate descends beneath another oceanic plate, the rising magma will form an "island arc," a chain of volcanic islands that parallel the convergent plate boundary (see Figure 6.7).

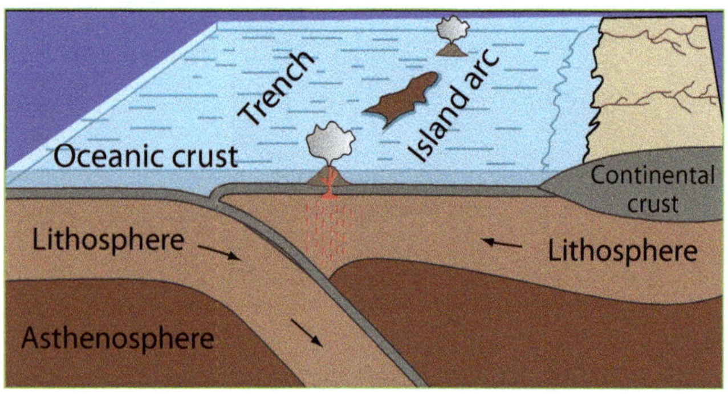

Figure 6.7 – Oceanic-Oceanic Convergent Boundary

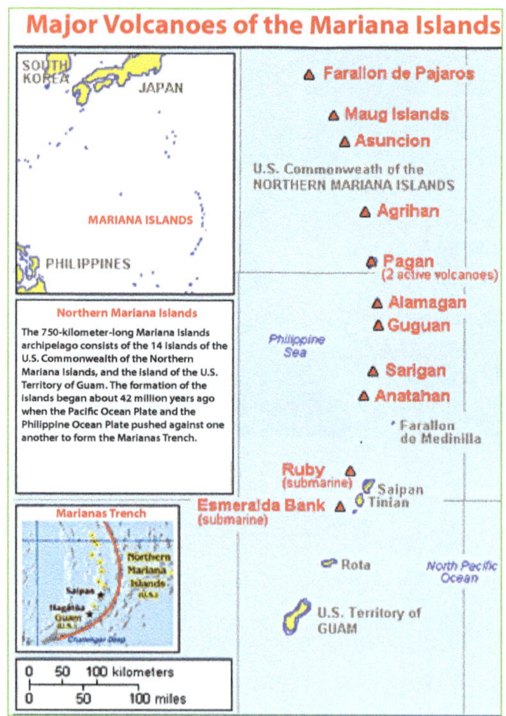

Oceanic-Oceanic Plate Boundary Example: Mariana Islands

8. Examine Sample A. Sample A is a rock that forms when basalt gets compressed in a high temperature, high pressure environment. Identify sample A. ______________________

9. Looking at Figure 6.7, this would be the rock that forms between the black arrows on the subducting slab. Notice the grains are all aligned in this rock. This is an example of foliation. Why are the grains arranged this way?
____
____

When an oceanic plate descends beneath a continental plate, the rising magma will float up and add to the existing continental crust, creating a chain of composite volcanoes on the overriding continent (see Figure 6.8).

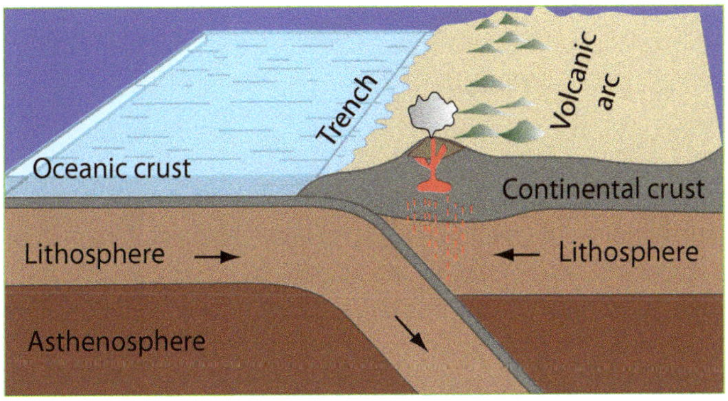

Figure 6.8 – Oceanic-Continental Convergent Boundary

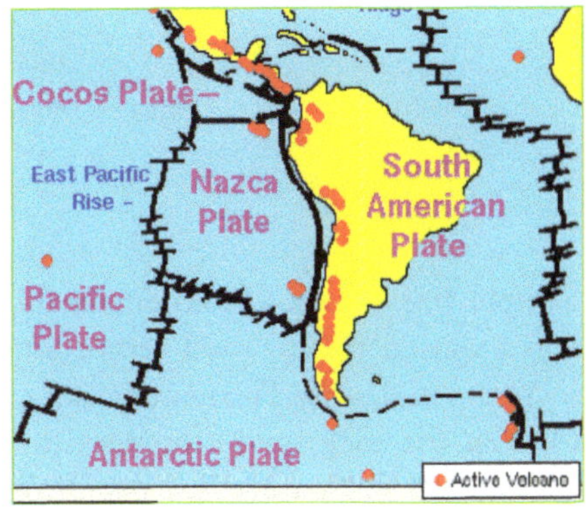

Oceanic-Continental Plate Boundary Example: Andes Mountain Range

As oceanic crust subducts beneath a continent, sediments that have been transported off the continent, as well as marine sediments scraped off the down-going slab, become wedged between the two colliding plates, forming a delta-shaped feature called an accretionary wedge/prism (see Figure 6.9). The wedge undergoes metamorphism due to higher temperatures and pressures associated with burial and compression resulting from the two colliding plates. As such, the accretionary wedge experiences a phenomenon called "tectonic foliation" (see Figure 6.10) in which the platy and elongate grains realign, and in some cases equant grains deform (see Figure 6.11) in addition to metamorphic changes in mineralogy.

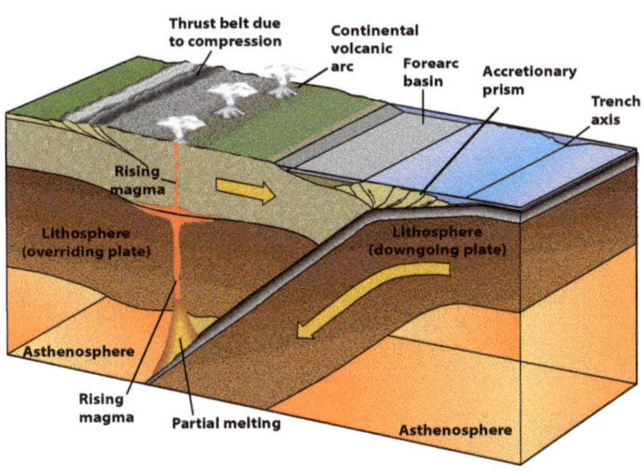

Figure 6.9 – Position of accretionary wedge

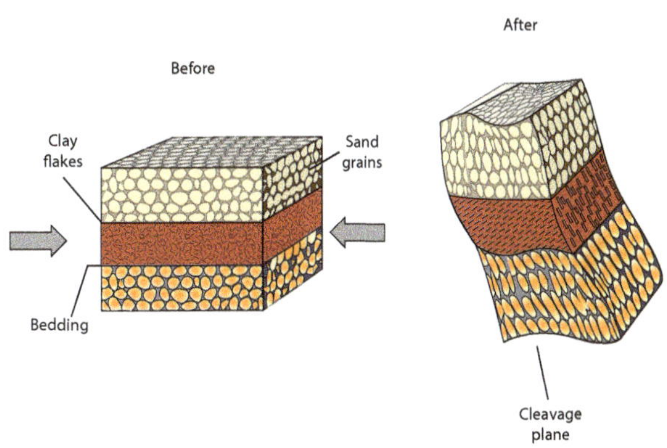

Figure 6.10 – Tectonic foliation

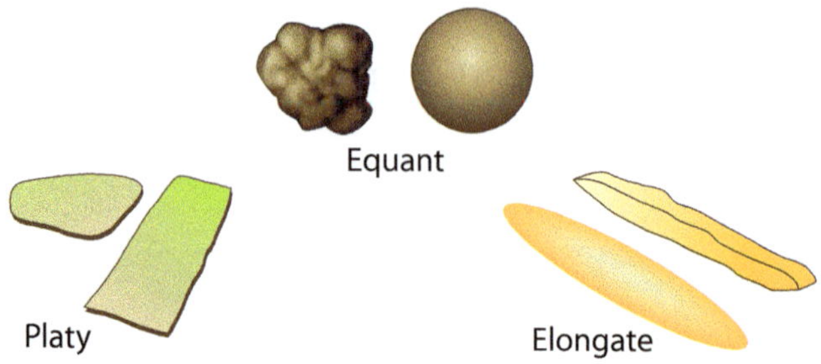

Figure 6.11 – Equant grains do not display foliation unless deformed, as there is no preferred orientation

Sample B is a shale, a sedimentary rock formed when mud-sized particles become lithified. When shale gets exposed to increased temperatures and pressures, as it does in an accretionary wedge, the rock increases in density. The new, denser rock has very fine ribbons running through it—a very subtle indicator that the rock was compressed in a high-temperature and high-pressure environment. As this rock gets buried further and subjected to even higher temperature and pressure conditions, muscovite, the primary mineral in the mud that composed the original shale, will become coarser and start to be visible to the naked eye. Eventually, the continually increasing stress will force the muscovite to develop clear, coarse crystals, and metamorphic minerals, such as garnet, begin to form. After this stage, any additional stress added to the rock will result in melting, and the rock will turn to magma, and return to the surface through volcanism in due time. As such, shale goes through three distinct metamorphic stages due to its being buried and compressed in an accretionary wedge environment.

Samples C, D, and E each represent one of the three metamorphic stages shale transforms to while stressed in an accretionary wedge environment.

10.

a. Which sample represents the first metamorphic stage? Sample C, Sample D, or Sample E? (Circle One)

b. What is your evidence? ____

c. What is the name of this rock? _____________________________________

11.

a. Which sample represents the second metamorphic stage? Sample C, Sample D, or Sample E? (Circle One)

b. What is your evidence? ____

c. What is the name of this rock? _____________________________________

12.

a. Which sample represents the third metamorphic stage? Sample C, Sample D, or Sample E? (Circle One)

b. What is your evidence? ____

c. What is the name of this rock? _____________________________________

Observe Samples F and G. Sample F is basalt, the same rock that makes up the oceanic crust, and sample G is the rock that forms when basalt is altered by hydrothermal fluids that are circulating throughout the subducting slab. Over time, slivers of metamorphosed oceanic crust are sheared off the down-going slab during subduction and become a part of the accretionary wedge. Notice the threadlike mineral in sample G. This mineral is asbestos, the very same material that is used to insulate homes and buildings, cushion brake pads, and make asbestos fabric, which is used as a fire retardant. Pay close attention to the crystal structure of the asbestos crystals. Notice they are shaped like tiny darts (see Figures 6.12a and 6.12b). It is this property that makes *airborne* asbestos dangerous in *large* quantities. Essentially, when these particles are freely floating in the atmosphere, they can be inhaled and will then treat the lungs like a dartboard, resulting in lung cancer after significant exposure. Have no fear here though. As you can see, these particles are not loose in the atmosphere, but are attached to the rock, there is an extremely small quantity, and your exposure is infinitesimal—so enjoy the view in a safe environment, as you likely will not get to see this again!

Figure 6.12a – Asbestos fibers

© EURICO ZIMBRES

Figure 6.12b – Scanning electron microscope image

USGS

13. Use Table 6.B to identify Sample G. What is the name of this rock? ______________________

Now refer back to Figure 6.9. Notice that the magma generated in the subduction zone from wet melting of the upper mantle must pass through the continental crust, and several layers of sedimentary rock before exiting from the volcano. In this part of the subduction zone, the shallow rocks are subjected to extremely high temperature, but no differential stress and significantly lower confining pressure in comparison to the accretionary wedge environment. For this reason, rocks that are affected by the upwelling magma do not show foliation, but instead display a texture called "fused-together grains," where the high heat instigated a reaction that fused together what were individual grains or crystals into an interlocking texture.

Compare Samples H and I. Sample H is a conglomerate, the same sedimentary rock you have kicked around outside thousands of times. Sample I is a rock formed when sedimentary conglomerate gets cooked by magma upwelling from a subducting slab. Look closely at Sample I's texture and notice how the grains have been "fused" together. Compare that texture to the way the sediment is "glued" together in the original conglomerate.

14. Identify Sample I: ______________________

15. Which do you think is stronger, i.e., less susceptible to breaking: the sedimentary conglomerate or Sample I? ______________________
    Explain: ______________________
    ____

When a rock is put under differential stress, it will only display the property of foliation if the grains are thin in one direction and elongate in another (i.e., flat, columnar, or dart-shaped). If, however, the rock is made from undeformed sand grains or calcite crystals, which have the same distance to all sides from the center of the grain, the rock might not appear foliated, even if it has been exposed to high heat and differential stress (refer back to Figure 6.11). Sample K is a quartzite, the rock formed when Sample J, a sandstone, is metamorphosed; Sample M is a marble, the rock formed when limestone, Sample L, is metamorphosed. Quartzite and marble can be found in environments of high temperature in conjunction with elevated confining pressure, like the area in a subduction zone where magma rises to the surface, or one of high temperature and differential stress like the accretionary wedge. Compare the parent rocks of sandstone and limestone, with their metamorphosed counterparts quartzite and marble. Notice that both the metamorphic rocks have a texture called fused-together grains, but no foliation. Because these rocks have such similar textures, it is often difficult to tell them apart with the naked eye.

16. What two diagnostic tools could you use to immediately distinguish these two samples? ______________________ and ______________________

17. Sample N is a rock formed when mudstone is exposed to contact metamorphism.

    a. Name this rock ____________________

    b. Is this rock **foliated or non-foliated?** (Circle one.)

18. Does this make sense with respect to how it was formed? Yes or No? Explain.
____
____
____

Exercise 3: Suture Zones

When a continental plate collides with another continental plate (see Figure 6.13), no subduction occurs, as the continental crust is too buoyant to sink. In this case, the two continents will "suture," forming a mountain range. As such, the mountain range is partially created from continental crust and overlying sedimentary layers that buckled due to the collision, and continues to grow higher as one plate plows into another.

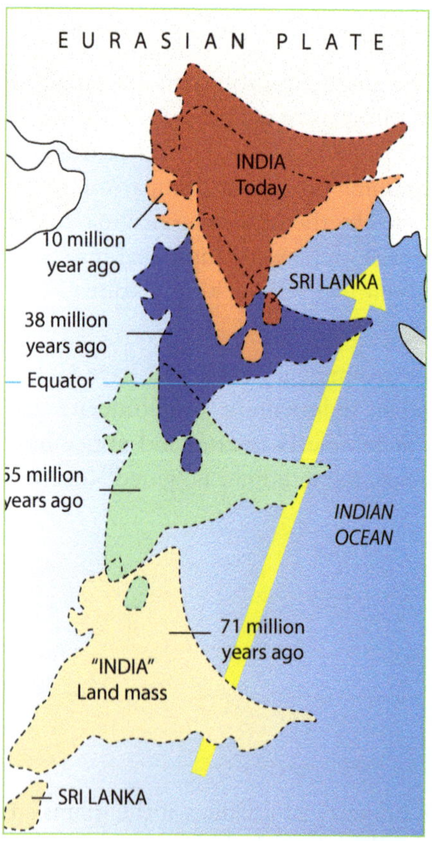

Continental-Continental Convergent Boundary Example:
Himalayas where India collided with Asia

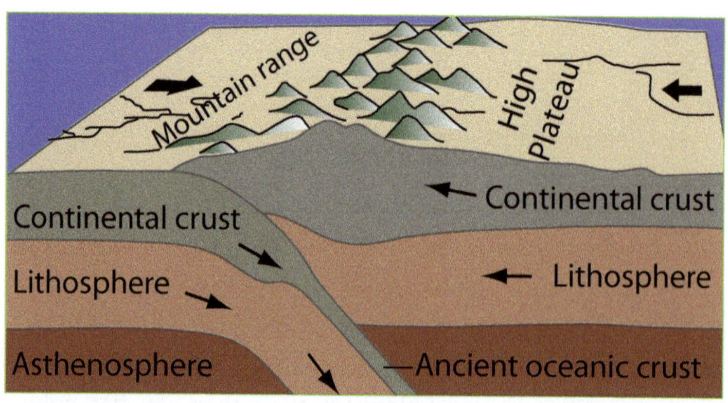

Figure 6.13 – Continental-Continental Convergent boundary

The differential stress that continental crust undergoes due to mountain building is astronomical, and will cause the elongate and platy grains in the granite comprising the continental crust to rotate such that their longer side is aligned in the direction of least stress. Samples O and P are rocks that formed from such a process.

19. What is the name of these rock samples? ______________________________

20. Do you see any resemblance between samples O and P and the granite (samples Q and R) that forms the continental crust? Yes or No? If yes, what?

    ____

    What's different? ______________________________

21. What kind of motion is taking place in all three of these collision zones?
    Tensional, Compression, or Shear.

22. What type of fault would you expect due to a plate collision? Normal, Reverse, or Strike-slip.

23. What hazards are these processes responsible for producing? Volcanism, Earthquakes, or both?

24. Are any of the hazards you circled in #23 a serious threat to humans? Yes or No?
    Explain: ______________________________

    ____

    ____

    ____

Exercise 4: Transform faults

Although transform faults are primarily located throughout spreading centers, where they are relatively harmless to humans, there are a few places on Earth where these faults develop as a result of entire plates grinding past each other in opposite directions. A classic example of this type of movement is the San Andreas Fault in California (see Figures 6.14a and 6.14b). If a continent is spanning across two plates that are grinding past each other, over geologic time, the continent will be ripped apart, each plate claiming its own piece.

25. What kind of motion is taking place here? **Divergent, Convergent, or Transform.**
26. What kind of stress is this fault under? **Tension, Compression, or Shear.**
27. What type of fault would you expect at this location? **Normal, Reverse, or Strike-slip.**
28. What hazards are these processes responsible for producing? **Volcanism, Earthquakes, or both?** (Circle all that apply)
29. Are any of the hazards you circled in #28 a serious threat to humans? **Yes or No?**
    Explain: ____

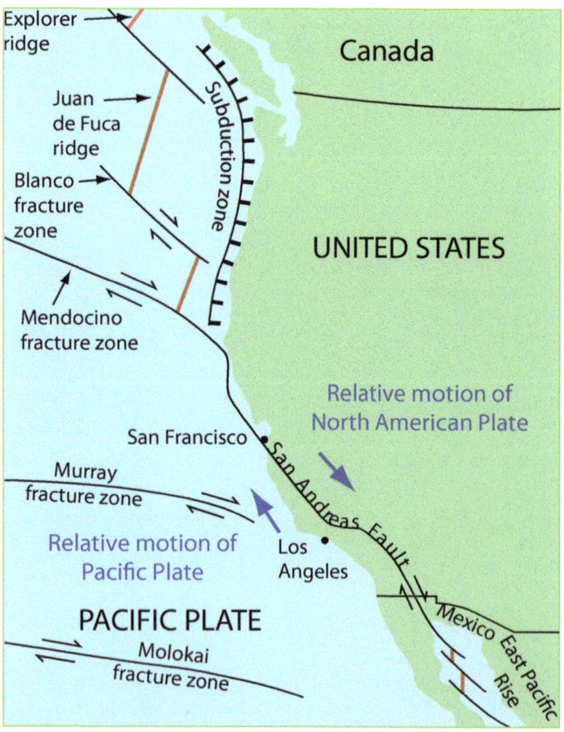

Figure 6.14a – Transform Fault Boundary

Figure 6.14b – Surface view of the San Andreas Fault

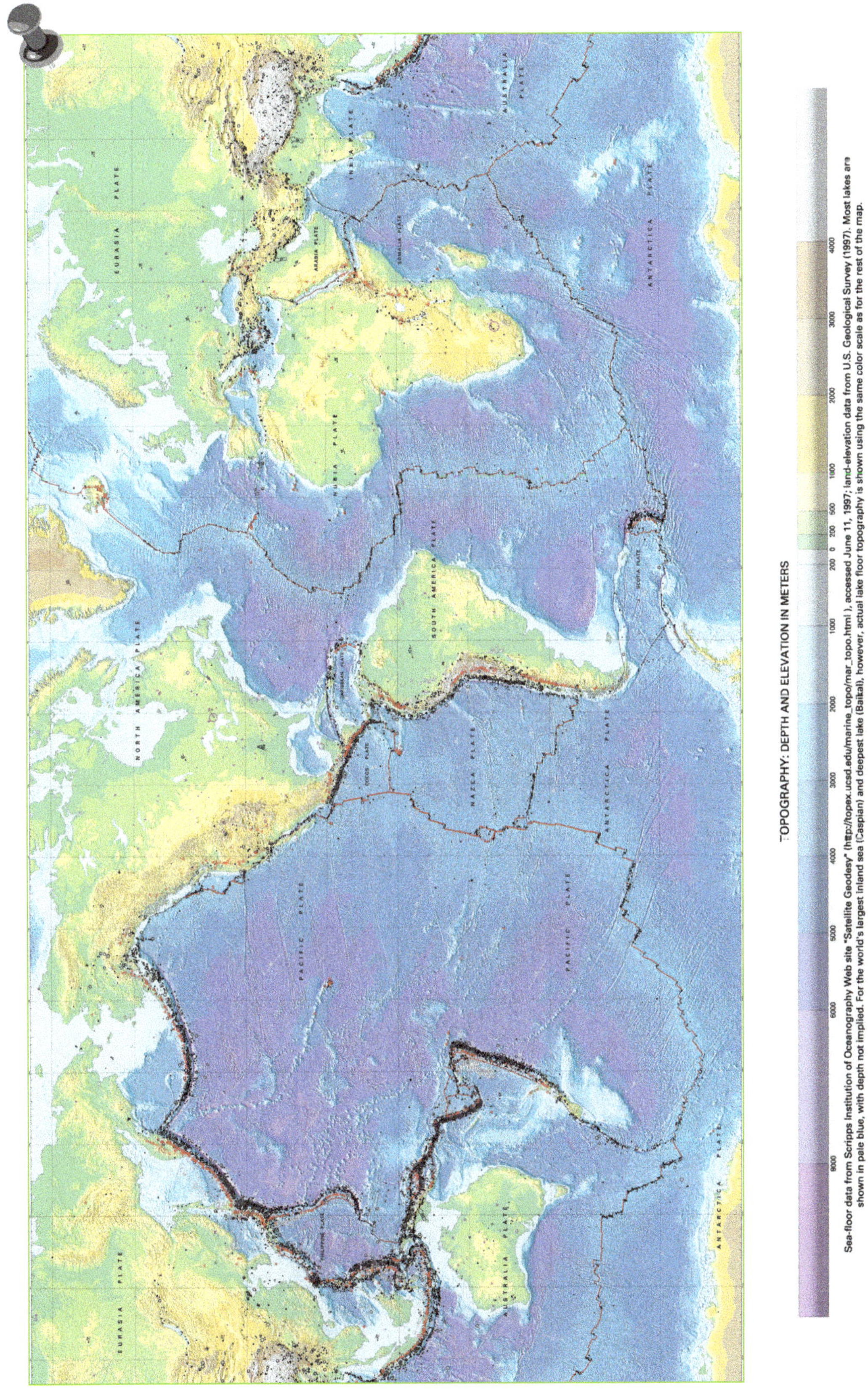

Figure 6.15: Relationship between plate boundaries, earthquakes, and volcanoes.

Legend for Figure 6.15

Volcanoes—Data from Global Volcanism Program, Smithsonian Institution, Washington, D.C.; accessed at http://www.volcano.si.edu/world/summary.cfm , March 16, 2005

- ▲ Erupted A.D. 1900 through 2003
- △ Erupted A.D. 1 through 1899
- △ Erupted in Holocene time (past 10,000 years), but no known eruptions since A.D. 1
- ▵ Uncertain Holocene activity and fumarolic activity

Impact Craters—Data from University of New Brunswick, Planetary and Space Science Centre, Earth Impact Database; accessed at http://www.unb.ca/passc/ImpactDatabase/, October 23, 2003 (also see Grieve, 1998). Geologic age span: 50 years to 2,400 million years. Crater diameter indicated below

- <10 km
- 10 to 70 km
- >70 km (shown at actual map scale)

Notable Events—Numbers next to a few symbols—of many thousands shown—denote especially noteworthy events, keyed to correspondingly numbered entries in tables found on the back of the map. These numbered events have produced devastating natural disasters, advanced scientific understanding, or piqued popular interest. They remind us that the map's small symbols may represent large and geologically significant events

- 19 Volcanoes
- 9 Earthquakes
- 23 Impact craters

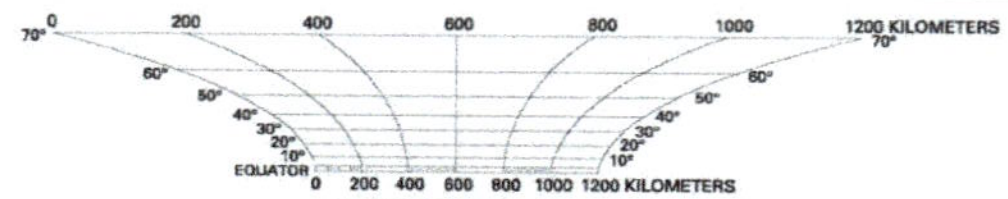

MERCATOR PROJECTION
Scale 1:30 000 000 at the Equator
One centimeter equals 300 kilometers (~186 miles) at the Equator
One inch equals 473 miles (~762 kilometers) at the Equator

Suggested citation: Simkin, Tom, Tilling, R.I., Vogt, P.R., Kirby, S.H., Kimberly, Paul, and Stewart, D.B., 2006, This dynamic planet; World map of volcanoes, earthquakes, impact craters, and plate tectonics: U.S. Geological Survey Geologic Investigations Series Map I–2800, 1 two-sided sheet, scale 1:30,000,000.

Plate Tectonics

- Divergent (sea-floor spreading) and transform fault boundaries—Red lines mark spreading centers where most of the world's volcanism takes place; thickness of lines indicates divergence rate, in four velocity ranges. White number is speed in millimeters per year (mm/yr) from DeMets and others (1994). The four spreading-rate ranges are <30 mm/yr; 30–59 mm/yr; 60–90 mm/yr; and >90 mm/yr. Thin black line marks the plate boundary, whether sea-floor spreading center or transform fault. On land, divergent boundaries are commonly diffuse zones (see interpretive map to the left); therefore, most are not shown. The only transform faults shown on land are those separating named plates
- 9 ← Plate motion—Data from Rice University Global Tectonics Group. Length of arrow is proportional to plate velocity, shown in millimeters per year. These approximate rates and directions are calculated from angular velocities with respect to hotspots, assumed to be relatively fixed in the mantle (see plate motion calculator at http://tectonics. rice.edu/hs3.html)
- 46 Plate convergence—More accurately known than "absolute" plate motion (above), convergence data are shown by arrows of uniform length showing direction and speed, in millimeters per year, relative to the plate across the boundary. Data from Charles DeMets (University of Wisconsin at Madison, written commun., 2003) and Bird (2003)

Earthquakes—Data from Engdahl and Villaseñor (2002). From 1900 through 1963, the data are complete for all earthquakes ≥6.5 magnitude; from 1964 through 1999, the data are complete for all earthquakes ≥5.0 magnitude. Most location uncertainties <35 km. Eleven more recent major or great earthquakes (magnitude ≥7.7) have been added for completeness through 2004; data from USGS National Earthquake Information Center at http://neic.usgs.gov/ , accessed January 4, 2005. An epicenter is the surface location of the first rupture on an earthquake fault. Symbols shown represent epicenters. For earthquakes larger than about magnitude 7.0, the size of the rupture zone, which can extend hundreds of kilometers from the epicenter, is larger than the symbols used on this map

Depth to earthquake, in km	Magnitude of earthquake			
	5.0–5.9	6.0–6.9	7.0–7.9	≥8.0
<60	•	•	○	○
60–300	•	•	○	○
>300	•	○	○	○
Global average occurrence[1]	1319/yr	134/yr	17/yr	1/yr

[1]Earthquakes of magnitude <5 (not shown on map) are much more frequent, with ~13,000/yr in the 4.0–4.9 range alone. Data from USGS National Earthquake Information Center.

- ○ Earthquakes that occurred from 1750 to 1963 within stable plate interiors on continents—Data from A.C. Johnston (Center for Earthquake Research and Information, University of Memphis, written commun., 2002). Even though these epicenters do not meet the precise location criteria of Engdahl and Villaseñor (2002), they are plotted here to remind readers of the potentially hazardous earthquakes that are distant from known plate boundaries. Size of symbol proportional to earthquake magnitude
- ○ Notable pre-1900 earthquakes—Nos. 1, 2, 3, 6, and 7 (see table 3, on back)

PART D: SUMMING UP

30. Complete the following summary sheet by placing an X in the appropriate box for each geologic process.

Geologic Process	Type of plate boundary		
	Divergent	Convergent	Transform fault
Creation of oceanic crust			
Destruction of oceanic crust			
Volcanism			
Creation of mountain ranges			
Compression			
Tension			
Shear			
Normal faulting			
Strike-slip faulting			
Reverse faulting			
Shallow EQs only			
Shallow and deep EQs			

31. Based on all the information above, what specific geographic location can be considered one of the most dangerous places on Earth? (*Hint*: See Figure 6.15.)

____

Part E: Tectonic Distribution of Rocks in the Landscape

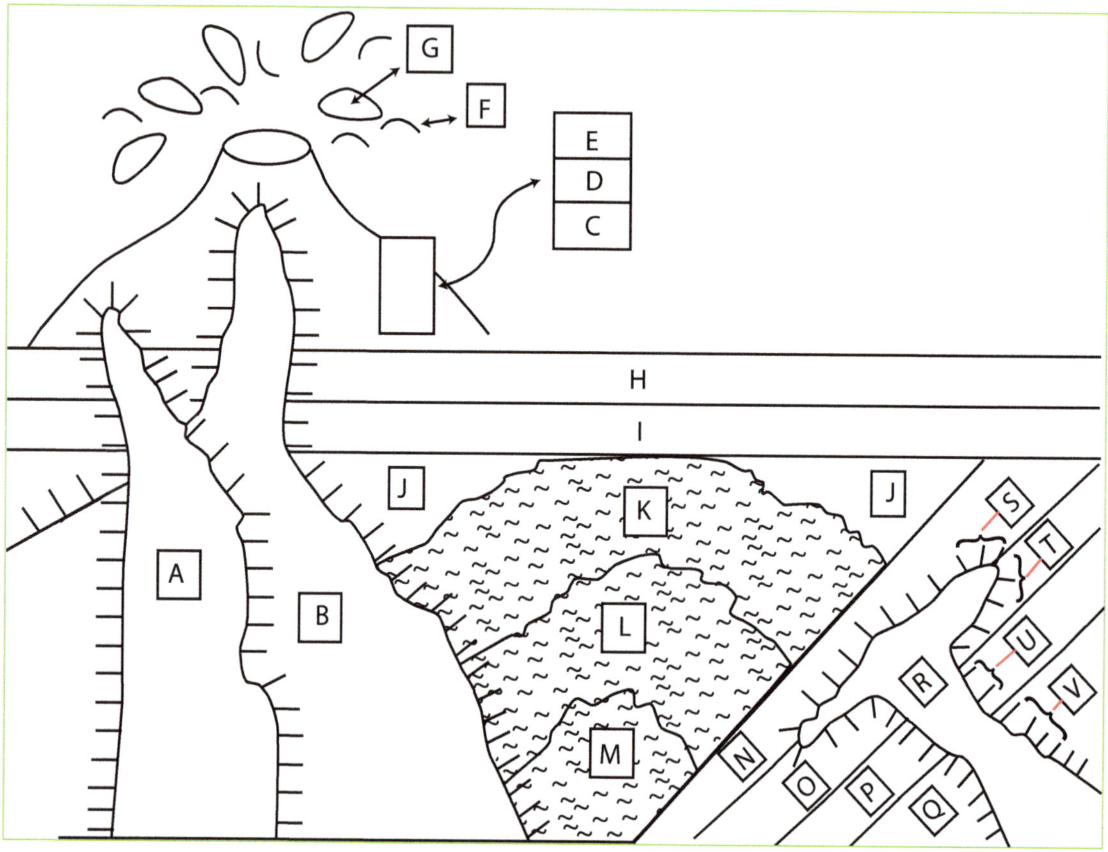

Figure 6.16 – Cross-section of three rock types in a tectonic environment

As demonstrated throughout the igneous, sedimentary, and metamorphic rocks laboratories, the type, distribution, and movement of any rock throughout the landscape is a direct result of tectonic activity. Figure 6.16 is a cross section representing the distribution of all three rock types in a tectonic environment. Throughout the lab are samples that correspond to the letters in the cross section. Use the hand sample in conjunction with the cross section and rock-naming charts from past labs to identify each rock in the diagram. You need to give each rock its specific name (e.g., granite, sandstone, gneiss etc.) and not just identify it as igneous, sedimentary, or metamorphic. As you identify each rock, fill in the chart on the following pages with the requested information.

Sample	Rock Classification	Environment of Formation	Name
A	Igneous, Sedimentary, or Metamorphic	If Igneous: Intrusive or Extrusive If Sedimentary: Detrital or Chemical If Metamorphic: Primarily Heat or Heat & Pressure	
B	Igneous, Sedimentary, or Metamorphic	If Igneous: Intrusive or Extrusive If Sedimentary: Detrital or Chemical If Metamorphic: Primarily Heat or Heat & Pressure	
C	Igneous, Sedimentary, or Metamorphic	If Igneous: Intrusive or Extrusive If Sedimentary: Detrital or Chemical If Metamorphic: Primarily Heat or Heat & Pressure	
D	Igneous, Sedimentary, or Metamorphic	If Igneous: Intrusive or Extrusive If Sedimentary: Detrital or Chemical If Metamorphic: Primarily Heat or Heat & Pressure	
E	Igneous, Sedimentary, or Metamorphic	If Igneous: Intrusive or Extrusive If Sedimentary: Detrital or Chemical If Metamorphic: Primarily Heat or Heat & Pressure	
F	Igneous, Sedimentary, or Metamorphic	If Igneous: Intrusive or Extrusive If Sedimentary: Detrital or Chemical If Metamorphic: Primarily Heat or Heat & Pressure	
G	Igneous, Sedimentary, or Metamorphic	If Igneous: Intrusive or Extrusive If Sedimentary: Detrital or Chemical If Metamorphic: Primarily Heat or Heat & Pressure	
H	Igneous, Sedimentary, or Metamorphic	If Igneous: Intrusive or Extrusive If Sedimentary: Detrital or Chemical If Metamorphic: Primarily Heat or Heat & Pressure	
I	Igneous, Sedimentary, or Metamorphic	If Igneous: Intrusive or Extrusive If Sedimentary: Detrital or Chemical If Metamorphic: Primarily Heat or Heat & Pressure	
J	Igneous, Sedimentary, or Metamorphic	If Igneous: Intrusive or Extrusive If Sedimentary: Detrital or Chemical If Metamorphic: Primarily Heat or Heat & Pressure	
K	Igneous, Sedimentary, or Metamorphic	If Igneous: Intrusive or Extrusive If Sedimentary: Detrital or Chemical If Metamorphic: Primarily Heat or Heat & Pressure	

Sample	Rock Classification	Environment of Formation	Name
L	Igneous, Sedimentary, or Metamorphic	If Igneous: Intrusive or Extrusive If Sedimentary: Detrital or Chemical If Metamorphic: Primarily Heat or Heat & Pressure	
M	Igneous, Sedimentary, or Metamorphic	If Igneous: Intrusive or Extrusive If Sedimentary: Detrital or Chemical If Metamorphic: Primarily Heat or Heat & Pressure	
N	Igneous, Sedimentary, or Metamorphic	If Igneous: Intrusive or Extrusive If Sedimentary: Detrital or Chemical If Metamorphic: Primarily Heat or Heat & Pressure	
O	Igneous, Sedimentary, or Metamorphic	If Igneous: Intrusive or Extrusive If Sedimentary: Detrital or Chemical If Metamorphic: Primarily Heat or Heat & Pressure	
P	Igneous, Sedimentary, or Metamorphic	If Igneous: Intrusive or Extrusive If Sedimentary: Detrital or Chemical If Metamorphic: Primarily Heat or Heat & Pressure	
Q	Igneous, Sedimentary, or Metamorphic	If Igneous: Intrusive or Extrusive If Sedimentary: Detrital or Chemical If Metamorphic: Primarily Heat or Heat & Pressure	
R	Igneous, Sedimentary, or Metamorphic	If Igneous: Intrusive or Extrusive If Sedimentary: Detrital or Chemical If Metamorphic: Primarily Heat or Heat & Pressure	
S	Igneous, Sedimentary, or Metamorphic	If Igneous: Intrusive or Extrusive If Sedimentary: Detrital or Chemical If Metamorphic: Primarily Heat or Heat & Pressure	
T	Igneous, Sedimentary, or Metamorphic	If Igneous: Intrusive or Extrusive If Sedimentary: Detrital or Chemical If Metamorphic: Primarily Heat or Heat & Pressure	
U	Igneous, Sedimentary, or Metamorphic	If Igneous: Intrusive or Extrusive If Sedimentary: Detrital or Chemical If Metamorphic: Primarily Heat or Heat & Pressure	
V	Igneous, Sedimentary, or Metamorphic	If Igneous: Intrusive or Extrusive If Sedimentary: Detrital or Chemical If Metamorphic: Primarily Heat or Heat & Pressure	

Laboratory Exercise VII

Rocks – The History Books of the Earth

Introduction

In one form or another, history has always been of interest to humanity. Since the dawn of civilization, mankind has gone in search of its roots, attempting to understand the sequence of events that led to its current circumstances. Such endeavors eventually resulted in the systemic study of socioeconomic, political, and cultural evolutions in the forms of art history, cultural anthropology, and archaeology, to name a few. From there, it was just a hop, skip, and a jump before we started asking questions about our physiological development, which led to the eventual organization of the biology, biochemistry, zoology, paleontology, and physical anthropology disciplines. As history in all these fields was uncovered, examined, and evaluated over the centuries, it became clear that the status of existing biota, now and at any time in the past, is a direct reflection of the era's global and local environmental circumstances—such as availability of resources, ease of transportation, and climatic conditions. Shortly thereafter, it was recognized that rocks preserve those past conditions, and a careful analysis of Earth's historical records could be used to investigate, describe, and interpret the physical, chemical, climatological, and biological conditions present at any stage in our planet's history. With the realization that the rocks beneath our feet are essentially Earth's diary, geology as a science was born. But before we begin investigating how geologists read the history written in the rock record, it's important to address why we would want to do that. What information is gained by understanding how Earth has behaved throughout its 4.6 billion–year lifetime? How is that information relevant to our modern, everyday lives?

Just like any sub-study of history, geology has both academic and practical applications. Because there are no limits to mankind's curiosity, many geologists spend their lives combing through Earth's archives trying to piece together the composition and formation of our planet; identify Earth's cycles, the rates at which they operate, and how they've changed over time; decipher the evolution of life; study environmental transformation; and understand how our planet compares to others in the universe. While certainly interesting and informative on an academic level, Earth's history can also be used to enhance our modern-day lives. It is the examination of Earth's past environments and their distribution throughout time and space that leads us to suitable locations for profitable mining of minerals, fossil fuels, and a myriad of other natural resources. By evaluating Earth's past behavior with respect to volcanic eruptions, earthquakes, landslides, flooding, and a plethora of other natural processes, we can understand why and where these changes come about, which leaves us better equipped to take advantage of the resources they deliver and protect ourselves from the dangerous side effects they produce. Finally, by charting how and at what rate Earth has changed prior to human evolution, we have a standard against which to measure the current changes. We can therefore more confidently assess whether the global systemic changes we are currently experiencing are just nature taking its evolving course, or whether anthropogenic alterations are interfering with Earth's natural rhythms and compromising the very processes that sustain us. As such, geology—just as any historical discipline—allows us to learn about our past in hopes of easing the path into the future.

Objectives

Keeping the above in mind, the threefold objective of this lab is to:

1) Familiarize you with the concept of geologic time and examine how the rates and frequencies of geologic processes differ from those of human time scales;

2) Introduce you to various techniques used by geologists to piece together Earth's history; and

3) Practice ordering, correlating, and interpreting stratigraphic data for both academic and practical applications.

Part A: A Geologist's Concept of Time

The first step in learning how to untangle Earth's 4.6 billion–year history is understanding the scales on which geologic time operates. As will become clear shortly, what is "slow or rare" on human time scales is "fast and frequent" when viewed over geologic time. To demonstrate this point, if we condensed all the time from Earth's birth to today into a 24-hour period, the time line would look something like this (see Figure 7.1):

- Earth was born at 12:00 am (4.6 billion years ago).
- We know only bits and pieces about what went on between 12:00 am and 6:00 pm, ¾ of Earth's history (the first 4 billion years of Earth's life).
- Between 6:15 pm and 6:45 pm, the Cambrian Explosion occurred (580–535 million years ago).
- 11:25 pm marks the extinction of the dinosaurs (65 million years ago).
- 11:59 and 55 seconds: Homo sapiens appear (200,000 years ago).
- All of recorded human history occurred in the last fraction of a second before the clock struck midnight.

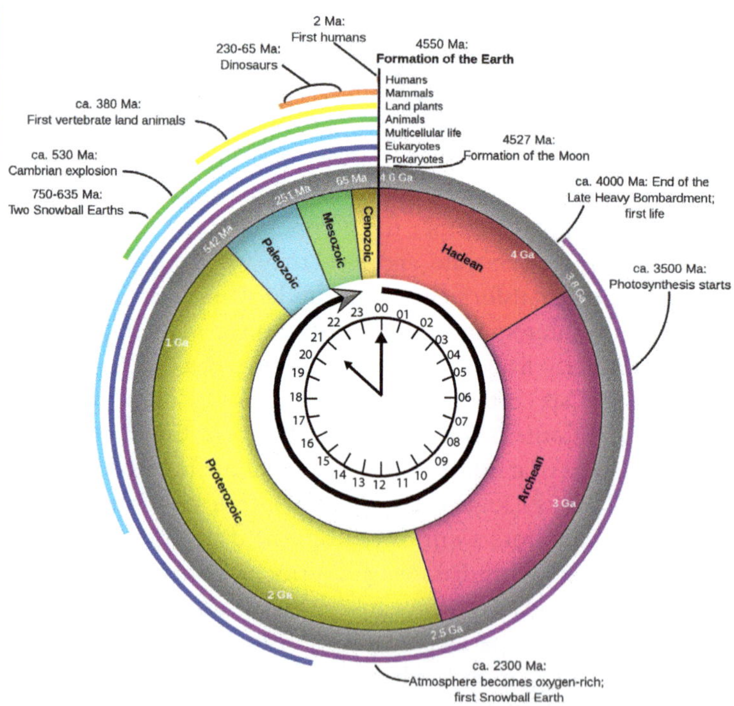

Figure 7.1 – Earth's Clock

As such, a few hundred thousand years is but the blink of eye in geologic time, and all geologic events must be scaled accordingly.

Now that you can conceptualize how we scale geologic time, next we need to examine the effect this scaling has on geologic rates and frequencies. The following exercise is designed to help you shift your perception of time from human standards to geologic standards.

Exercise: Getting Acquainted with Geologic Time

1. When working with the 4.6 billion years that comprises Earth's history—as opposed to the decades and centuries that concern mortal man—the first important question that has to be evaluated is "What constitutes fast?" Consider the following two examples, which highlight the vast difference between "fast" in geologic time and "fast" in human time.

   a. Rates in Geologic Time

      - 12,300 years ago, the waters from Lake Erie overflowed onto the Niagara Escarpment, and formed a 27-mile-long river that currently drains into Lake Ontario. As the Niagara River flowed over the escarpment into lower-elevation Lake Ontario, the falls began eroding the underlying substrate, which ultimately formed the Niagara Gorge. Over the past 12,300 years, Niagara Falls has eroded the gorge 7 miles to the south toward Lake Erie, at a rate of approximately 3 feet per year (see Figure 7.2). However, because significant amounts of water along the Niagara River have recently been diverted for hydroelectric power, the current erosion rate has been reduced to approximately 1 foot per year. At present, there are approximately 20 miles between the location of Niagara Falls and the end of the gorge where Lake Erie is located. Hypothetically speaking, if the natural erosion rate of 3 feet per year were to continue for the remaining 20 miles, in how many years will the falls reach Lake Erie? (Show your work.) ____________ years

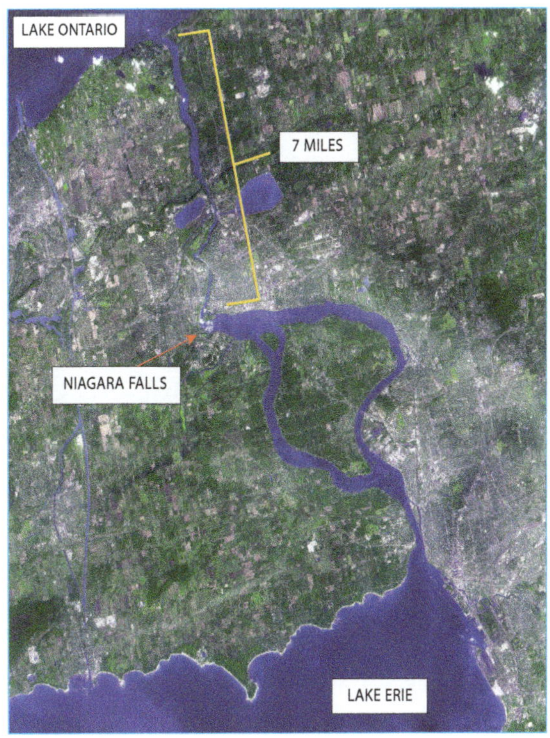

Figure 7.2 – Erosion of Niagara Gorge

      - How many years total will it have taken to erode all 27 miles of the gorge?
        ______________ years

      - Using your answer from the question above, calculate what percentage of Earth's history the eroding of the gorge represents. (Show your work.) ________________________ %

- Within the context of geologic time, would you consider the Niagara erosion event to be happening "quickly" or "slowly?" ____________________

- Within the context of a human lifetime, would you consider the Niagara erosion event to be happening "quickly" or "slowly?" ____________________

b. Rates in Human Time

- According to a 2009 report by the World Bank, average life expectancy in the United States is approximately 78 years. If someone who lived to the age of 78 built a home, and the construction took the same percentage of their lifetime as the eroding of the Niagara Gorge took of Earth's lifetime, how long would the construction take? (Show your work.) ____________days

c. Note that in both the above cases, the same percentage of time is being discussed, and therefore both events are happening at the same relative rate. This being the case, why don't these two events mark the passing of the same amount of time?

____

____

2. Now that it has been established that the term "fast" takes on a whole new meaning when viewed in the context of 4.6 billion years, the next concept that requires some adjustment when evaluating geologic time is frequency. Do events that are considered rare and unlikely on human time scales appear more frequently when viewed over much longer stretches? In order to answer this question, consider the scenarios below.

a. Based on geologic evidence, there is a virtual certainty that a mega-flood will occur in a particular region sometime within a 500-year period.

- What is the likelihood that this catastrophic mega-flood will occur in any given year? (Show your work.) ______% chance

- Given the odds you calculated in the question above, would you be concerned about buying land in this region? **Yes** or **No**? Explain.

____

____

- How many times did this mega-flood occur over the past 500,000 years? (Show your work) __________times

b. Yellowstone National Park is home to one of the world's "super-volcanos," which produces the kind of eruptions that have the potential to cause mass extinctions and change global climate. Geologic evidence has shown that the Yellowstone Volcano erupts approximately every 700,000 years.

- How many times has this eruption taken place over the past 35 million years? __________times

- Last eruption is believed to have occurred around 640,000 years ago. In approximately how many years is the next eruption expected? ______________

- From the standpoint of geologic time, is the next eruption "any day," or way off into the future? ________________

- From the standpoint of human time, is the next eruption "any day," or way off into the future? ________________

c. Given your answers to the questions above, would you agree or disagree with the following statement: "By human standards, 'catastrophes' are rare events, but must be considered frequent and normal on a geologic time scale." **Agree or Disagree**

3. With the understanding that rates and frequencies are viewed differently on a geologic time scale, there remains one more issue that must be addressed before examining how geologists uncover and interpret Earth's history. Because we are dealing with such large stretches of time and are investigating processes no human was or ever will be around long enough to witness, the only data we have to work with concerning the rates at which long-term events occurred is when the process started and when it ended. For example, the Indian Plate is believed to have broken off from Gondwanaland approximately 90 million years ago and collided with the Asian Plate between 55 and 50 million years ago—but we have no knowledge of whether it moved at a steady pace or if its velocity fluctuated during its journey. As such, when evaluating the rates of Earth's past processes, more often than not the process will be discussed in terms of its "average" rate. Does this number accurately depict the real rate at which the process occurred? Consider the following examples.

a. You and your friends decide to go on a cross-country trip. Over the course of three weeks, you drive from Los Angeles, California, to Washington, D.C., a total of approximately 3000 miles.

- Using exclusively the total travel time and distance, calculate the average speed at which the car was traveling in mi/hr (miles per hour). (Show your work.) _______mi/hr

- Is the number you calculated above an accurate representation of the real rate the car was moving? **Yes** or **No**?

- Assuming the car stayed within about 15 miles per hour of a 70 mile per hour speed limit, within what range was the "real rate?" ______________ to ______________ mi/hr

- Was the car moving nonstop throughout the entire three-week period? **Yes** or **No**?

- What circumstances are causing the real rate to be so drastically different from the average rate? ____
____
____

b. Humans, like all biological beings, go through three distinctive stages between birth and death: growing, sustaining, and degrading. Based on the average lifespan of ~ 80 years,

   The Growing Phase: 0–18 years
   The Sustaining Phase: 19–35 years
   The Degrading Phase: 36–80 years

- How many years does the growth phase consist of? ____________ years

- How many years does the sustaining phase consist of? __________ years

- How many years does the degrading phase consist of? __________ years

- Given that the average human lifespan is 80 years, and each person undergoes three life phases between birth and death, what is the average number of years each phase lasts? __________ years

- Is the average an accurate representation of the rate at which this process progresses? Yes or No?

Now that you've become acquainted with a geologist's concept of time, the next step is to investigate how geologists decipher the sequence of events that created the various features comprising a locale's landscape, as well as when each event occurred. Techniques used for this task are categorized into two main groups: Relative Dating Techniques and Numerical Dating Techniques. Relative dating uses the relationships between geologic features to establish the order in which the events took place (see Table 7.A). Numerical dating permits an actual age to be assigned to an event, and therefore specifies the precise time it took place (see Table 7.B). This differs significantly from relative dating, which can only disclose whether an event happened before or after something else.

As will become clear shortly, both methods have their strengths and limitations and compilation of a complete history requires the joint use of relative and numerical dating. Let's begin by examining how relative dating is accomplished, then follow up with a discussion on techniques involved with numerical dating.

Table 7.A – Relative Dating

Age	Period
Stone Age	Paleolithic
	Mesolithic
	Neolithic
Bronze Age	Copper Age
	Bronze Age
Iron Age	Iron Age

Table 7.B – Numerical Dating

Age	Period	Dates
Stone Age	Paleolithic	2.5 million–10,000 B.C.
	Mesolithic	10,000–5500 B.C.
	Neolithic	5500–3000 B.C.
Bronze Age	Copper Age	3500–2300 B.C.
	Bronze Age	3300–1200 B.C.
Iron Age	Iron Age	1200–550 B.C.

*Note: Dates are subject to fluctuation depending on the region.

PART B: RELATIVE AGE DATING

SECTION I—PRINCIPLES OF RELATIVE AGE DATING

Back when geology was more of a hobby than a science, some extremely astute observers, such as Steno, Hutton, Lyell, and others, correctly discerned that nature adheres to a defined set of rules when creating past and present landscapes. With the realization that there is clear "method in the madness," Charles Lyell made an attempt to advance geology from an amateur's activity to an epistemological science by formally establishing the basic principles that govern age relationships within the geologic record. In doing so, Lyell unified the observations and interpretations of his predecessors and contemporaries, thereby providing a systematic approach to sequencing events throughout geologic history. The following principles are the result of that effort and provide the guidelines used by geologists to understand the relationships between geologic features in the rock record and enable them to sequence them accordingly.

- *Principle of Original Horizontality (see Figure 7.3)*: Sediments are deposited from a fluid under the influence of gravity, and therefore accumulate in comparatively flat areas (sediment on a slope will mass waste to the bottom before lithification takes place). Therefore, it can be confidently stated that sediment, when first deposited, results in a relatively horizontal layer. As such, it stands to reason that if a sediment bed is folded, tilted, or otherwise distorted, the source of that deformation was a process that took place after the horizontal layer lithified.

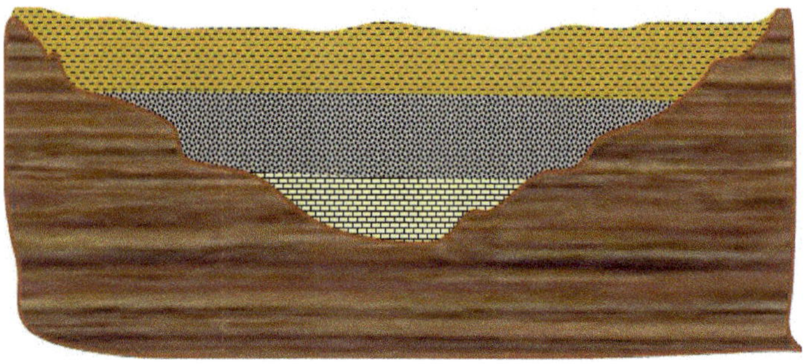

Figure 7.3 – Original Horizontality

- *Principle of Lateral Continuity (see Figure 7.4)*: Anything deposited in layers, including sediments, lava flows, and ash falls, will form large sheets that extend laterally in all directions throughout a defined region. As such, if there is a substantial gap in the strata within the sheet(s), it can be assumed that the strata were once intact, contiguous layers, and the missing interior was eroded after deposition.

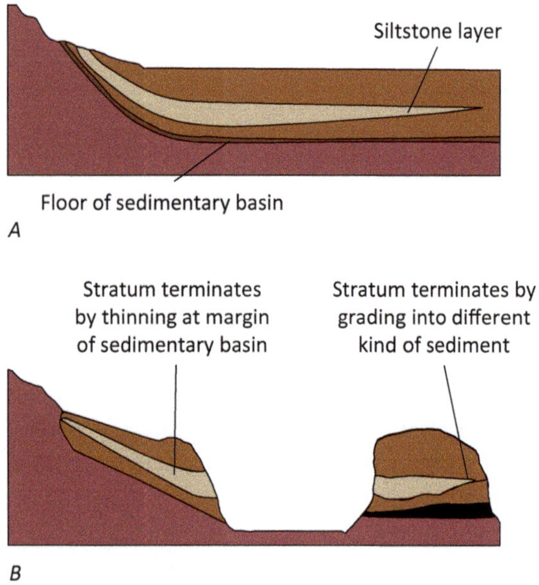

Figure 7.4 – Lateral Continuity

* Note that depositional sheets will laterally extend until they either terminate at the margin of the basin or until they thin to zero and grade into a different kind of sediment.

- *Principle of Superposition (see Figure 7.5)*: In any sequence of strata that has NOT been disturbed, youngest material is on the top and each subsequent layer must be sequentially older than the one above it and sequentially younger than the one below it.

Grand Canyon Strata Sequence

Figure 7.5 – Superposition

⋆ Note: Strata can be deformed by tilting, folding, or faulting. Many times, the strata maintain their original depositional order and other times it becomes so distorted that older rock ends up on top of younger rock. When any of the following deformation features are present, the area must be thoroughly examined to ensure the strata have not been overturned or older rock thrust up on top of younger rock. See Figure 7.6.

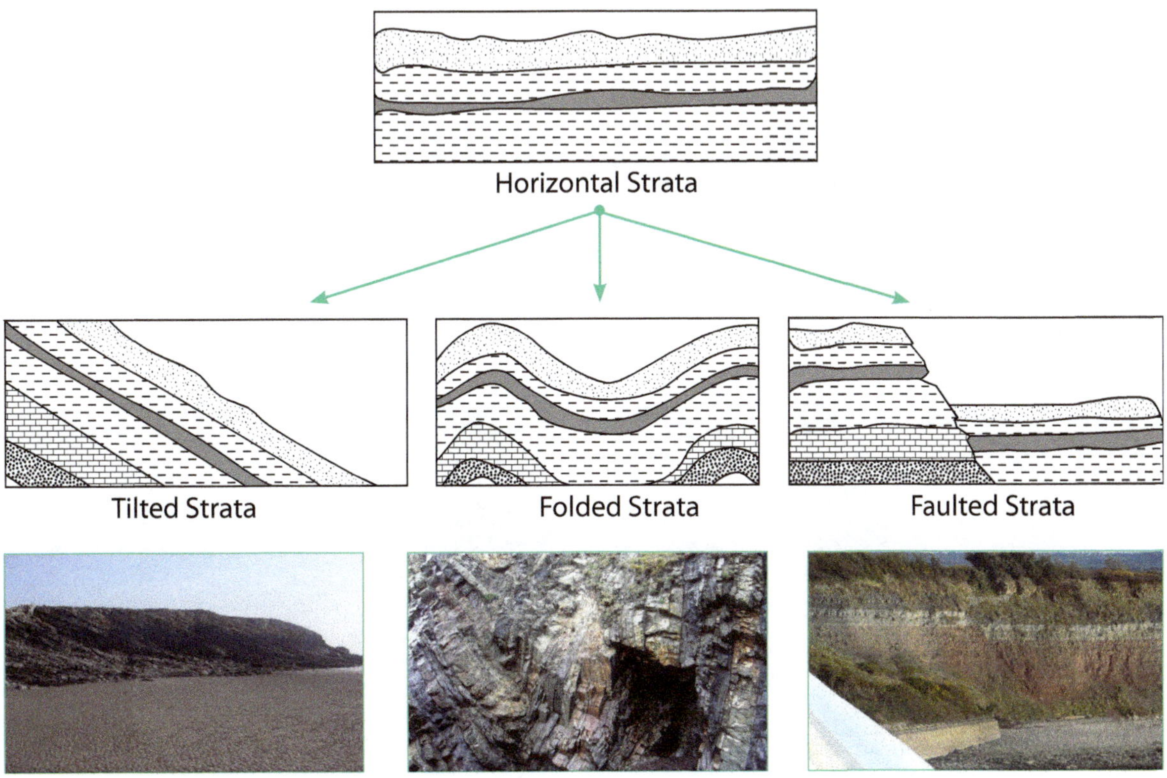

Figure 7.6 – Tilts, Folds, and Faults

- *Principle of Cross-Cutting Relationships (see Figure 7.7)*: If one geologic feature is cross-cutting another, whatever is doing the cross-cutting has to be younger then whatever is being cross-cut. Examples of cross-cutting bodies include dikes, sills, massive intrusions, and faults (see Figures 7.8a–7.8c). This assumption is valid because igneous intrusions and faults are not supported in midair, and therefore could only have occurred in rocks already present to support them.

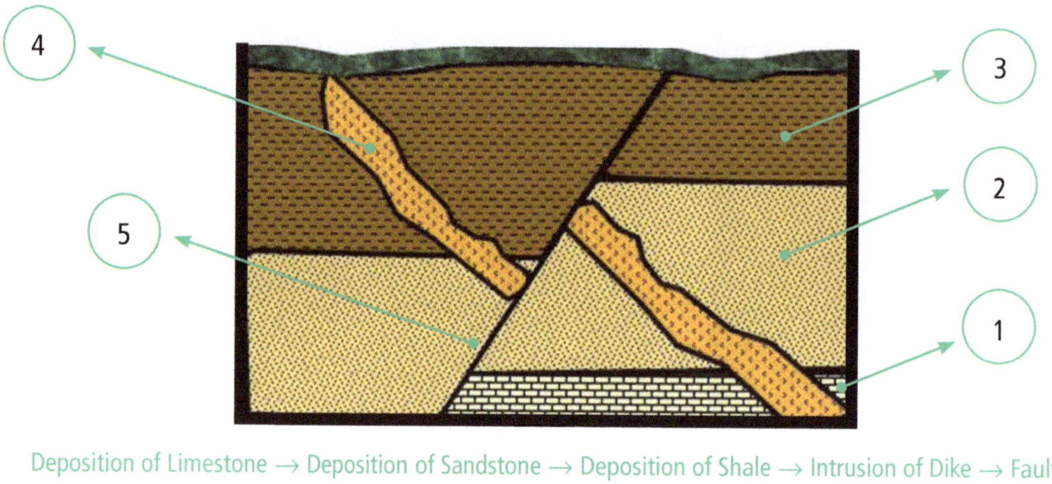

Figure 7.7 – Cross cutting relationships

Figure 7.8a – Dikes and Sills

Figure 7.8b – Massive Intrusions

Figure 7.8c – Faults

- *Principle of Inclusions (see Figure 7.9)*: If foreign particles (inclusions) are included in a rock body, the rock body from which the inclusions came has to be older than the rock body in which they're included. This assumption is valid, because the inclusion process occurs due to fresh, mobile material incorporating pieces of preexisting material during the lithification or cooling process. Rock that has already lithified or cooled is rigid and immobile and cannot resituate to incorporate freshly arrived material. See Figure 7.10 for examples of inclusions.

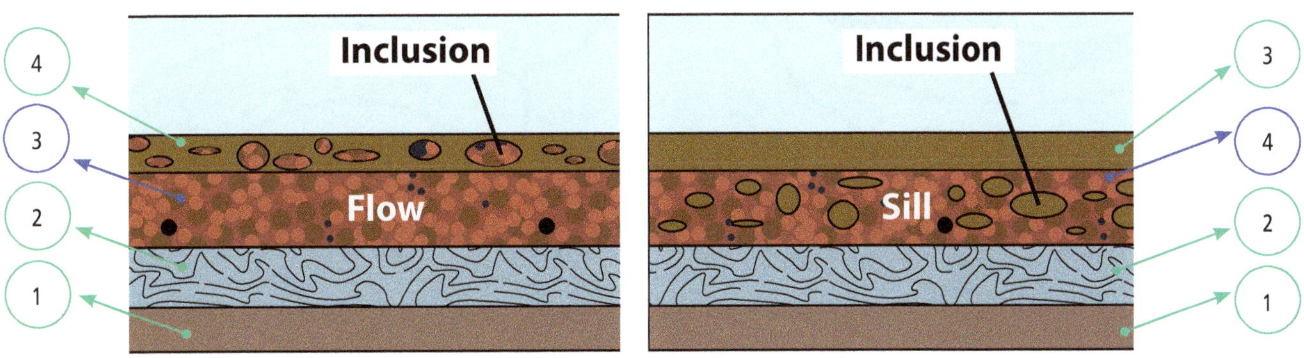

Figure 7.9 – Principle of Inclusions

Figure 7.10 – Inclusions

It should also be noted that with the addition of some rudimentary geology, you can infer events such as sea level rises and falls—by identifying transitions between chemical and clastic rocks; regional metamorphic events—by the presence of foliated rock; and the timing of sill intrusions—by observing the baked contacts.

Figure 7.11a is a hypothetical cross-section sketched from a geologists observations in the field. Figure 7.11b is an example of how to develop a geologic history of this area using the principles described above.

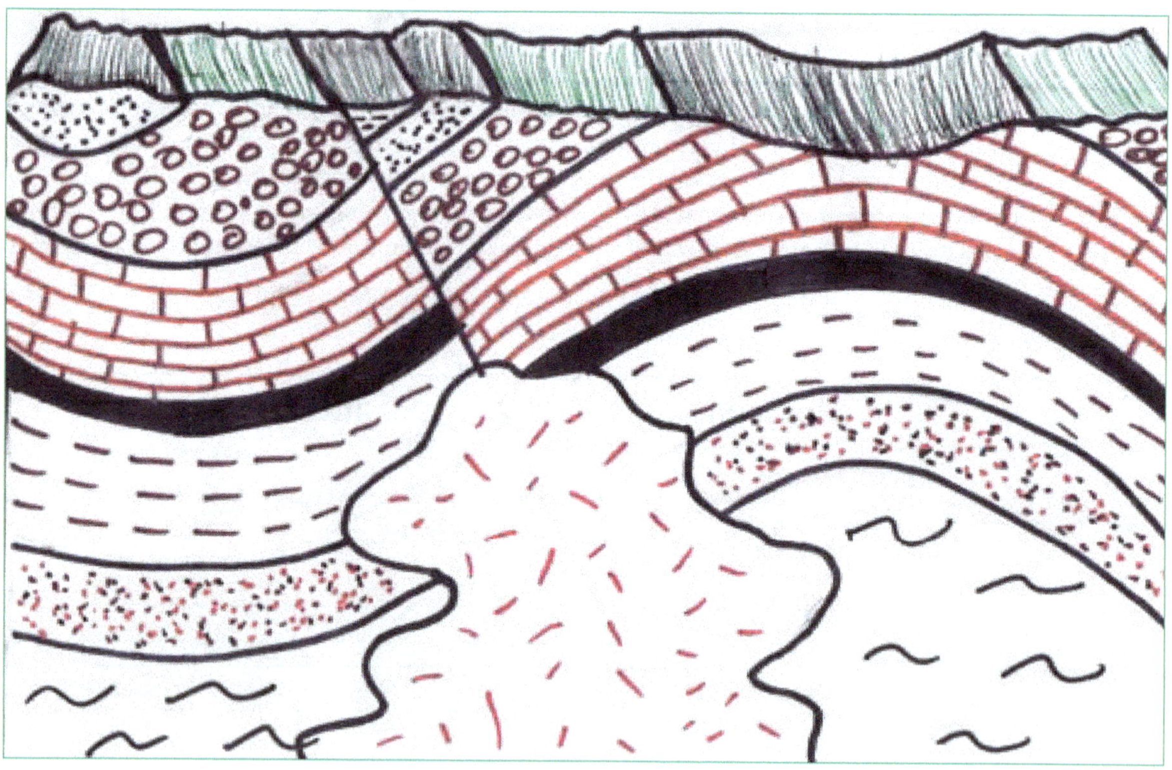

Figure 7.11a – Hypothetical cross-section

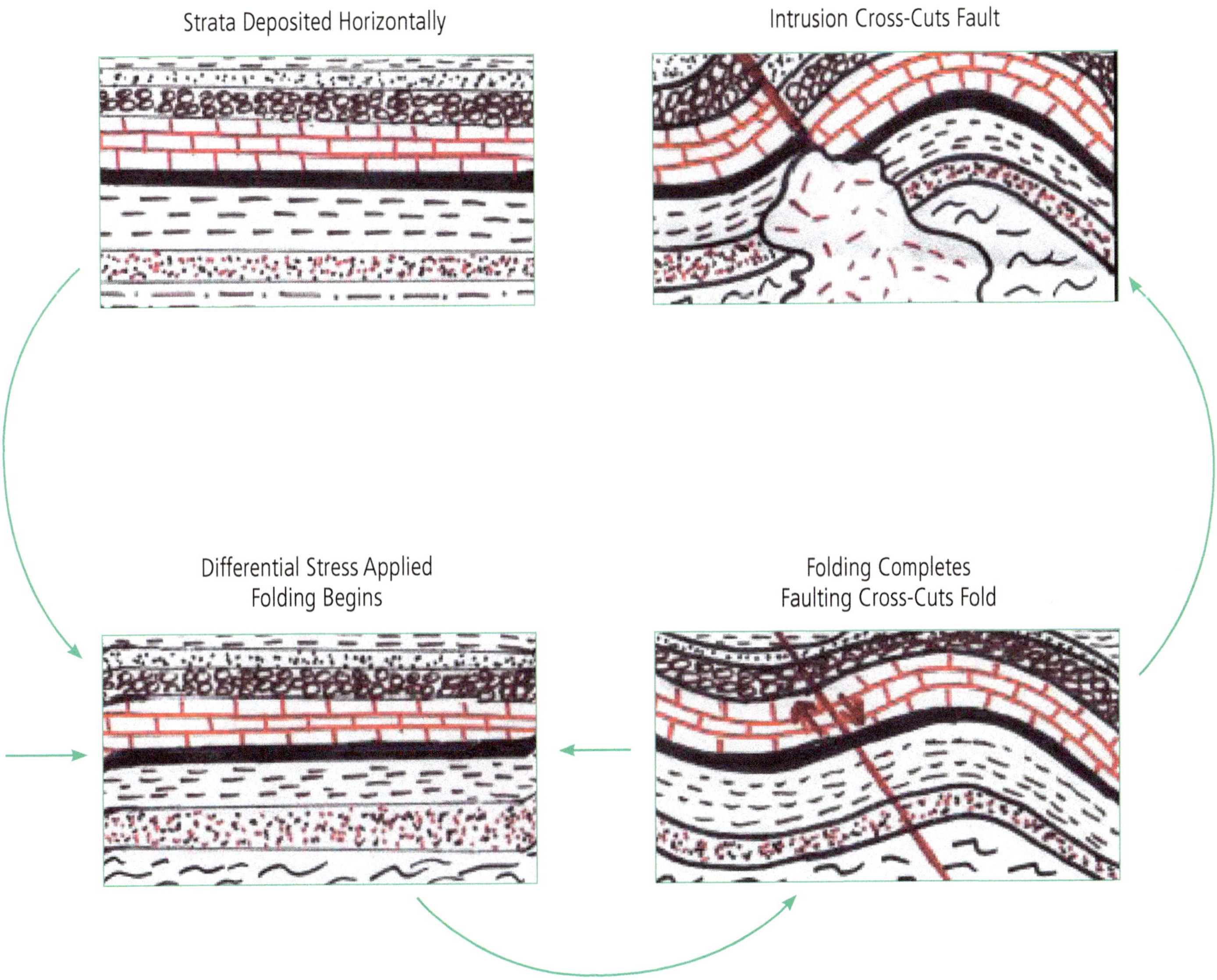

Figure 7.11b – Relative sequence of events that formed Figure 7.11a

Exercise 1: Learning to use relative dating techniques

Figures 7.12a–7.12e are visual depictions of how a landscape looked at various times throughout its history. Put the story of this landscape in order by placing a 1 under the oldest landscape, a 5 under the youngest landscape, and sequentially order the three in between.

Figure 7.12a

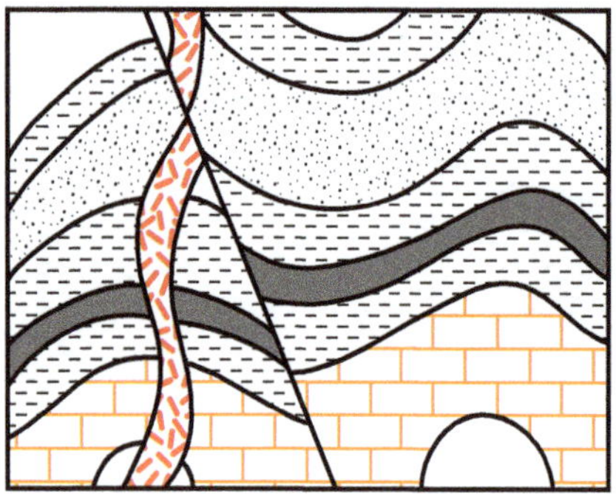

____

Figure 7.12b

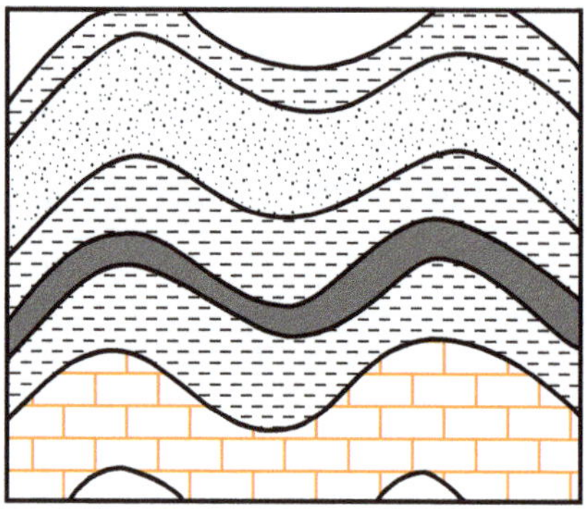

____

Figure 7.12c

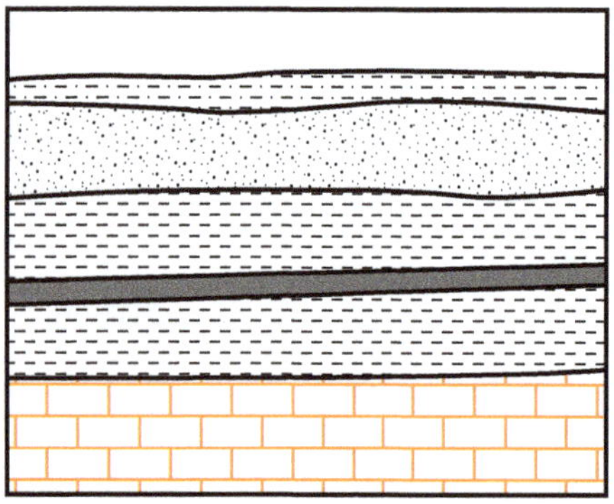

____

Figure 7.12d

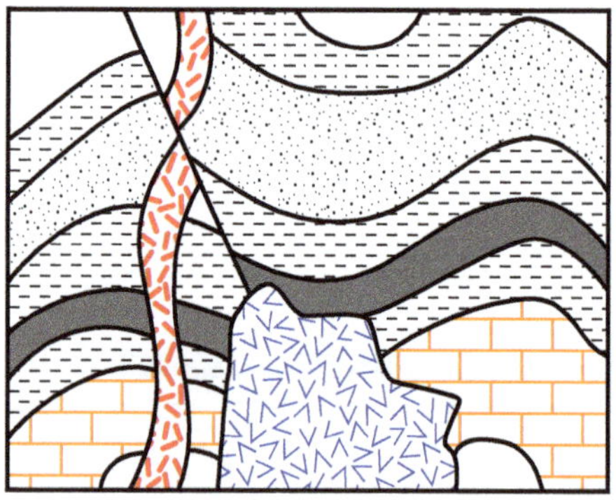

____

Figure 7.12e

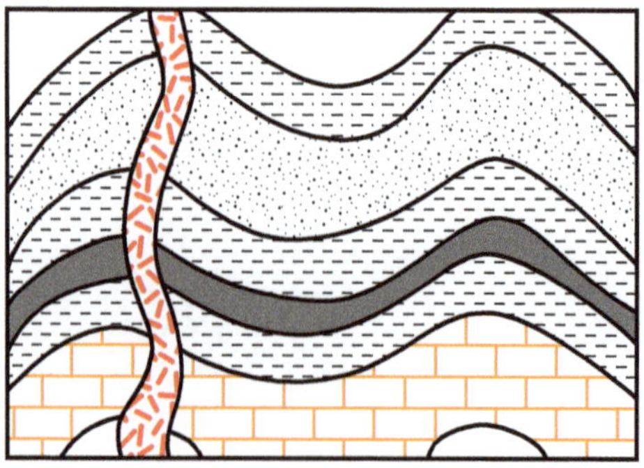

____

Exercise 2: Building a History from Relative Sequencing

Below is a cross-section and corresponding key (Figs. 7.13a and 7.13b), recording a series of geologic events within a region. Using the relative dating principles described, put the events in their proper sequence on the lines provided to the right. Several events have been filled in for you.

Note: You will find it considerably easier to start with the oldest event and work toward the youngest.

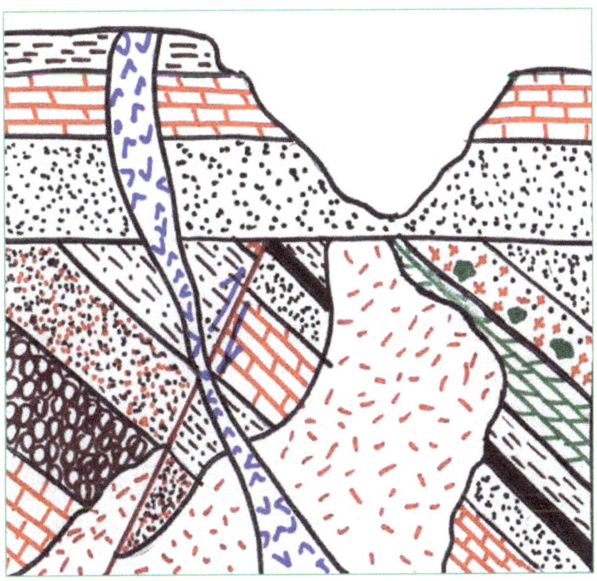

Figure 7.13a

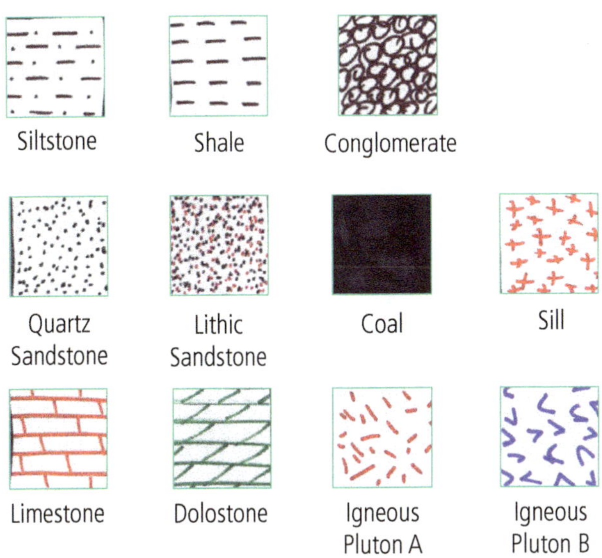

Figure 7.13b

Youngest

19. ______________________

18. Deposition of ______________________

17. Deposition of ______________________

16. Deposition of ______________________

15. Erosion

14. ______________________

13. ______________________

12. Tilting

11. ______________________

10. Deposition of ______________________

9. Deposition of ______________________

8. Deposition of ______________________

7. Deposition of ______________________

6. Deposition of ______________________

5. Deposition of ______________________

4. Deposition of ______________________

3. Deposition of ______________________

2. Deposition of ______________________

1. Deposition of ______________________

Oldest

Section II—Unconformities

While it would make life much easier if landscape evolution happened from start to finish without interruption, thereby leaving us a complete and perfectly documented history of the region's geologic processes, unfortunately nature does not work in this manner. As such, like any historian, it is the job of a geologist to identify where the historical record is missing documentation or has been altered by later events, and then attempt to fill in the missing gap with information obtained from another source. Geologists now know that there are two reasons why the rock record might demonstrate gaps in time: 1) If during that time period, conditions were such that deposition did not occur, thereby leaving no record of what went on in that time interval; or 2) a later period of intense erosion removed previously deposited material, thereby effectively erasing any evidence left from that time period. In either case, there will be a gap (in many cases, multiple gaps) in the rock record, meaning that the historical record is incomplete. Geologists refer to time periods that are not represented in the rock record because of non-deposition or erosion as an "unconformity." Unconformities are classified into three categories (see figures 7.14a–7.14c):

- *Disconformity*: Missing time between parallel layers such as strata or lava flows.

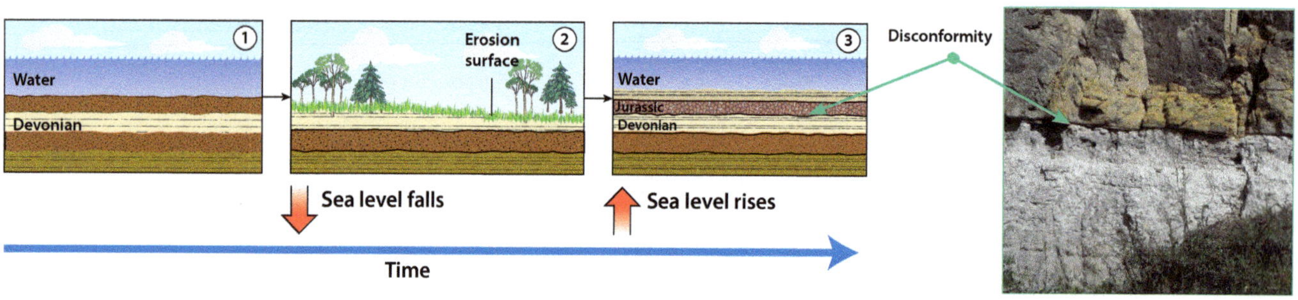

Figure 7.14a – Disconformity

- *Angular Unconformity*: Younger, parallel strata sit on top of older folded or tilted strata that have been eroded down to a flat surface. Results in areas where rock layers have been tilted due to folding or faulting, then exposed to the surface and eroded down before deposition recommenced.

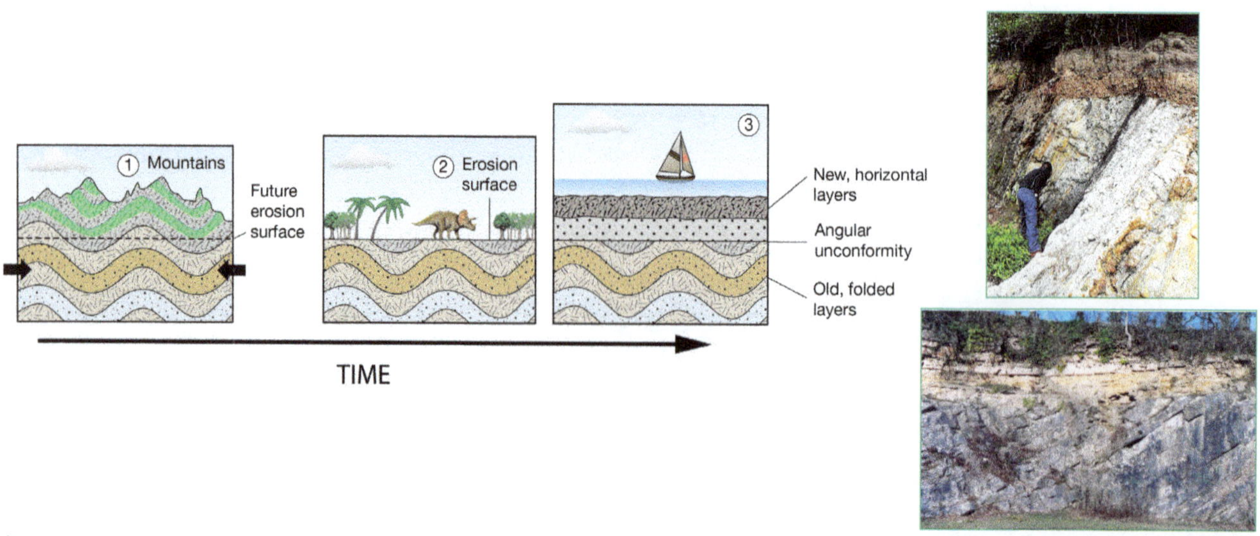

Figure 7.14b – Angular Unconformity

- *Nonconformity*: Younger, parallel strata sit on top of older, igneous or metamorphic rocks that have been eroded down to a flat surface. Results from exposure of the metamorphic or igneous rock to surface weathering after cooling, then subsequent erosion before deposition recommenced.

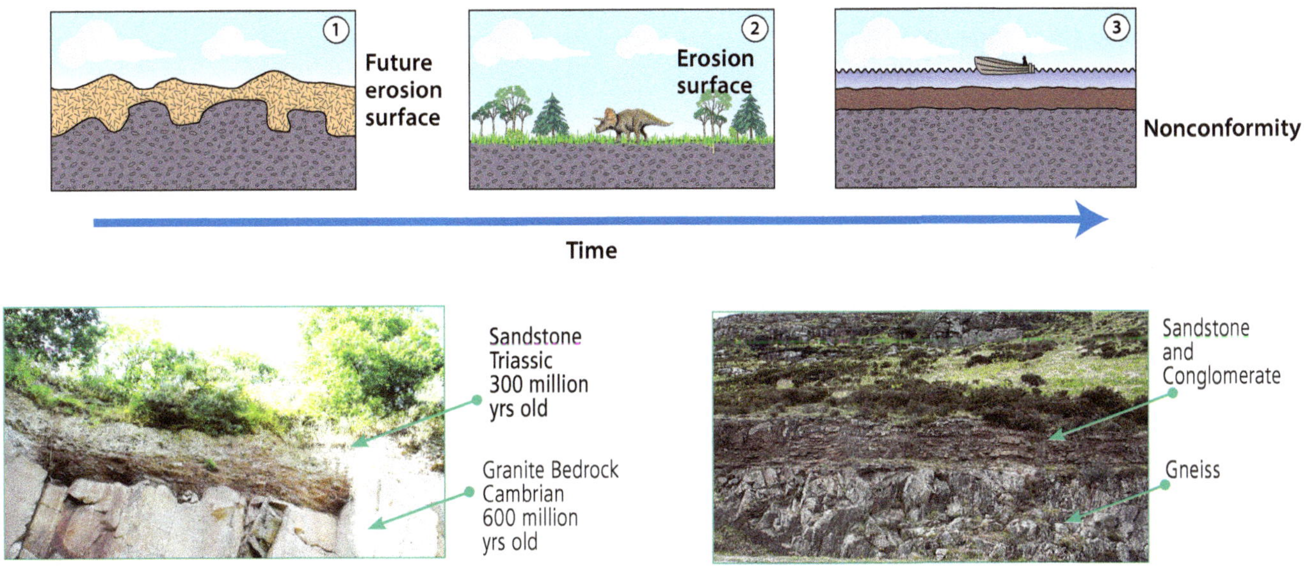

Figure 7.14c – Nonconformity

Now that you have been introduced to the types of unconformities erosion and non-deposition generate, the next thing that needs to be addressed is how we recognize them in the geologic record. In other words, if the record of a defined time period has been removed, exactly how do we know that something is missing, and how do we figure out how much time the gap represents?

Some unconformities in the geologic record can be recognized simply by observing an exposed outcrop. Angular unconformities and nonconformities are visually detectable to a trained geologist and can be identified by the presence of horizontal strata on top of tilted strata in the first case, and horizontal strata on top of relatively flat, much older igneous or metamorphic rock in the second case. Disconformities are rarely readily apparent, but sometimes leave signs that geologists are able recognize. However, difficulty arises if a disconformity has left no signs of its presence, and particularly problematic is when there is a disconformity between the same type of rock. In addition, in all three cases, landscape observation is of little use in determining the span of time the unconformity represents. Luckily, a solution presented itself in the form of fossils.

It has been observed and confirmed throughout millions of locations around the world that specific sedimentary layers can be recognized and established as having been deposited during a particular period of time using the fossil assemblages preserved in the rock layer. Fossils are able to serve as timekeepers, because examination after examination has shown that certain fossils will be found only within a restricted interval of strata known as a range zone. In essence, all species come into being, exist for a finite expanse of time, and then go extinct, allowing the evolutionary process to fill those empty niches with newly evolved species. It has also been conclusively demonstrated that once a species goes extinct, (i.e., disappears from the strata), it never reappears at a later time. These observations led to the formal establishment of the Principle of Fossil Succession, which states that there is a documented, evolving sequence of fossils throughout geologic history and it can therefore be used to date the layers relative to each other and help identify the presence of unconformities. See Figure 7.15 and subsequent discussion to appreciate how this principle works in practice.

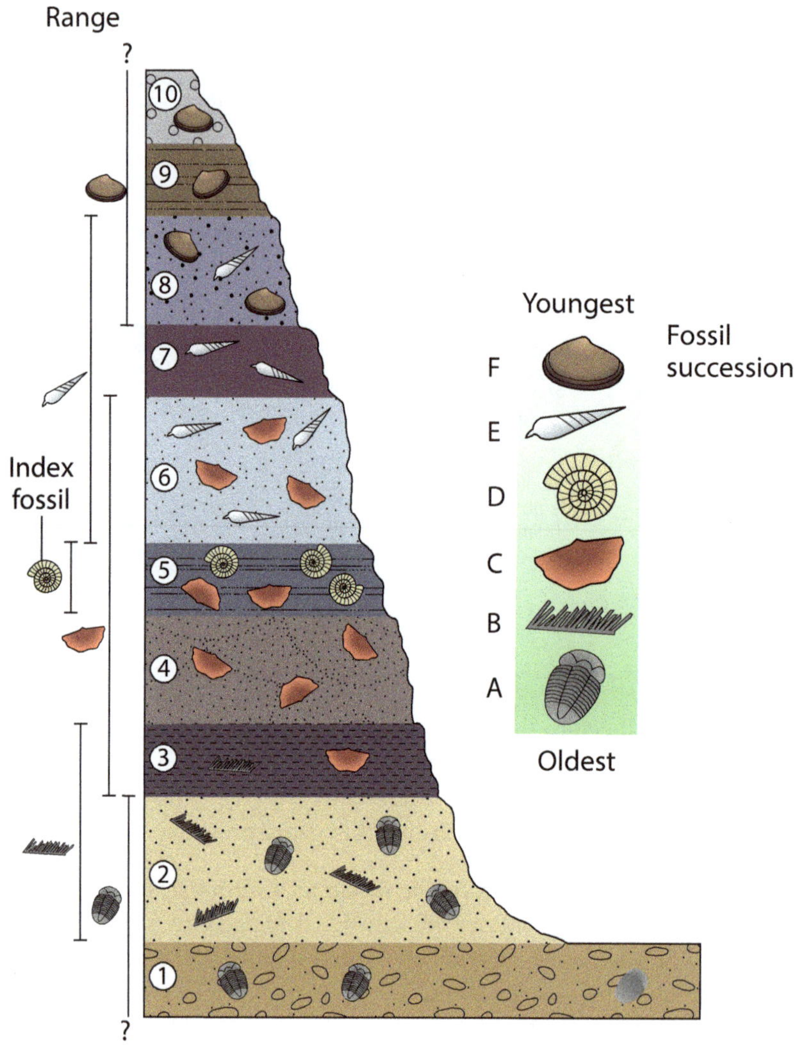

Figure 7.15 – Principle of Fossil Succession

Above is a hypothetical sequence of fossil-containing strata. The range in which a specific fossil is present in the sequence has been defined for each of the fossils. Note that while the ranges of some fossils overlap with others, there is a distinct succession of fossils throughout this sequence (A, B, C, D, E, and F). This enables geologists to use the Principle of Fossils Succession for sequencing sedimentary strata in the following ways:

- Strata containing fossils earlier in the succession must be older than strata containing fossils found later in the succession
    - Ex: A bed containing fossil B has to be older than a bed containing fossils D, E, and F. Note that no conclusion can be drawn with regard to the relationship between a bed containing

fossil B and a bed containing fossil C, as their ranges overlap, which means there was a time when they coexisted. This comparison of one fossil earlier in the succession to another fossil later in the succession is only valid if there is no overlap in the ranges of the two fossils in question.

- In order to circumvent the limitation described in the previous point, assemblages of fossils can be considered, as opposed to individual fossils. Just as in the previous case, strata containing assemblages of fossils earlier in the succession have to be older than strata containing assemblages later in the succession.
  - Ex: A bed containing fossils A and B must be older than a bed containing fossils B and C. Again, this conclusion is valid only because the range of fossils A and C do not overlap.
- Fossil studies have shown there are certain species that existed for an unusually short duration, and therefore demarcate a specific time interval. Fossils that are diagnostic of a specific time period are called index fossils; a fossil assemblage that defines a specific time period is known as an index fossil assemblage. See Figure 7.16 for select examples of index fossils.

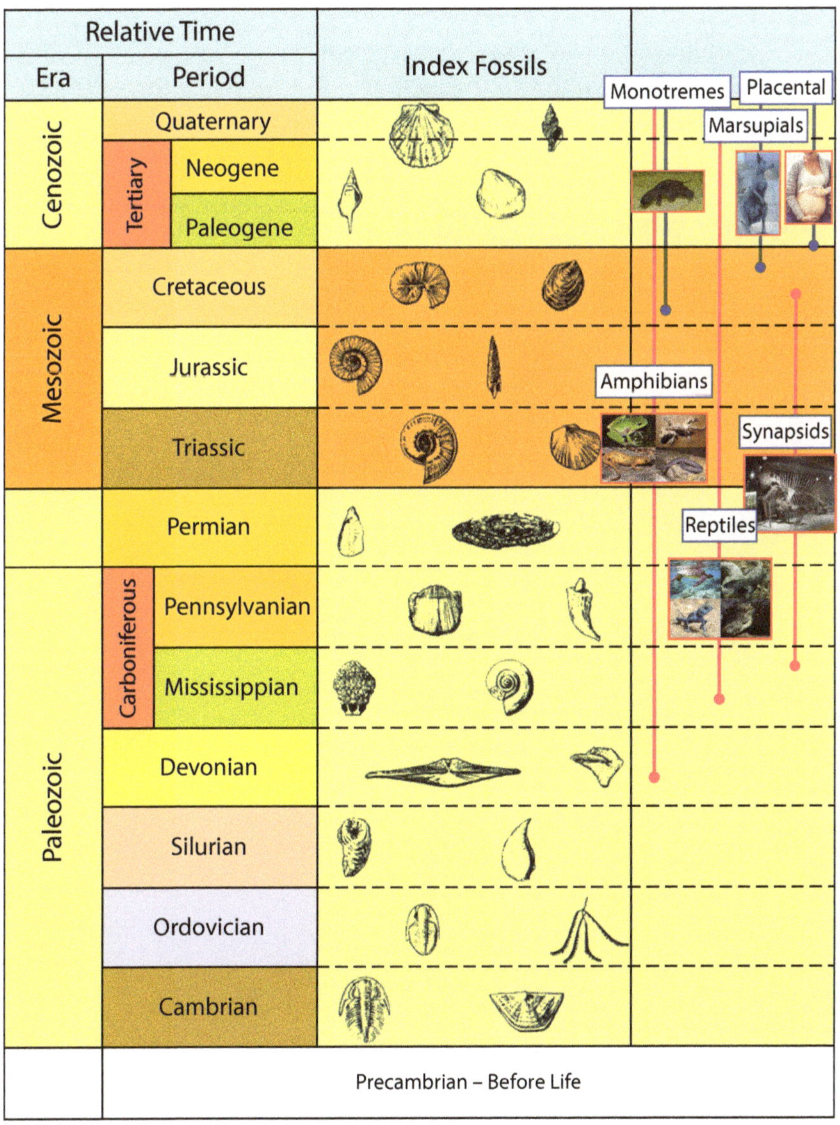

Figure 7.16 – Index Fossil and the Geologic Time Column

Figure 7.16 depicts two index fossils for each period in the geologic time column. Using that figure and what you know about unconformities, answer Questions 1–3 in the Identifying Unconformities exercise.

Exercise: Identifying Unconformities

1. Examine Figure 7.17a.
    a. Using a visible marker, highlight the unconformity in this figure.
    b. Which period(s) are missing from this sequence? ____________________
    c. What type of unconformity is this? ____________________
    d. Now examine Figure 7.17b. This sequence is a duplicate of Figure 7.17a, minus the fossils. Is there any indicator that an unconformity is present in this image? Yes or No?
    e. Is yes, what? ____________________

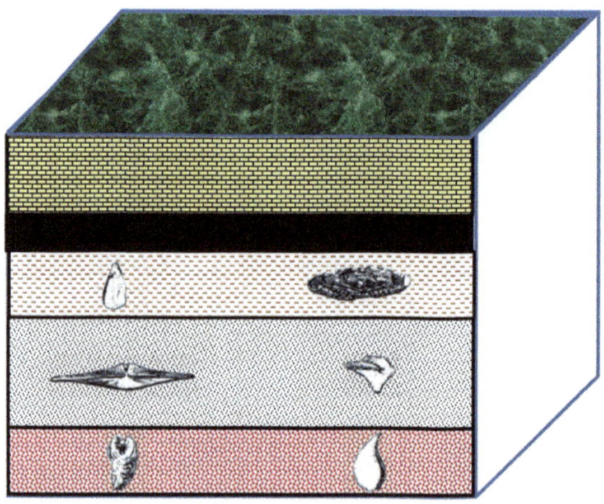

Figure 7.17a

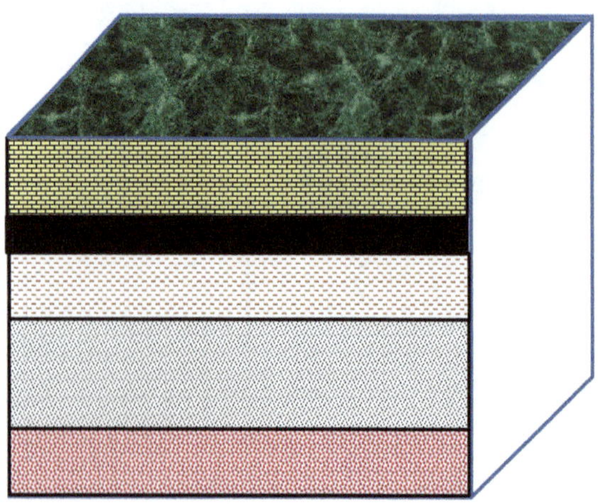

Figure 7.17b

2. Examine Figure 7.18a

   a. There are two unconformities in this figure. Using a visible marker, highlight both unconformities.

   b. Draw an arrow from each unconformity and label the type.

   c. Now examine Figure 7.18b. This sequence is a duplicate of Figure 7.18a, minus the fossils. Is there any indicator that an unconformity is present in this image? Yes or No?

   d. Is yes, what? ______________________

   e. If yes, which type of unconformity is detectable? ____________________

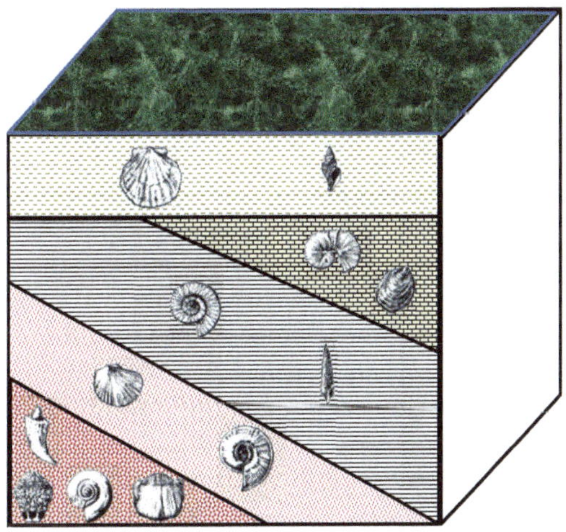

Figure 7.18a

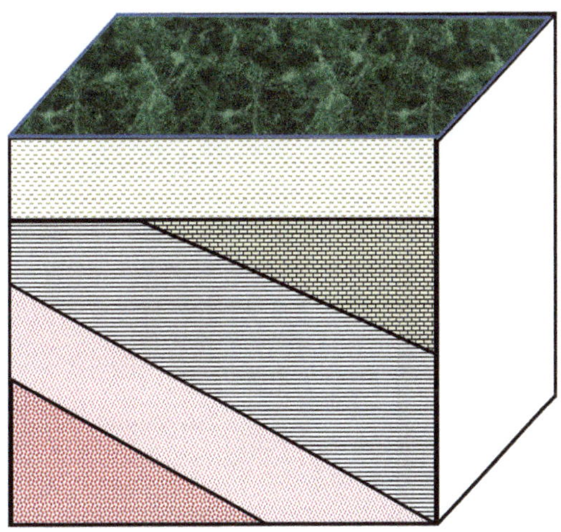

Figure 7.18b

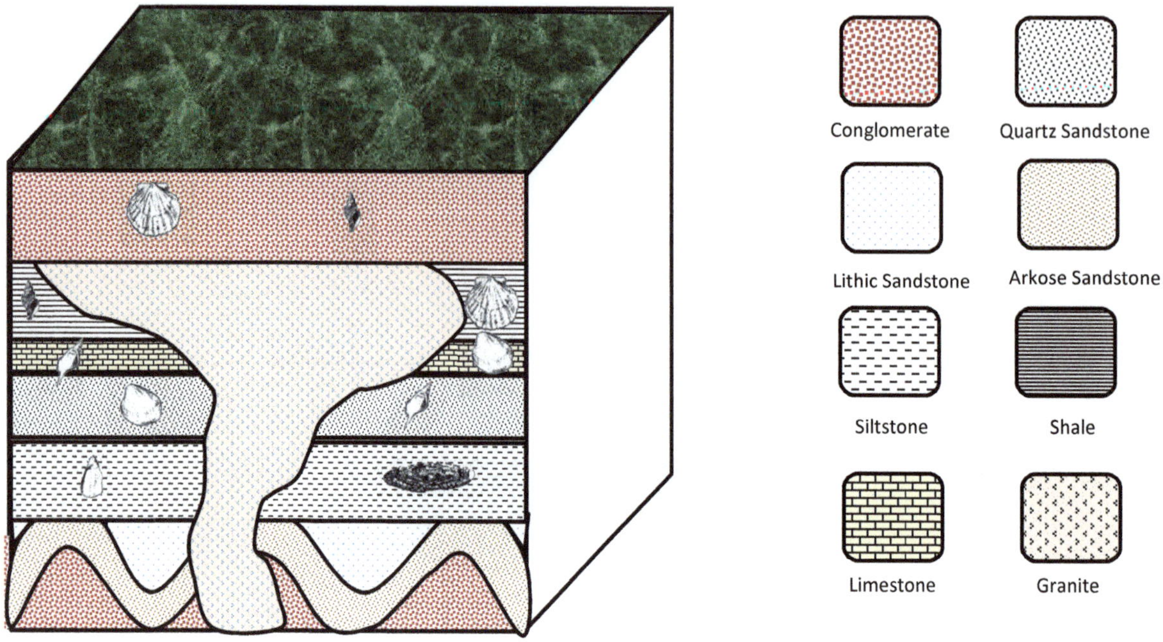

Figure 7.19

3. Examine Figure 7.19

   a. Figure 7.19 contains all three types of unconformities. Using a visible marker, highlight all three unconformities.

   b. Using the lines provided on the right, put the events that formed this landscape in their proper sequence. For the three lines representing an unconformity, write the following information:

      - "erosion"
      - Type of unconformity

   c. What entire era is missing from this strata?

      ____

Youngest

13. Deposition of ______________________

12. ______________________

11. ______________________

10. Deposition of ______________________

9. Deposition of ______________________

8. Deposition of ______________________

7. ______________

6. Deposition of ______________________

5. ______________

4. ______________

3. Deposition of ______________________

2. Deposition of ______________________

1. Deposition of ______________________

Oldest

Section III—Filling in the Gaps: Stratigraphic Correlation:

As stated earlier, it is not only the job of a geologist to identify unconformities, but they must also attempt to find another source that can help fill in the missing gaps in order to complete the story. One technique used for accomplishing this goal is called correlation. Correlation is a method by which geologists define a relationship between strata at various seemingly unconnected localities. There are two types of correlation: Lithostratigraphic Correlation and Biostratigraphic Correlation. Lithostratigraphic correlation identifies the same layer(s) across long distances based on similarities in rock type and sequence. To simplify the process, geologists often specify "formations"—an entire sequence of rocks that can be traced over long distances. Because this method of correlation relies on the Principle of Lateral Continuity, it is generally reserved for correlating strata within a given region. Biostratigraphic correlation relies on the Principle of Fossil Succession, and uses the age of fossils to match layers of the same age in two unconnected sequences. This type of correlation is helpful for both nearby and broad areas. Lithostratigraphy is often not useful over large areas, because the depositional environment is expected to change over long distances; therefore, the two areas will have very different rock types deposited during the same period of time. However, the organisms whose lifespans coincide with that time interval will remain the same.

How is correlation done? In order to organize the history of a particular area, geologists sketch scaled drawings called stratigraphic columns, which accurately represent the thickness and position of the layers as seen at that location. Once a stratigraphic column has been obtained for various localities, the columns are compared, and distinctive sequences of strata can be correlated throughout the area in question. Generally speaking, when correlating within a region (a relatively close area), geologists rely on lithostratigraphy to match up distinctive rock sequences from one area to another. In addition, many times there will be one or several key beds, an unusually unique layer, such as a coal seam, that can be traced from location to location and thereby serve as a base for the correlation. While biostratigraphic correlation is primarily used for long-distance correlation, it can assist in identifying unconformities at a specific location, and then lithostratigraphic correlation can be used to fill in the gap. Once correlation is complete, it must be concluded that any layers that stand alone—i.e., cannot be matched to other localities—were either eroded out of or were never deposited at the localities in which they are not present, and therefore represents a missing gap of time. If there is enough data from the various localities to fill in all the unconformities (and often there is not), a complete stratigraphic column representing the area's entire, uninterrupted history can be built.

Examine Figure 7.20 for a demonstration on how this is done.

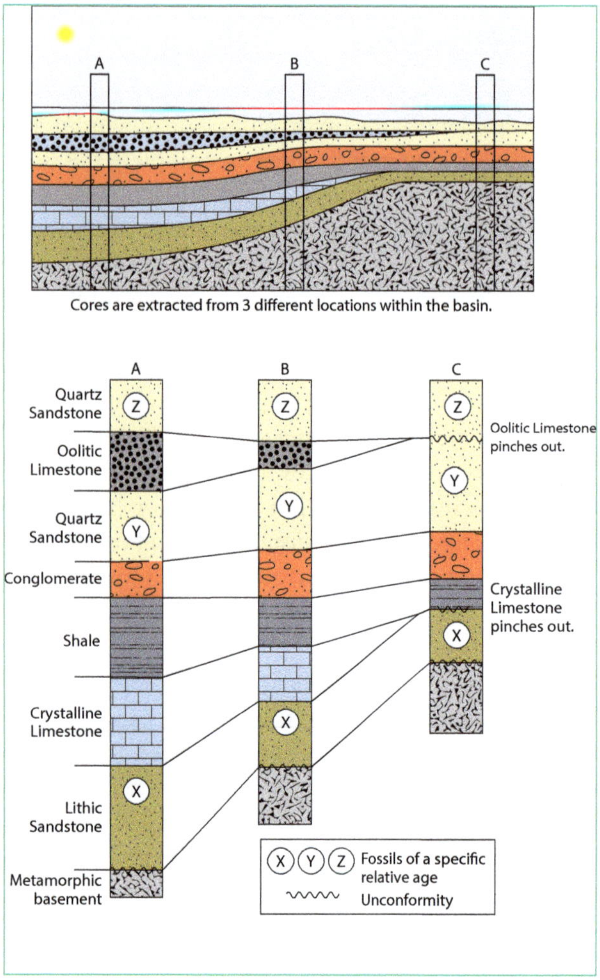

Figure 7.20 – Principles of Correlation
Notice how an unconformity can be identified either by fossil evidence, using the Principle of Fossil Succession, or by strata that are "*pinched out*," shown to be missing after lithostratigraphic correlation.

Exercise 1: Building the Picture

Imagine you are a geologist who has been employed by a major hotel and casino chain to investigate an as-yet undeveloped island that is being considered for the site of an enormous health and wellness facility. You have been asked to document and describe the layers beneath the surface along the southern coast, and assess whether this area is suitable for a development of this magnitude, with respect to both strength and long-term sustainability. Given that it is a coastal environment that is prone to rapid change, you correctly surmise that you need to thoroughly investigate the area's current landscape as well as its geologic history, in order to forecast how that coast is expected to evolve in the future. As such, you decide to drill multiple cores along the coast, and then use correlation to map out the subsurface layers. In this way, you can build a history and accurately evaluate whether this project is a worthy investment. On the following page is a map of the southern coast of the island with the cores you had drilled superimposed on the map in their respective places (Figure 7.21). Using lithostratigraphy and biostratigraphy where appropriate, correlate the cores. When you have completed drawing your lines, use colored pencils to fill in each layer, thereby creating a complete cross-section. Note: You may need to refer back to figure 7.16 in order to complete this exercise.

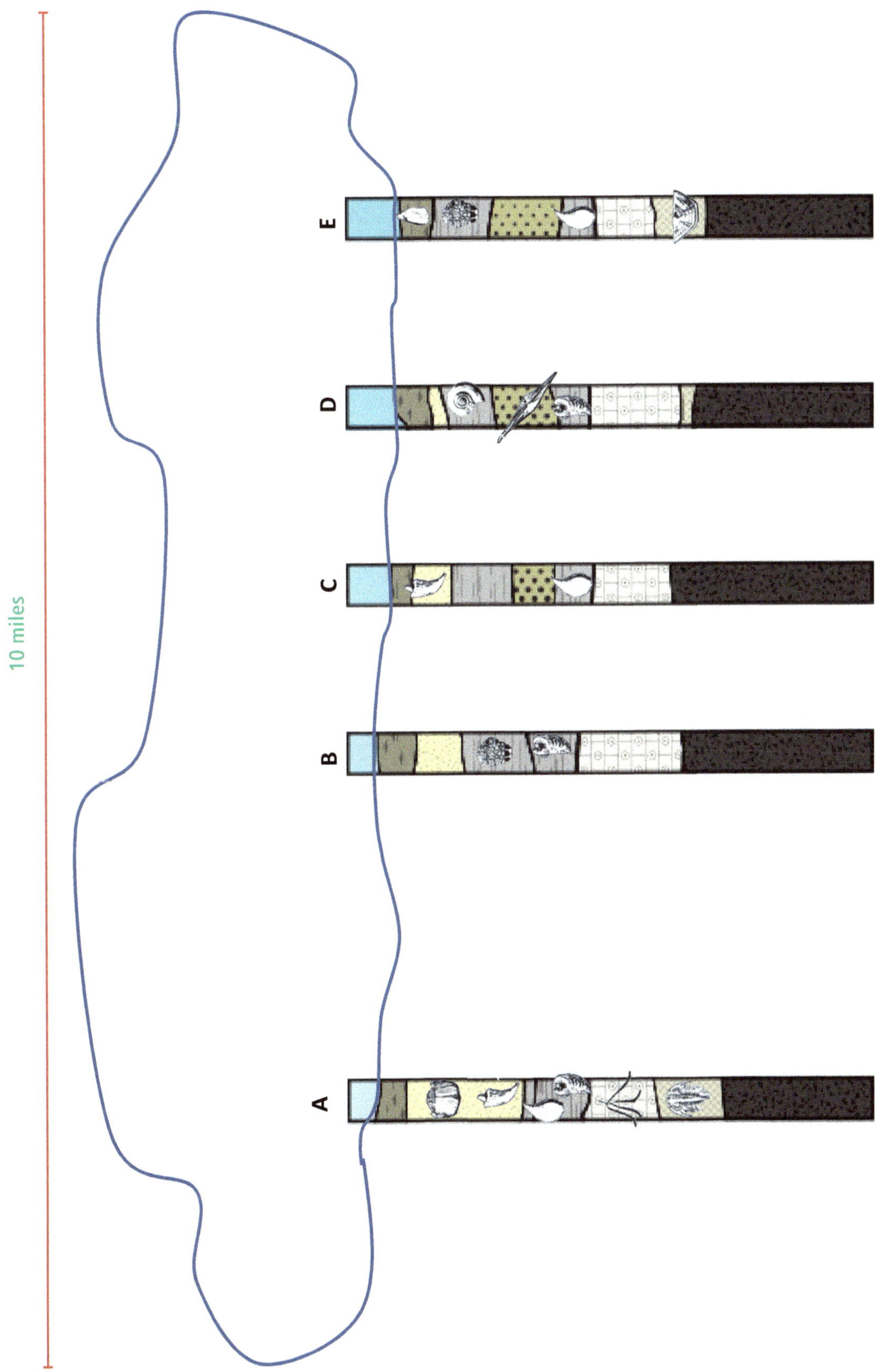

Figure 7.21

Exercise 2: Assessing the History

Exercise 1 demonstrates how lithostratigraphic and biostratigraphic correlation can be used to reconstruct the geologic history within a localized region. When applied properly, these tools can also be helpful in reconstructing large scale processes, such as mountain-building events, ocean basin rifting, and the past position and movements of the continents.

Figure 7.22a depicts six stratigraphic columns throughout the interior of the European Alps, a mountain chain that resulted from a plate collision that began in the late Mesozoic. While there is a tremendous amount of complexity throughout this region, the multiple rock types have been grouped together into five distinct formations that have been identified and color coded for you: 1) African Basement Rock; 2) African Sedimentary Rock; 3) Mixed Continental and Oceanic Sediments; 4) Oceanic Crust; and 5) European Basement Rock. Correlate the formations throughout these six stratigraphic columns. Remember to connect contact to contact—and if there is no matching contact, then the layer has been "pinched out" of that area.

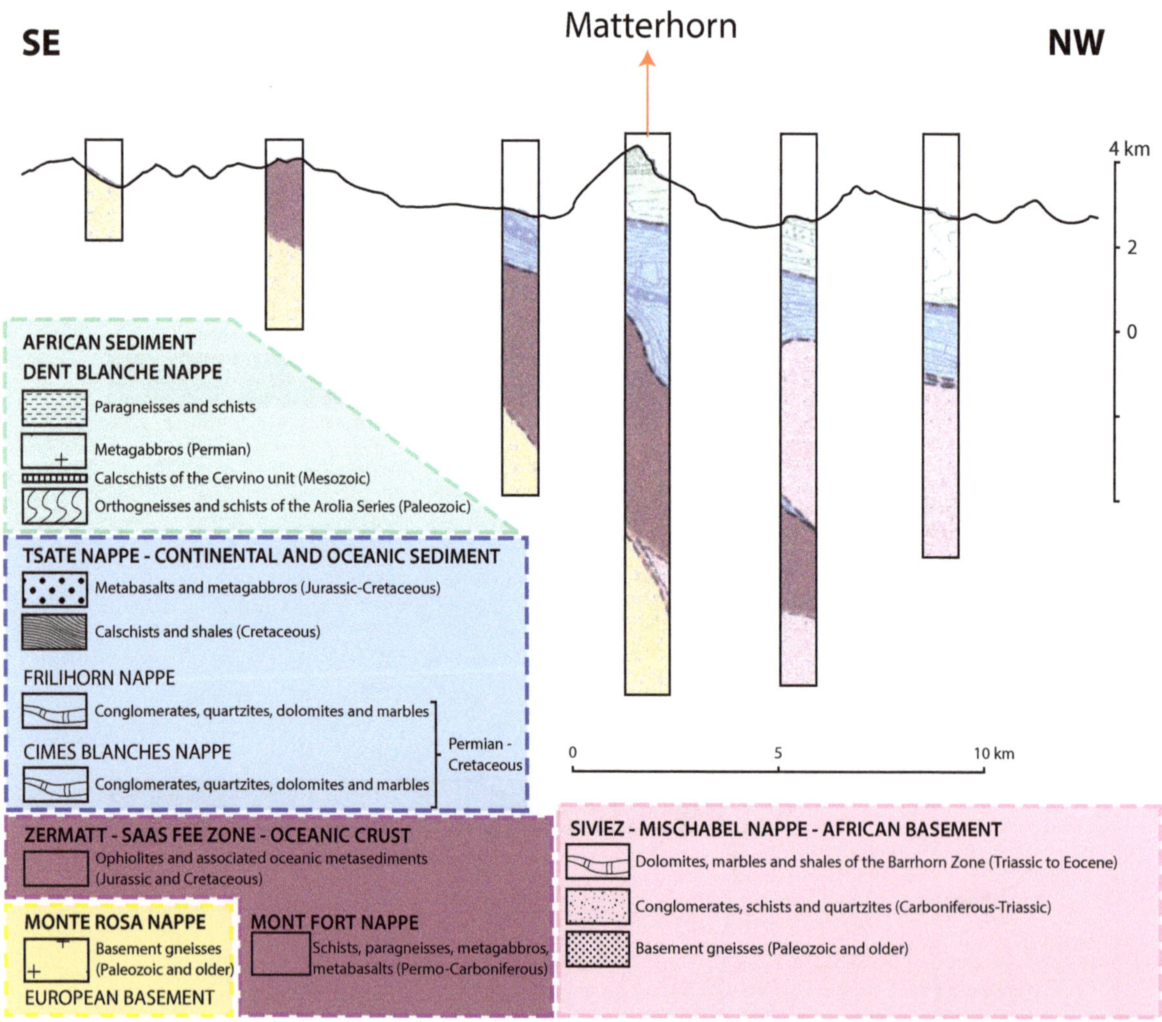

Figure 7.22a – Stratigraphic Columns in the European Alps

Now that you have completed your own correlation, look below to view a correct and complete simplified cross-section of the internal Alps. Use this figure (Figure 7.22b) to answer the questions that follow and untangle the history of this region.

1. Which two continents collided to form the Alps? ______________ and ____________

   How do you know? ____

2. Which continent was moving faster (in other words, who hit whom?)
   ______________________________ hit ______________________________

   How do you know? ____
   ____

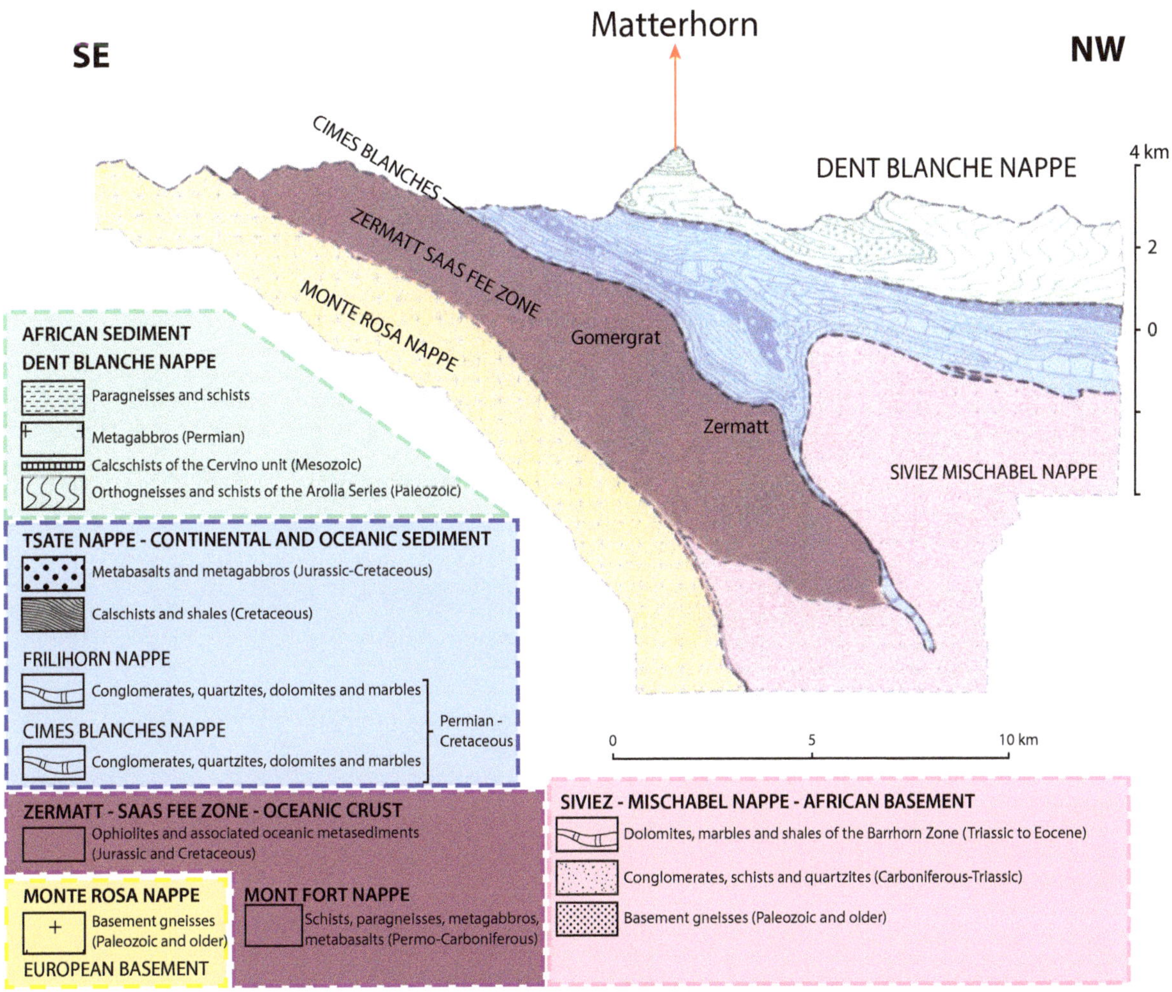

Figure 7.22b – Completed Correlation of European Alps

3. Name an example present in this cross-section for each of the features listed below that indicates this area underwent extreme compressional stress:

   - Landscape features ______________________
   - Structural features ______________ and ______________
   - Rock type ____________________________

4. Even if the mountains were eroded down, what evidence would there be indicating this was an ancient collision zone? ____

5. Notice that there is a large wedge of oceanic crust (pink) and oceanic and continental sediments (blue) sandwiched between the European basement and the African sediment. Based on what you've discovered about the geologic history of this region, explain how that got there.
____
____
____

6. What types of faults are present in this region: normal, reverse, or strike-slip? Circle all that apply.

7. Using a high-lighter, high-light the locations of the main faults.

8. By filling in the blanks below, give a step-by-step account of how the internal Alps were formed. Start from the bottom (older processes) and work up.

Youngest

- ____________________
- Uplift creating the ____________
- ____________ sediments thrust on top of ______________
- Continental and oceanic sediments mixed and __________________
- ______________ thrust up onto _________________
- Africa approaches Europe and pinches ____________along with the ___________ on top of it in between the two continents.
- Tethys seaway, located between the African Plate and the European Plate begins to close.

Oldest

Section IV—Putting the Pieces Together

Figure 7.23 is a cross-section that represents a 100-mile region, which was correlated using cores drilled at 10-mile intervals. Using what you know about all the techniques discussed above and the relative geologic time column (Figure 7.16), sequence the events in this cross-section on the lines provided. Make sure to include events such as folding and unconformities. For each unconformity, write the type of unconformity and what period of time is missing if the evidence permits you to deduce that information. To help keep you on track, several events have already been filled in for you and others have hints supplied. When you have completed this task, answer the follow-up questions.

For the initial chronology, DO NOT INCLUDE THE SILL. You will be asked to do that later with the follow-up questions.

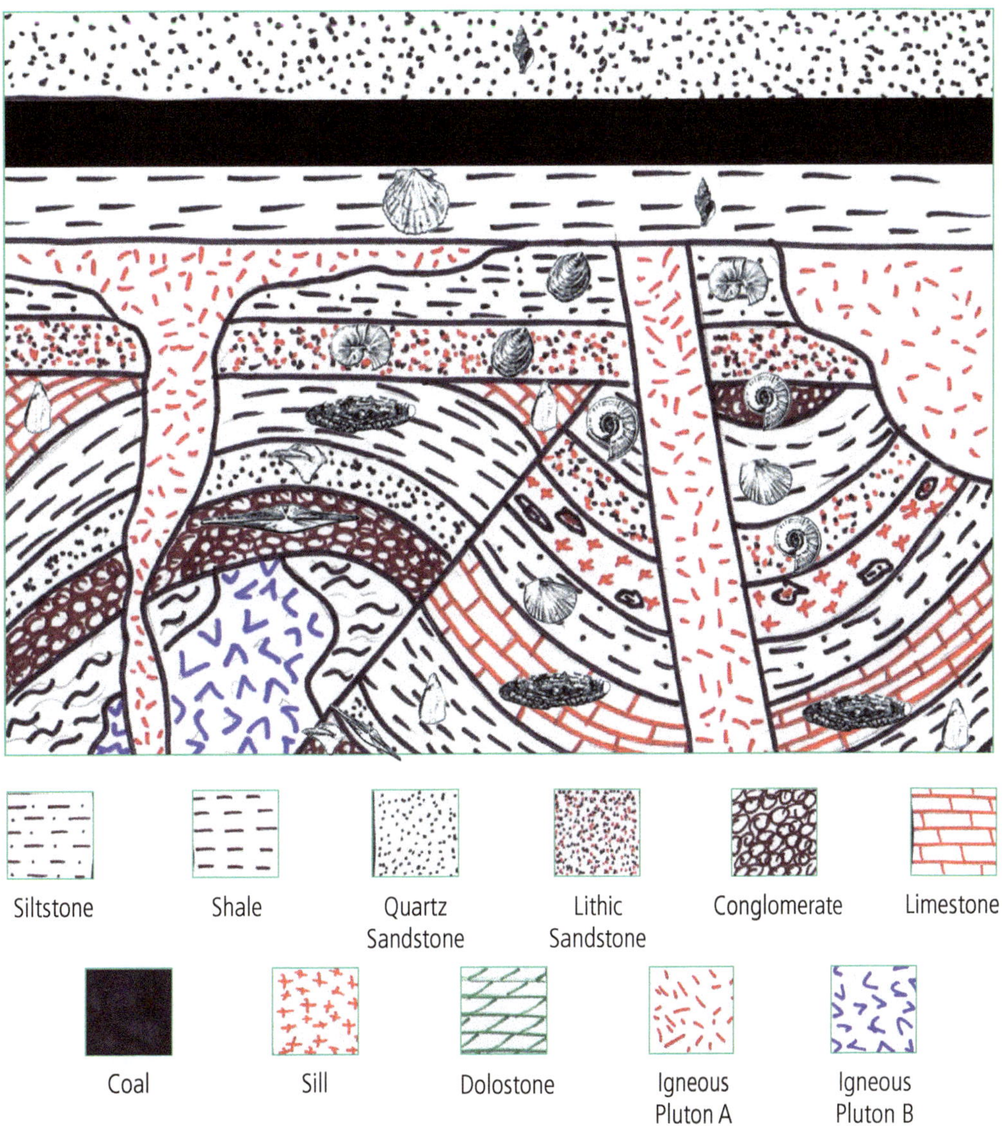

Figure 7.23

Youngest ↑

23. Deposition of ______________________ in the _______________ period

22. Deposition of ______________________ in the _______________ period

21. Deposition of ______________________ in the _______________ period

20. ____________ and _______________ (unconformity types), depending on the spot, due to erosion of strata from the __________ period

19. Three Igneous Type A plutons intrude

18. Deposition of ______________________ in the _______________ period

17. Deposition of ______________________ in the _______________ period

16. ________________________________ (Unconformity type) due to erosion of strata during the _______________ period(s)

15. ____________________ (Event)

14. ____________________ (Event)

13. Deposition of ______________________ in the _______________ period

12. Deposition of ______________________ in the _______________ period

11. Deposition of ______________________ in the _______________ period

10. Deposition of ______________________ in the _______________ period

9. Deposition of ______________________ in the _______________ period

8. Deposition of ______________________ in the _______________ period

7. ________________________________ (Unconformity type) due to erosion of strata during the _______________ period(s)

6. Deposition of ______________________ in the _______________ period

5. Deposition of ______________________ in the _______________ period

4. ________________________________ (Unconformity type)

3. ____________________ (Event)

2. ____________________ (Event)

1. Deposition of shale or mudstone

Oldest ↓

Follow-up Questions

1. Return to the lines you used to sequence this cross-section and put a bracket around the events that were occurring during the time when the sill could have intruded. Explain why you feel the sill intrusion was restricted to the time frame you selected (i.e., why it could only have come after a particular event and before something else). ______________________________
2. In an attempt to create a complete stratigraphic column, geologists went to a local outcrop approximately 50 miles away in hopes of finding an area that can help fill in some of the gaps represented by the unconformities in the previous cross section. Figure 7.24 is the column they were able to obtain from the second area. Correlate this second column with the previous cross-section, which has been reproduced below.
3. What rock type was deposited during the second unconformity? ______________________________
4. Does the geologic record at location 2 extend as far back in time as that of location 1? Yes or No? If no, why not? ______________________________
5. At what period does the geologic record end at location 2? ______________________________
6. Are you able to get a full and complete column using a combination of both these locations? Yes or No?
7. What does this imply about the likelihood that any one location or correlating just a few locations will permit you to compile a complete history? ______________________________
8. Can you deduce from this cross-section the order in which the three Igneous A intrusions occurred? Yes or No? If yes, how? If no, why not? ______________________________

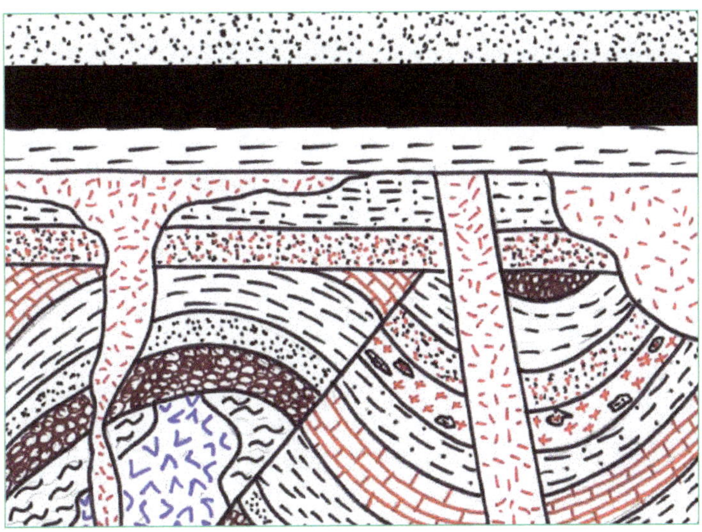

Figure 7.23 – Cross-section from location 1

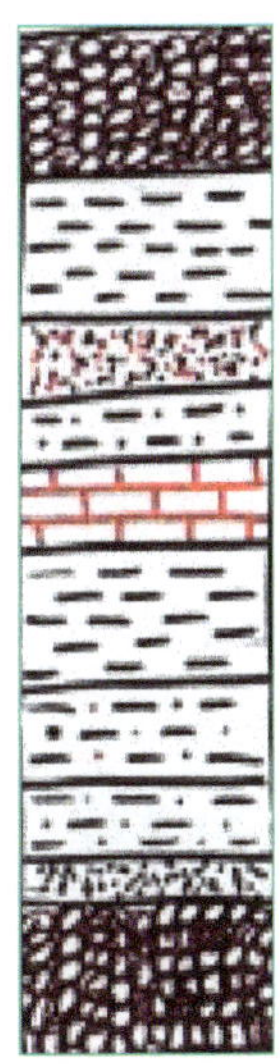

Figure 7.24 – Column from Location 2

PART C: NUMERICAL AGE DATING

As was demonstrated in the last exercise, sequencing events can only get you so far, as some events do not depict relationships with surrounding strata that are useful when applying the principles of relative dating. Furthermore, in order to completely reconstruct any historical record, actual dates must be assigned to major events so that the time span being represented can be put into context. With the advent of radiometric dating, the problem of assigning actual ages to certain types of geologic events was solved. As such, when used within its limitations, radiometric dating allows for a numerical age to be assigned to igneous and metamorphic events, therefore giving sedimentary events between these numerical ages a bracketed age range.

Radiometric dating is accomplished through the use of isotopes. Recall that an isotope is a group of atoms that have the same number of protons, but a different number of neutrons. For example, there are three isotopes of carbon:

	# Protons	# Neutrons	Protons + Neutrons
^{12}C	6	6	12
^{13}C	6	7	13
^{14}C	6	8	14

All three atoms have six protons, meaning they are all carbon, but they each have a different number of neutrons, making these three atoms isotopes.

Isotopes can be categorized as either "stable" or "unstable." Stable isotopes are comfortable in their configuration and therefore maintain their atomic structure permanently. Unstable isotopes, aka radioactive isotopes, are unhappy and will spontaneously undergo a process known as radiometric decay, in which constituents of the atom are released. This results in the original atom (parent) transforming into a new atom (daughter) that is stable and happy in its new form. A parent isotope and its subsequent daughter are jointly known as a "decay pair."

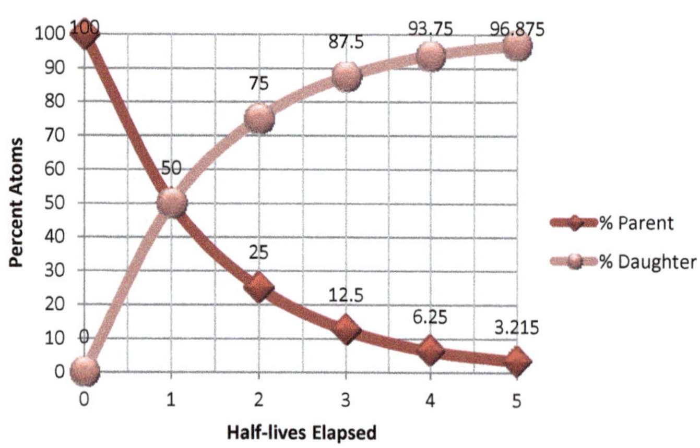

Figure 7.25 – Half-life Relationship for all Decay Pairs

While the exact time at which any one unstable parent will decay to its daughter product is unpredictable, statistical testing has demonstrated that a group of radioactive parent atoms will decay exponentially at a constant rate described as a half-life. A half-life is the amount of time it takes one half of the unstable parent sample to decay into stable daughter product. While we don't know which specific atoms will decay or precisely when the decay will take place, we do know that after a half-life has passed, 50% of the remaining parent sample will have decayed into daughter products. Figure 7.25 depicts the <u>*half-life relationship*</u> for all decay pairs.

As discussed above, all parent atoms will decay to daughter products at a rate described as a half-life. However, each decay pair has its own individual half-life, thus the actual decay rate differs from one decay pair to the next. Half-lives of some of our most common decay pairs have been supplied for you in Table 7.C.

Decay Pair	Half-life (Yrs)	Isotope Containers
$^{235}U \rightarrow {}^{207}Pb$	713 million	Minerals with Uranium (Zircon & Uranite)
$^{238}U \rightarrow {}^{206}Pb$	4.5 billion	
$^{40}K \rightarrow {}^{40}Ar$	1.3 billion	Minerals with Potassium (Feldspar & Mica)
$^{87}Rb \rightarrow {}^{87}Sr$	48.8 billion	
$^{147}Sm \rightarrow {}^{143}Nd$	106 billion	Micas and Garnet

Table 7.C Half-lives of common decay pairs

Because the half-life of each decay pair is known to be constant, it is the equivalent of a clock that started ticking when the parent atom was locked into the crystal structure of its mineral housing. This occurs when the cooling magma/lava reaches the "blocking temperature," the temperature at which it locks the atoms into the crystal lattice for that particular mineral. As such, the age of a mineral and/or rock can be calculated by using the parent/daughter ratio in the mineral to determine the number of half-lives elapsed, then multiplying that number by the decay rate (half-life) for that specific decay pair.

# of half-lives elapsed x half-life of the decay pair = numerical age

While radiometric decay unquestionably revolutionized the historical geology discipline, it too has its limitations. Each situation must be carefully evaluated to see if radiometric decay is capable of delivering an accurate date under the circumstances at hand. Table 7.D lists a number of scenarios, where radiometric decay is either completely inappropriate or where the age divulged indicates something other than the date of crystallization, and therefore must be interpreted based on the surrounding geology.

Limitations	Ramifications
Must have a closed system	If during the lifetime of the mineral in question, parent or daughter atoms have been gained or lost, the sample has been contaminated and the date is unreliable.
In non-metamorphosed minerals, the date marks the date of crystallization.	Radiometric dating of minerals in a sedimentary rock will convey when the mineral crystallized in the magma, not when it was incorporated into the sedimentary rock.
In metamorphosed minerals, if during metamorphism the temperature is raised above the blocking temperature, the atoms are released from the lattice and the clock is reset.	Radiometric dating of metamorphic minerals dates the last metamorphic event, not the time of original crystallization.
The differing decay rates of each system impose limitations on the amount of time each is useful for.	Ex: ^{14}C decays so rapidly that after ~60,000 years, there is not enough of it left to be measured by modern techniques. • Good for archeology, climatology and studies of groundwater movement. Ex: The long half-life isotopes are of no use dating minerals < 100 million years old, because no perceptible change will have occurred in such a short time. • Applicable to Precambrian rocks (rocks over 700 million years old).
Inherent Error A. Because we are dealing with a start and finish time with no knowledge of how rates proceed in between, under the best of circumstances, only the average decay rates are determinable, which inherently introduce some uncertainty. B. Laboratory analyses of isotopic ratios come jointly with a small percent error that contributes to some inaccuracy. C. Events such as leakage, contamination, delayed cooling, and thermal interruption all introduce the potential for significant inaccuracy in the results.	• There is no such thing as an "absolute date." • It should be understood that the numbers produced by radiometric dating inherently have varying degrees of error. • This does not mean these numbers are not overall accurate, merely that they should be interpreted in the context of the geological history of the source material.

Table 7.D – Limitations of the Radiometric Decay Dating Technique

Exercise: Crunching Some Numbers

The following exercise is designed to acquaint you with the use of radiometric decay as a tool for numerical dating. Examine Figure 7.26a, then answer the corresponding questions.

Igneous Sill

Sandstone

Figure 7.26a

1. Isotopic analysis of zircon crystals from the igneous sill in Figure 7.26a showed the ratio of $^{235}U/^{207}Pb$ to be 75%/25% respectively. Calculate the age of the igneous sill and show your work. _______________ years

2. Other types of minerals from the igneous sill were also radiometrically dated. They divulged ages significantly different, both from the age of the zircons and the age of each other. Does this mean something has happened in this area that has contaminated the isotopic ratio, and therefore renders unreliable dates? Yes or No? Explain ____

3. If various minerals from a surface lava flow were analyzed for their isotope ratios, would you expect the dates to be significantly different, as with the sill, or more or less the same? Explain. ________

4. Does the sandstone beneath this sill have to be older or younger then the sill? _________ How do you know? ____

5. Isotopic analysis of zircon crystals from the sandstone beneath the igneous sill in Figure 7.26a showed the ratio of $^{235}U/^{207}Pb$ to be 50%/50%, respectively. Does this mean that the sandstone was lithified approximately 713 million years ago? Yes or No? Explain.

____

____

____

6. We know that the age you calculated for the sill in question number 1 is analogous to the carboniferous period. This being the case, are there any unconformities in this outcrop? Yes or No?

7. If yes, which period(s) are missing? ____________________

8. Examine the close-up of Figure 7.26b taken from Figure 7.26a. Notice how the sandstone has been lifted upward. Suggest how this likely happened.

____

____

Figure 7.26a

Figure 7.26b

9. Using the available fossil and numerical data, on the lines provided, put together a brief history of the events that took place to form this landscape.

Youngest

D. ______________ approximately ______________ years ago.

C. ______________

B. Hiatus from the ____________ to the __________ period

A. Deposition of ____________ in the ____________ period

Oldest

PART D: TYING IT ALL TOGETHER

It has been demonstrated throughout this laboratory assignment that both relative and numerical dating have stringent limitations when used independently. However, when used jointly, they each compensate for the other's deficiencies by telling different parts of the story. Therefore, attainment of a complete picture requires the use of both dating techniques. It was through the combined use of both relative and numerical dating that we now have a universal geologic time scale, which gives us both the sequence of major events since the birth of the Earth, as well as numerical dates at which these monumental events occurred. In order for you to gain an appreciation for the joint use of geologic dating techniques, as well as the scale of geologic time and the events contained therein, the following exercise will guide you through the same processes geologists used to create our modern geologic time scale model.

Exercise: Obtaining the Modern Perception of Geologic Time

Because interest in compiling a large-scale history of the planet developed long before the advent of radiometric dating, initial attempts at organizing Earth's history were limited to relative sequencing. As such, scientists needed some kind of tracer that marked the passage of time by constantly evolving in a unilateral manner, such that prior conditions were never repeated—and which could be traced worldwide, thereby representing conditions on the planet as a whole. With the discovery of the Principle of Fossil Succession, fossils proved to be the timekeeper geologists were searching for.

However, the scientists were hardly on easy street, as the next major hurdle was that no one location on the planet could provide a complete record of Earth's history, due to the presence of unconformities. As such, it would be necessary to draw local stratigraphic columns for millions of places around the globe, then correlate them to create a baseline worldwide stratigraphic column.

This process began with thousands of local investigations and correlations that produced columns representative of each major continent. The stratigraphic columns representing each major continent have been reproduced for you in Figure 7.27. (Note: Any organism depicted that is not an invertebrate represents a large group of similar organisms, not a specific species).

1. Your job is to take this data and create a worldwide stratigraphic column by applying biostratigraphic correlation. In order to simplify your task, the columns have been lined up horizontally in Figure 7.28. Cut out each column from Figure 7.28, then mix and match them as necessary to effectively correlate all nine continental columns.

2. Once you have completed your correlation, build one composite stratigraphic column representing Earth's entire geologic history by cutting out the necessary strata from your correlated columns and taping them in the correct sequence on the space provided in Figure 7.29. If there are several squares representing one stratum—as there will be if that stratum was found on multiple continents—cut out the square with the most information and use it to build your complete column.

3. Now observe the stratigraphic column you have just compiled, and compare it to the geologic time column next to it.

    a. What did geologists use to make the division between the two eons? ____________ (Hint: Phanerozoic means "visible life.")

b. Following that first division, geologists next set out to subdivide the complex life forms that had evolved since the onset of the Phanerozoic. In doing so, they identified three distinct eras: "Paleozoic," which translates to "ancient life"; "Mesozoic," which translates to "middle life"; and the "Cenozoic," which translates to "recent life."

- Which organisms dominated the Paleozoic: Invertebrates, reptiles, or mammals?
- Which organisms dominated the Mesozoic?: Invertebrates, reptiles, or mammals?
- Which organisms dominated the Cenozoic: Invertebrates, reptiles, or mammals?
- How does nature go about downsizing or completely eliminating existing, thriving organisms so that developing organisms can diversify, fill the open niches, and dominate for a fixed time span? ____________________
- Based on your answer to the question above, and your knowledge of which groups of organisms dominated each era, what event do you think defines the Paleozoic/Mesozoic boundary? ____________________
- How about the Mesozoic/Cenozoic boundary? ____________________

4. Subdivisions of the three eras are called "periods." Boundaries between the periods are dictated by the appearance and extinction of specific organisms within the dominant group of that era. The smallest subdivisions are known as "epochs" and are based on fossil percentages of organisms that have living counterparts. As demonstrated, the major biological transitions in Earth's history supplied a framework for sequencing geologic events, and therefore allowed us to produce what is called "the geologic time column." How does this differ from the "geologic time scale"? ________
____

5. In order to create a geologic time scale from the geologic time column, Earth has to have a birthdate. It is believed that Earth formed from bombardment of meteorites; the oldest meteorites found on Earth have 50%/50% $^{238}U/^{206}Pb$ ratio. This being the case, how old is Earth? ____________________ (Show your work.)

6. For what percentage of Earth's history has the planet been home to complex life? ____________% (Show your work.)

7. What percentage of Earth's history has little to no fossil evidence? ______ %

8. Based on your answer to the question above, do we have much information about what went on for the bulk of Earth's history? ____________________

9. Each boundary on the geologic time scale is now associated with a numerical age. What type of rock was dated in order to obtain those numbers—igneous, sedimentary, or metamorphic?

10. As mentioned previously, the best material for correlation involves a component (such as a fossil) that is widespread throughout the globe. Suggest an igneous material that can be carried far and wide and therefore traced long distances. This material was dated and traced at many of the boundaries on the time scale. Thus, it serves as our primary numerical timekeeper for the geologic time scale. ____________________

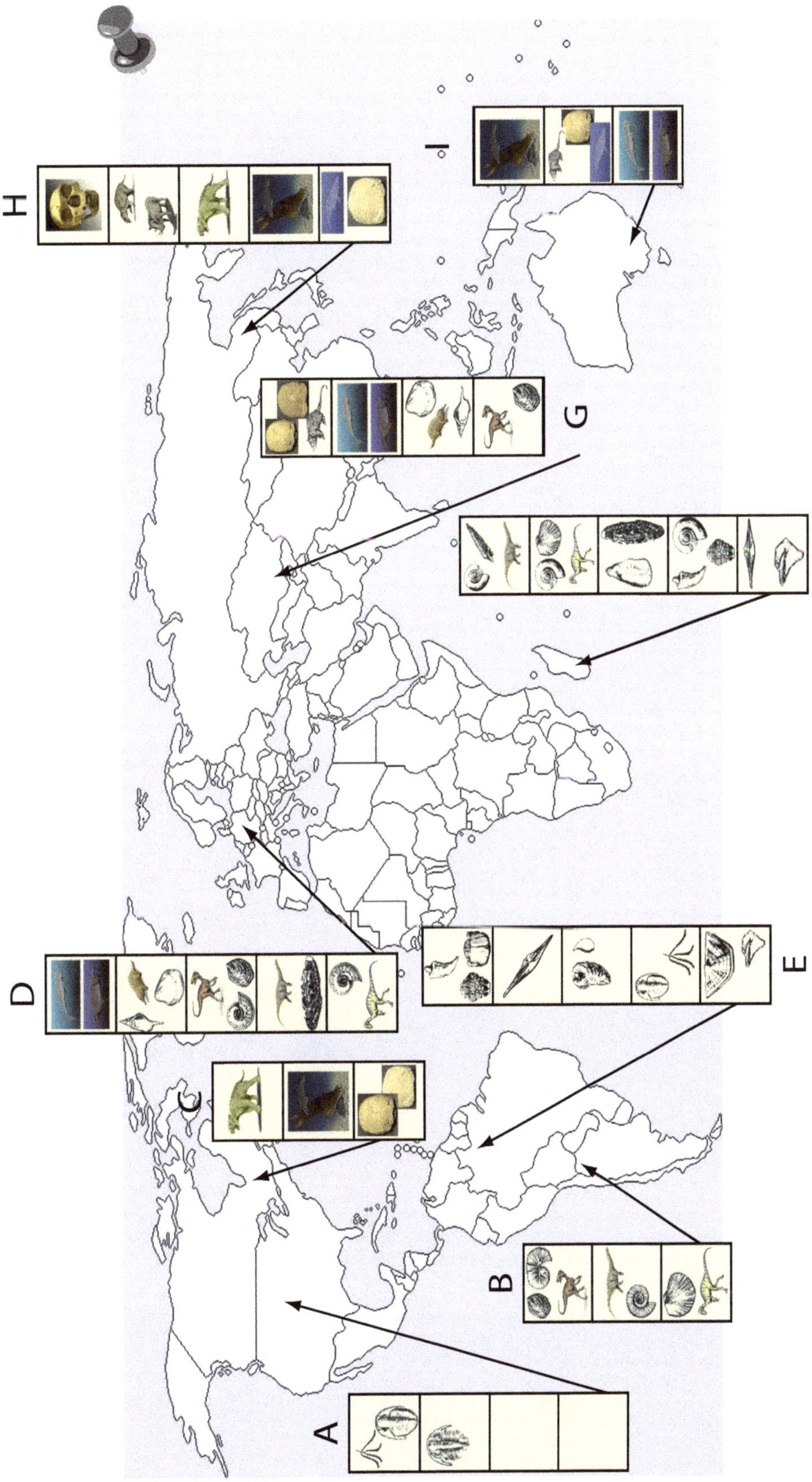

Figure 7.27

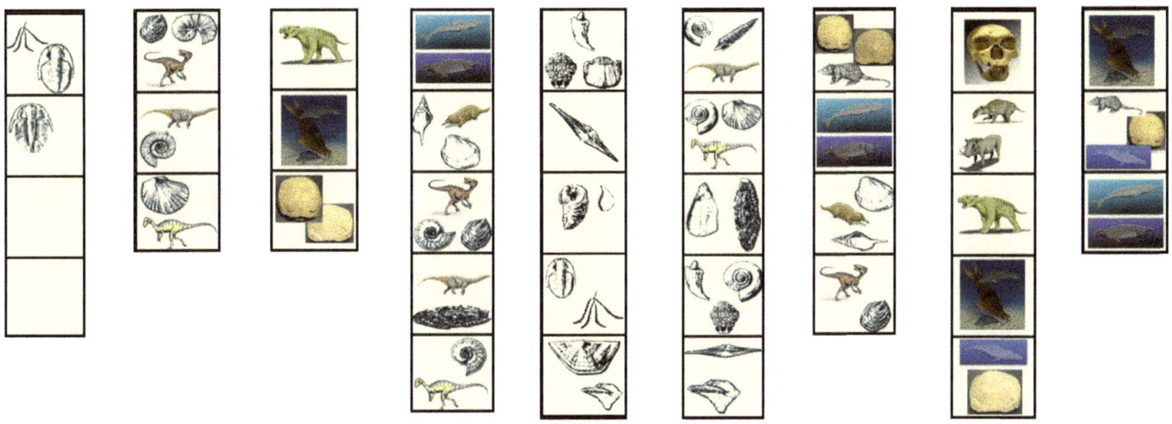

Figure 7.28

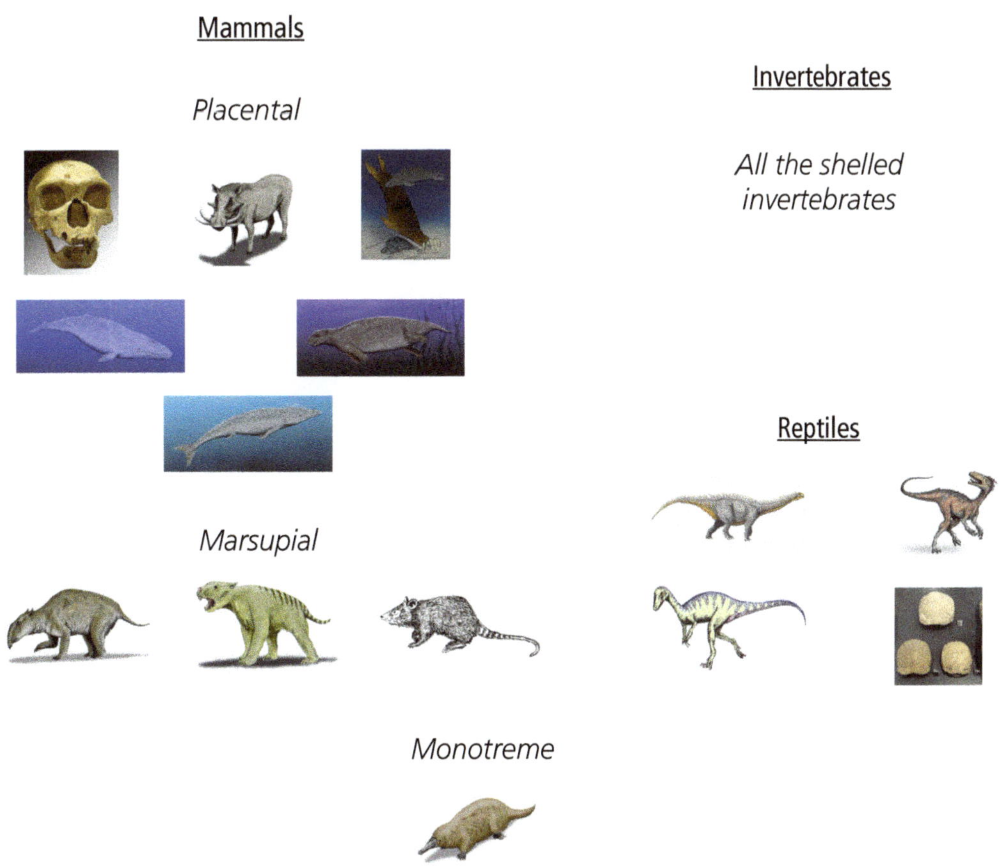

Legend for Figures 7.27 and 7.28

Eon	Era	Period		Epoch	Major Geologic and Biologic Events	Millions of Years Ago
Phanerozoic	Cenozoic	Quaternary		Recent or Holocene	Ice Age ends	0–0.01
				Pleistocene	Ice Age begins	0.01–1.6
		Tertiary	Neogene	Pliocene	Earliest humans	1.6–5.3
				Miocene		5.3–23.7
			Paleogene	Oligocene		23.7–36.6
				Eocene	Formation of Himalayas Formation of Alps	36.6–57.8
				Paleocene		57.8–66
	Mesozoic	Cretaceous			Extinction of dinosaurs	66–144
		Jurassic			First birds Formation of Sierra Nevada	144–208
		Triassic			First mammals First dinosaurs	208–245
	Paleozoic	Permian			Mass Extinction Formation of Pangaea Formation of Appalachians	245–286
		Carboniferous	Pennsylvanian		Abundant coal-forming swamps	286–320
			Mississippian		First reptiles	320–360
		Devonian			First amphibians	360–408
		Silurian				408–438
		Ordovician			First land plants	438–505
		Cambrian			First fish	505–545
Cambrian Explosion						
Proterozoic	Pre-cambrian				Earliest shelled animals	545–2500
	Archean				Earliest fossil record of life	2500–4000
	Hadean					4000–4600

Figure 7.29 – Earth's Geologic Time Scale.
Note: not to scale

Laboratory Exercise VIII

Earthquakes

Introduction

An earthquake is an episode of ground shaking caused by elastic waves propagating in the Earth that were generated by a sudden release of slowly accumulated stress in rigid bedrock. This sudden release of stored energy results from an abrupt slippage of rock masses along a fault. The place where the slippage occurs is known as the hypocenter, or focus, of the earthquake. The point on the surface vertically above the hypocenter is the epicenter. See Figure 8.1.

Objective

The threefold objective of this lab is to:

1) explore the physics of earthquake waves and their relationship to magnitude and intensity;
2) apply those concepts to an actual case study of the 1989 Loma Prieta earthquake; and
3) investigate the large-scale ramifications of earthquakes with respect to their tectonic environments.

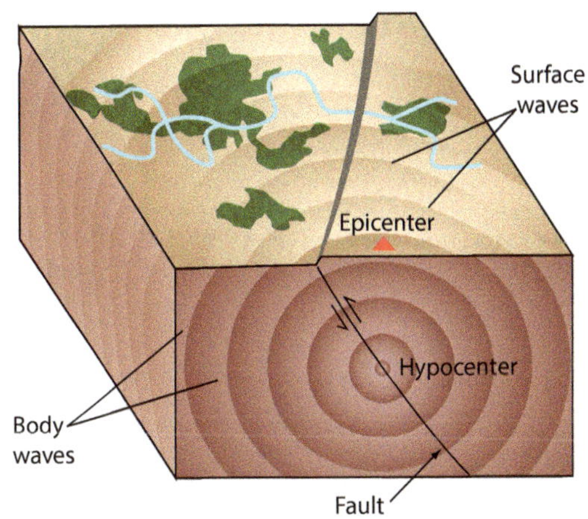

Figure 8.1 – Types of Earthquake Waves

Part A: Earthquakes and Waves

The energy released from the hypocenter of an earthquake travels as several types of vibrations, or waves, that are transmitted through the Earth in all directions. Some waves travel through the Earth to the surface and are known as "body waves." Others travel along the Earth's surface and are known as "surface waves." See Figure 8.1.

During any earthquake, there are two types of body waves that are produced. The first is known as the "P-wave," or "primary wave," and the second is known as the "S-wave," or "secondary wave." P-waves propel themselves forward by compression and extension; this motion can be demonstrated by thumbtacking a Slinky to the wall and repeatedly pulling the Slinky backward and forward. As such, P-waves are said to travel in a horizontal motion. S-waves, unlike P-waves, have the additional component of vertical motion. Their method of travel can best be demonstrated by thumbtacking a jump rope to the wall, then taking hold of the loose end and moving your hand up and down. See Figure 8.2. *Note: Both the P-wave and the S-wave are generated and released from the hypocenter at exactly the same time, but reach their destination at different times due to the S-wave traveling at approximately half the speed of the P-wave.*

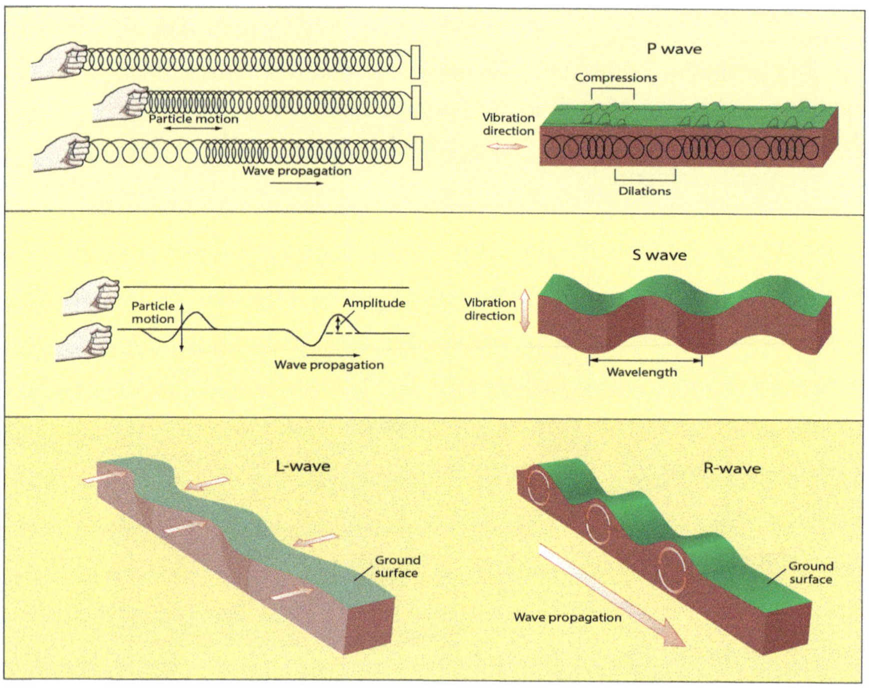

Figure 8.2 – Motion of Earthquake Waves

Another class of seismic waves is surface waves. These waves are generated as a result of the body waves smashing into Earth's surface. Surface waves consist of "Love waves" (travel forward but shake sideways) and "Rayleigh waves" (travel forward via a rolling motion). So the combined effect of both surface waves creates a very aggressive motion in both a horizontal and vertical direction. See Figure 8.2.

Figure 8.3 is a typical seismogram—a record of the ground's motion (y-axis) as a function of time (x-axis)—marking the arrival of the P-waves, S-waves, and surface waves (L-waves and R-waves) into a measuring station called a seismograph. Use the seismogram in Figure 8.3 to answer the following questions.

1. Based on the seismogram, which of these waves is the fastest? P-wave, S-wave, or Surface wave (L/R-wave) ____________________

2. Based on the seismogram, which of these waves is the slowest? P-wave, S-wave, or Surface wave (L/R-wave) ____________________

3. Which of the three waves would you expect to cause the most damage? P-wave, S-wave, or surface wave (L/R-wave) ____________________

   What is your evidence? ____________________

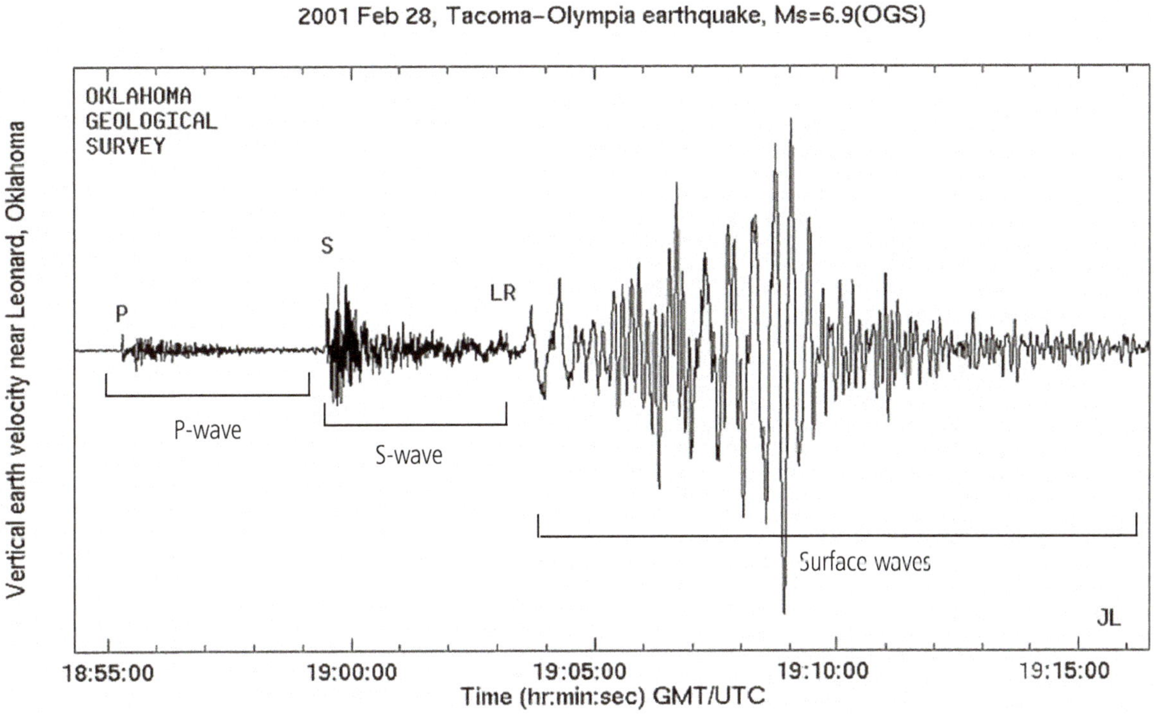

Figure 8.3 – Earthquake seismogram. Source: The Iris Consortium

PART B: EPICENTER, MAGNITUDE, AND INTENSITY

Epicenter

After an earthquake, seismologists are faced with the task of finding when and where the shaking began, meaning they must locate the epicenter. Keeping track of epicenter locations in earthquake prone regions is essential to understanding the fault behavior, and therefore is a central part of earthquake forecasting, land-use assessment, and building code requirements. Since most earthquake-prone areas are currently outfitted with seismographs to record the shaking during an earthquake, the location of an epicenter is usually determined in a matter of seconds, as the seismographs transmit their data to seismic laboratories in real time. However, it is important to understand the physics behind this process, as epicenter identification, along with anything linked to earthquakes, is directly correlated to wave behavior.

It is now known that earthquake waves travel at a fixed speed, depending on the type of rock the wave is traveling through. The denser the material, the more confined the wave and the faster the wave can propel forward. Laboratory experiments have carefully documented the speed at which each of the different waves travel through the various rocks making up both the surface and internal Earth. Those experiments permitted the development of a universal graph that demonstrates the distance each of the waves can travel throughout Earth per unit time, based on the rock types comprising the planet (Figure 8.4). As such, if the travel time of a wave from its generation to the time it reaches a seismograph is known, the universal graph can be used to determine the distance from the epicenter to the seismograph. Note that since the creation of this universal graph, it has been used to accurately identify the epicenter locations of thousands of earthquakes around the world and is therefore supported by field data.

Exercise 1: Locating the Epicenter of an Earthquake

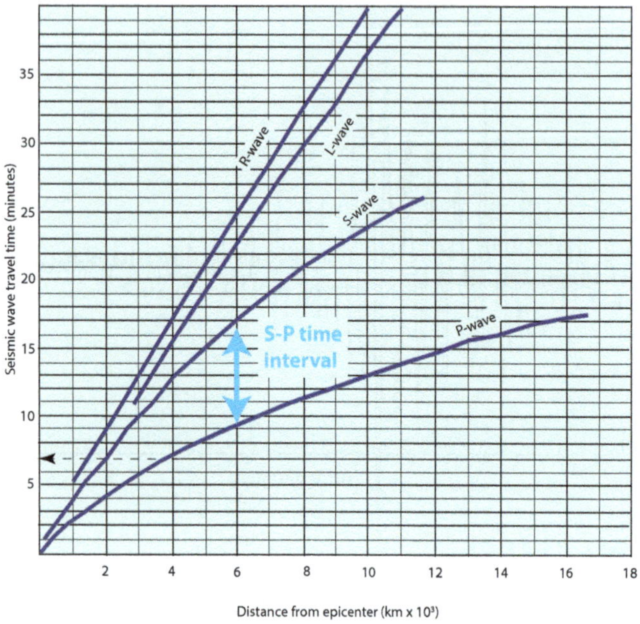

Figure 8.4 – Wave Distance and Travel Time Curves

1. Examine Figure 8.4. Why are the lines of the P- and S-waves curved, whereas the line for the surface waves is straight?
____
____
____
____
____
____
____

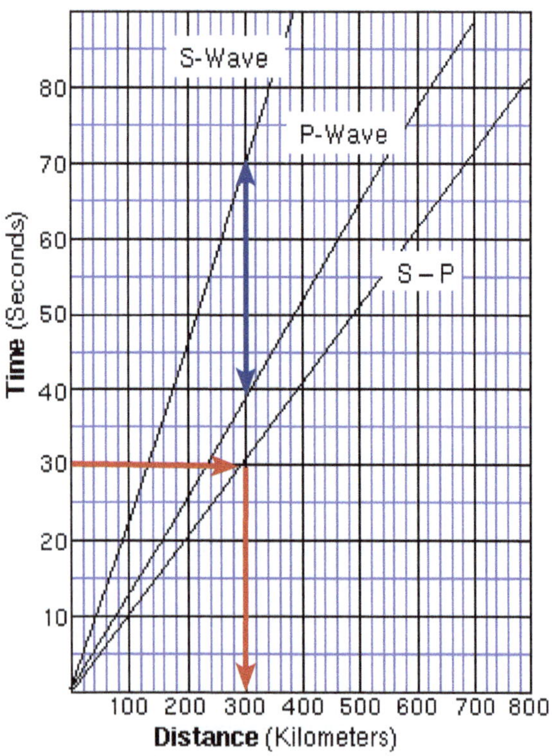

Figure 8.5 – Simplified Wave Distance and Travel Time Curves Directly Showing S-P Time Interval

As previously discussed, in order to use this graph, the travel time of the wave must be known. Yet when the P-wave arrives at a seismograph, the only information available is the arrival time of the wave—which does not tell us how long the wave has been traveling. In order to rectify this problem, scientists use what is known as the S-P time interval, which is the amount of time between the arrival of the P-wave and the arrival of the S-wave. Using this relationship, we have a known wave travel time interval and can therefore use the graph to calculate the distance of the seismograph from the epicenter. Figure 8.5 is a simplified version of the universal graph in which the S-P time interval has been depicted on one curve. For example, if it was determined that 30 seconds passed between the arrival of the P-wave and the arrival of the S-wave, on a traditional graph it would be necessary to find the spot between the P and S wave curves that represents 30 seconds (see blue line on Figure 8.5), then read down to the Y-axis to determine the distance from the epicenter (300km). However, the addition of the S-P time interval curve allows you to start at 30 seconds on the Y-axis, read across till you reach the S-P time interval curve, then read down to the X-axis; again getting 300km as the distance from the epicenter (see red lines on Figure 8.5). Using the simplified S-P time interval curve will allow you to quickly determine the distance from the epicenter for any seismogram.

2. What happens to the S-P time interval as the distance from the epicenter increases? ______________________________

3. Refer back to Figure 8.3.

   a. What is the arrival time of the P-wave? _______

   b. What is the arrival time of the S-wave? _______

   c. What is the S-P time interval? _____________ minutes

   d. Using Figure 8.4, determine how far this seismograph was from the epicenter? _______________km

4. While going through the process described above will tell you the distance of the seismograph from the epicenter, it does not tell you the actual location of the epicenter. Why not? ______________________________ ______________________________

5. What additional information do you need to determine the actual location of the epicenter? ______________________________

Once you obtain the distance of three different seismographs surrounding the epicenter, you can triangulate to find the actual epicenter location. This can be done by following the steps below:

I. Obtain a map of the region you are working with.

II. Plot the locations of the three seismographs on your map and label them accordingly.

III. Using the bar scale on the map, open a compass the width of the distance from seismograph 1 to the epicenter.

IV. Put the sharp end of your compass on the location of seismograph 1, and draw a full circle around that location. The radius of that circle is the distance of seismograph 1 from the epicenter.

V. Repeat steps III and IV, using the distances from seismographs 2 and 3, respectively.

The place where all three circles intersect is the location of your epicenter! See Figure 8.6.

Figure 8.6 – Example of Triangulation

6. Use the three seismograms that follow, the corresponding map (Figure 8.7), and the simplified S-P time interval and distance graph (Figure 8.5) to locate the epicenter of an earthquake.

Step 1: Locate the distance of the 3 seismographs from the epicenter

Station	S–P Interval (seconds)	Epicentral Distance (km)
Chichuahua		
Mazatlan		
Rosarito		

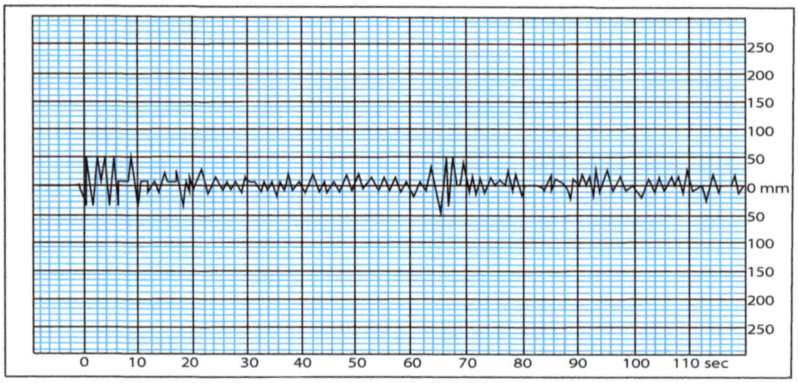

Chihuahua Seismic Station

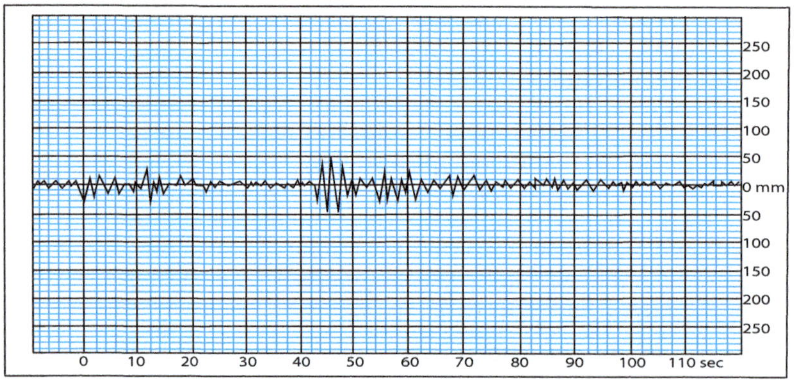

Mazatlan Seismic Station

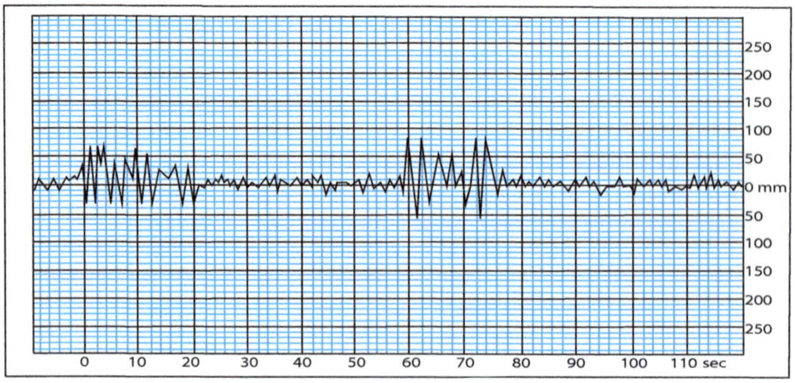

Rosarito Seismic Station

Figure 8.7 – Triangulation Map

Step 2: Triangulate

Where is the epicenter located?

____

Magnitude

Magnitude is the term used to describe the highest amplitude of an earthquake wave produced during an earthquake. The Richter Magnitude Scale is a commonly known standardized system of amplitude measurement, and allows for comparison of different earthquakes around the world. However, due to inherent limitations associated with the Richter Scale, scientists now use a more quantitative and accurate scale called Moment Magnitude, which involves examining the amplitude, the amount of slip, and the size of the area over which the slippage occurred. Both the Richter and Moment magnitude scales are logarithmic. This means that for every whole number increase in magnitude, there is a 10-fold increase in amplitude and a 32-fold increase in the amount of energy released by the earthquake. See table 8.A. It is important to note that the magnitude of an earthquake is the same for all areas affected by the earthquake. In essence, an earthquake will have *only* one magnitude.

Δ Magnitude	Δ Ground Displacement	Δ Energy
1	10x	~32x
0.5	3.2x	~5.5x
0.3	2x	~3x
0.1	1.3x	1.4x

***Note: Δ = "change"**

Table 8.A – Earthquake magnitude, ground displacement, and energy release—they're all connected

7. If an earthquake in Australia had a magnitude 5.0 and an earthquake in Indonesia had a magnitude 7.0, how much bigger was the amplitude of the second quake than the first? ________ x larger—How much more energy was released in the second quake than the first? ________ x more energy

Intensity

In addition to measuring the size of an earthquake by means of seismographs, we also survey people who felt the earthquake and interpret their recollection of the level of intensity using the Modified Mercalli Intensity Scale (see Table 8.B). The intensity of an earthquake at a site is based on the observations of individuals during and after the earthquake. It represents the severity of the shaking, as perceived by those who experienced it, as well as observations of damage to structures, movement of stationary objects, and change in the Earth's surface as a result of geologic processes during the earthquake. As such, unlike magnitude—where there can only be one per earthquake—one earthquake can have many different intensities.

Isoseismal (intensity) maps, like the one shown in Figure 8.8, demonstrate the distribution of seismic intensities associated with an earthquake. The greatest impact of earthquake waves is usually in the epicentral region, with progressively lower intensities occurring in nearly concentric zones outward from the epicenter. It must, however, be kept in mind that there are exceptions to this general rule, as the quality of construction and variation of geologic conditions affect the distribution of intensity.

Exercise 2: Understanding Magnitude and Intensity

8. Given that a high percentage of intensity determination is based on interviews with survivors, as well as observations of damage to structures, name a few non-geologic factors that could significantly impair the geographic accuracy of the contour lines.

- ____
- ____
- ____

Modified Mercalli Scale		Magnitude Scale
I	Detected only by sensitive instruments	1.5
II	Felt by few persons at rest, especially on upper floors; delicately suspended objects may swing	2
III	Felt noticeably indoors, but not always recognized as earthquake; standing autos rock slightly, vibration like passing truck	2.5
IV	Felt indoors by many, outdoors by few, at night some may awaken; dishes, windows, doors disturbed; autos rock noticeably	3 3.5
V	Felt by most people; some breakage of dishes, windows, and plaster; disturbance of tall objects	4
VI	Felt by all, many frightened and run outdoors; falling plaster and chimneys, damage small	4.5
VII	Everybody runs outdoors; damage to buildings varies depending on quality of construction; noticed by drivers of autos	5
VIII	Panel walls thrown out of frames; fall of walls, monuments, chimneys; sand and mud ejected; drivers of autos disturbed	5.5
IX	Buildings shifted off foundations, cracked, thrown out of plumb; ground cracked; underground pipes broken	6 6.5
X	Most masonry and frame structures destroyed; ground cracked, rails bent, landslides	7
XI	Few structures remain standing; bridges destroyed, fissures in ground, pipes broken, landslides, rails bent	7.5
XII	Damage total; waves seen on ground surface, lines of sight and level distorted, objects thrown up in air	8

Figure 8.8 – 1989 Newcastle, Australia Intensity Map. Illustration: Bidgee

Left: Table 8.B – Modified Mercalli Scale

9. Use Figure 8.9 to create an isoseismal map for the December 1872 Pacific Northwest earthquake by drawing boundaries between the intensities. Then use the isoseismal map you created to answer the questions that follow.

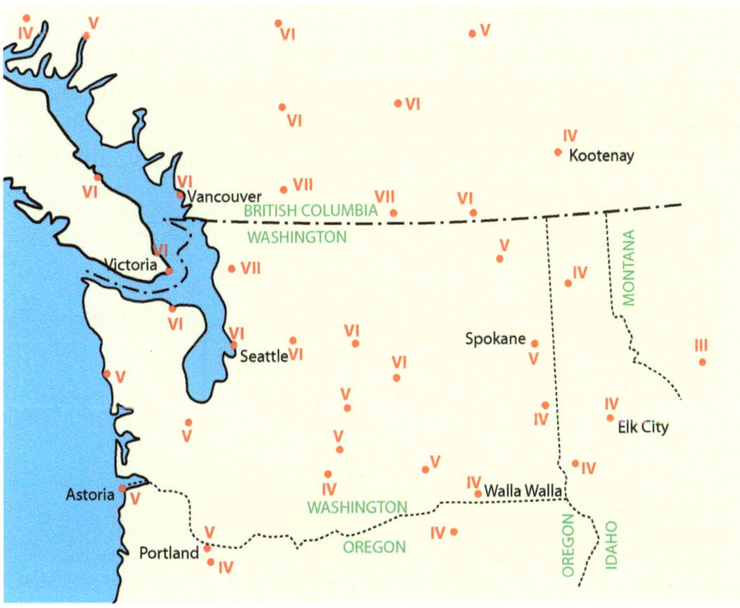

Figure 8.9 – Intensities for the December 1872 Pacific Northwest Earthquake

a. What was the maximum intensity for the 1872 Pacific Northwest earthquake?

____

b. Shade in the area that represents the epicenter of the 1872 earthquake.

c. While the intensity map in Figure 8.9 depicts no irregularities, often the expected bulls-eye pattern is disrupted with intensities that are "out of place." An example of this can be seen in Figure 8.10, which is an isoseismal map of the 1971 San Fernando Valley earthquake. Notice there are pockets of intensity VIII that are nearly on the boundary between the areas delineated as intensity VI and intensity VII. What are the two most likely causes for this localized increase in intensity? (Hint: One issue is geologic and the other is engineering.)

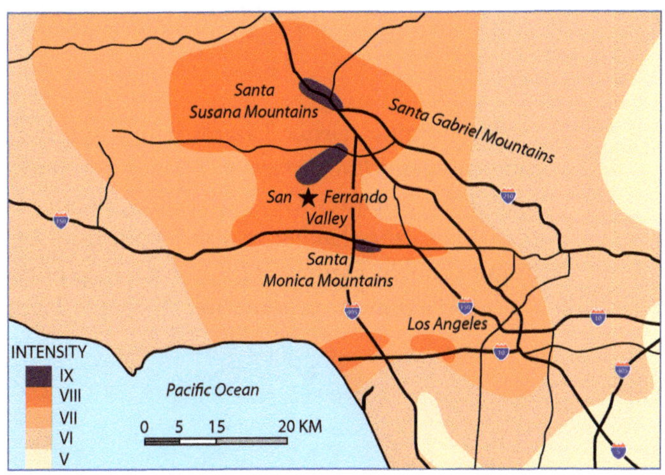

Figure 8.10 – Intensity Map of the 1971 San Fernando Earthquake, USGS

- _______________________
- _______________________

Material Amplification

Earth materials of various types behave differently during earthquakes and the difference is related to their degree of consolidation and saturation. Seismic waves move forward faster in consolidated bedrock than through unconsolidated sediment or soil. They slow even further if the unconsolidated material has a high water content. When the front part of the wave slows down, but the middle and back continue forward at the original speed, the wavelengths become compressed. As the wavelengths shorten, the amplitude increases, and the energy that was once directed forward becomes directed upward, effectively increasing the intensity of the shaking. This effect is known as material amplification. See Figure 8.11.

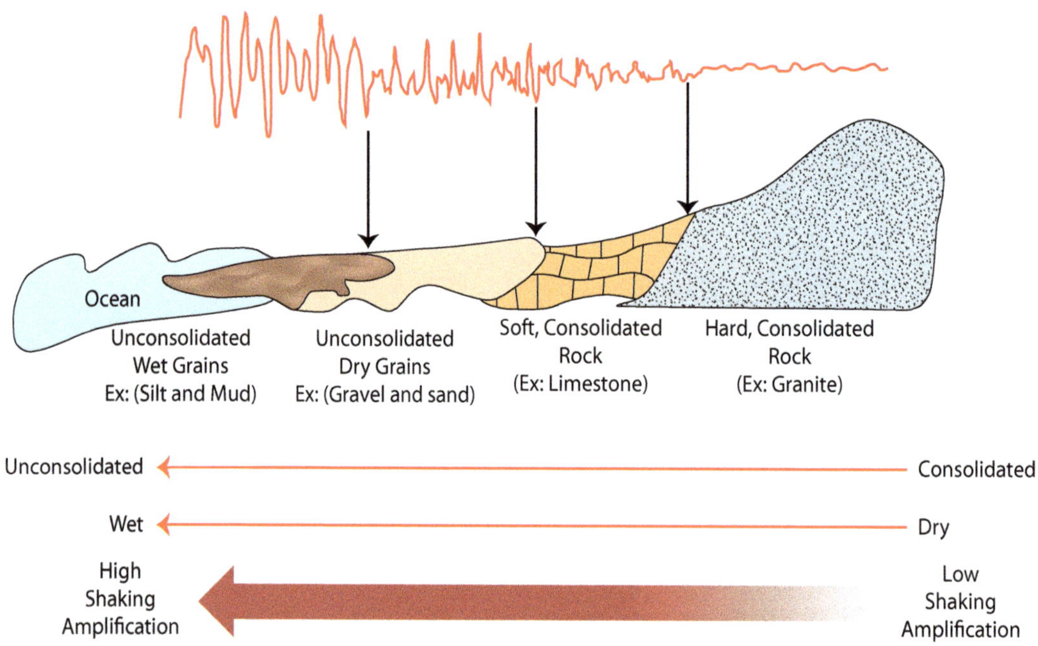

Figure 8.11 – Effect of material amplification on ground motion.

In determining seismic hazards within specific areas, geologists consider the potential for material amplification in local surface and shallow subsurface geology, and the effects of the amplified shaking on manmade structures. Of particular importance is the potential for liquefaction, where the ground "liquefies" due to the shaking, and slides out from under the structures it supports, leaving the city or town a heap of rubble. Figure 8.12 is a shake map of the San Francisco region. Figure 8.13 depicts the potential for liquefaction in the same area. Examine Figures 8.12 and 8.13, then answer the questions below.

Exercise 3: Material Amplification

10. On Figure 8.13, areas depicted by white are primarily underlain by bedrock, areas depicted by yellow and orange are primarily underlain by alluvium (sand, gravel, and cobbles), and areas depicted by pink and red are primarily underlain by mud.

    a. Comparing Figures 8.12 and 8.13, what type of shallow, subsurface geology is most susceptible to liquefaction? ______

    b. What type is least susceptible to liquefaction? ______

    c. Explain the reason for this difference: ______

    ______

11.

    a. Comparing Figures 8.12 and 8.13, which type of geology typically experiences the greatest shaking amplification? ______

    b. Which type of geology typically experiences the least shaking amplification? ______

12.

    a. Compare areas designated by red and yellow in Figure 8.13. What do you notice about the degree of development in those two areas with respect to each other?

    ______

    b. Why do you think there is less development in one area than in the other area?

    ______

    ______

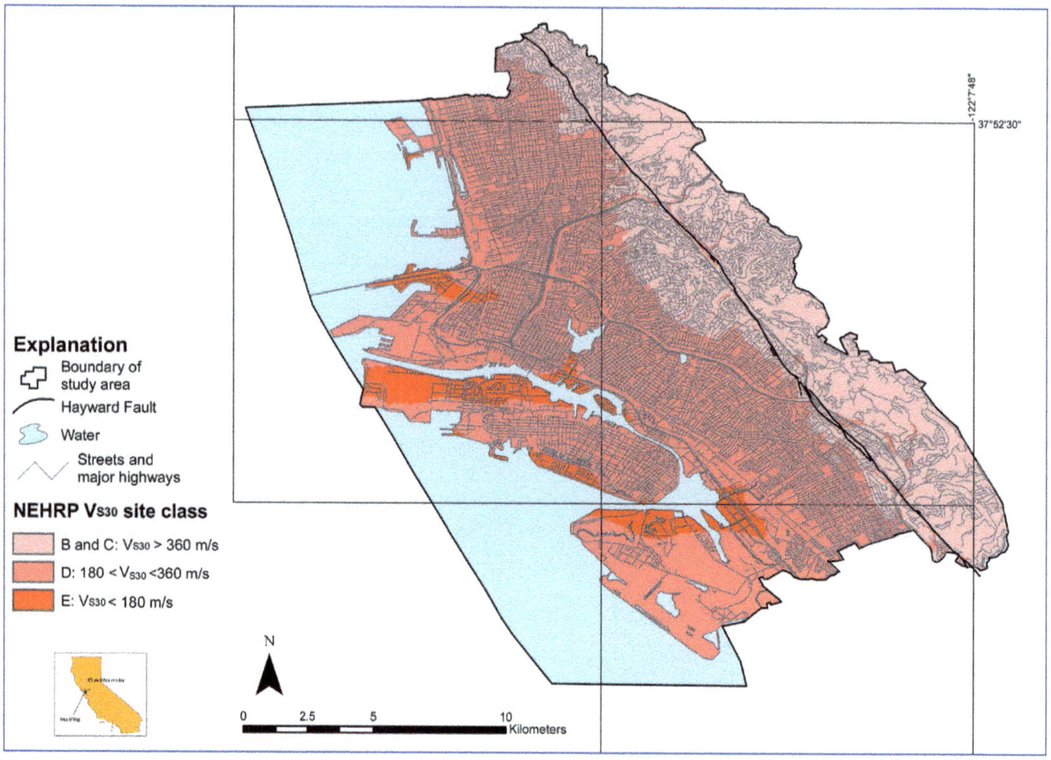

Figure 8.12 – Shaking Hazard Map for Northwestern Alameda County

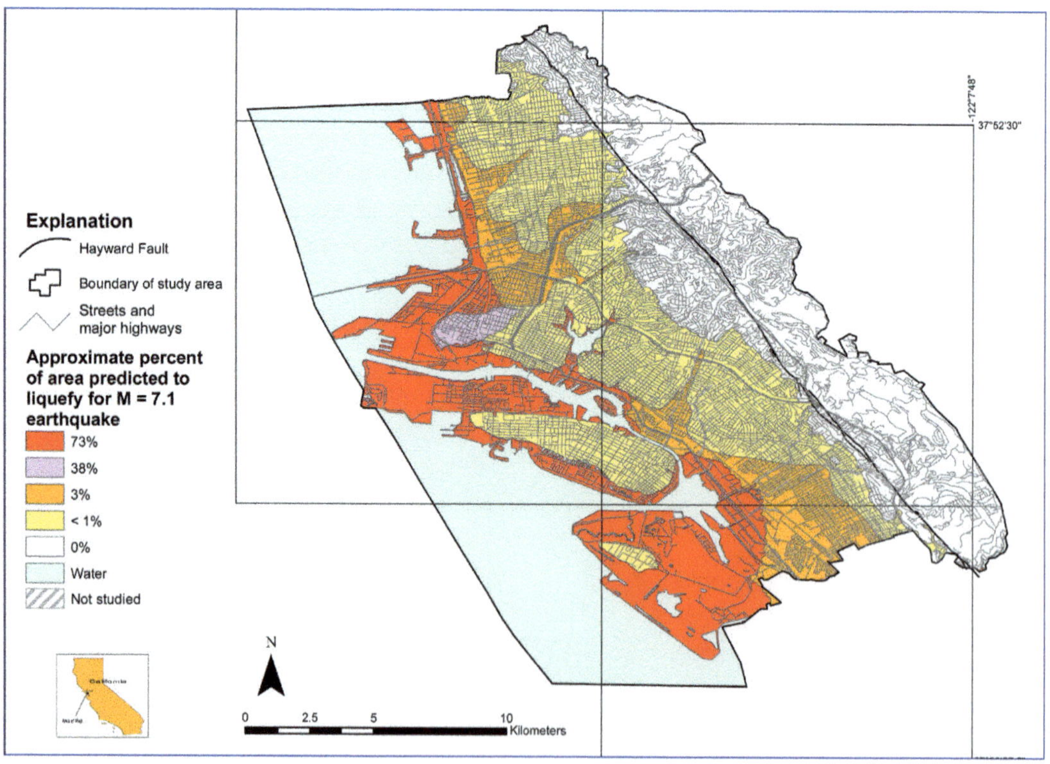

Figure 8.13 – Liquefaction Hazard Map for Northwestern Alameda County

PART C: LOMA PRIETA CASE STUDY

On October 17, 1989, the San Francisco Bay region was shaken up by the Loma Prieta Earthquake, named after the small community located near the epicenter. San Francisco and Oakland were the focus of media attention because of their size and the amount of damage they sustained, but it's important to remember that they were 100 km away from the epicenter. Other smaller, but closer communities, such as Watsonville and Santa Cruz, also suffered great damage and fatalities. This exercise looks at the geology of the earthquake and the geologic reasons for the building failures in the San Francisco and Oakland region.

Geology of the San Andreas Fault System

Use Figure 8.14 to answer the following questions:

1. What is the geographic orientation (compass direction) of the faults shown in Figure 8.14?

   ____

2. What large-scale geological, tectonic process is occurring in this location that accounts for the fault orientation and motion? (Hint: What type of fault is this, and in what direction is each side of the fault moving with respect to the other side?) See Figure 8.15 for a visual aid.

   ____

   ____

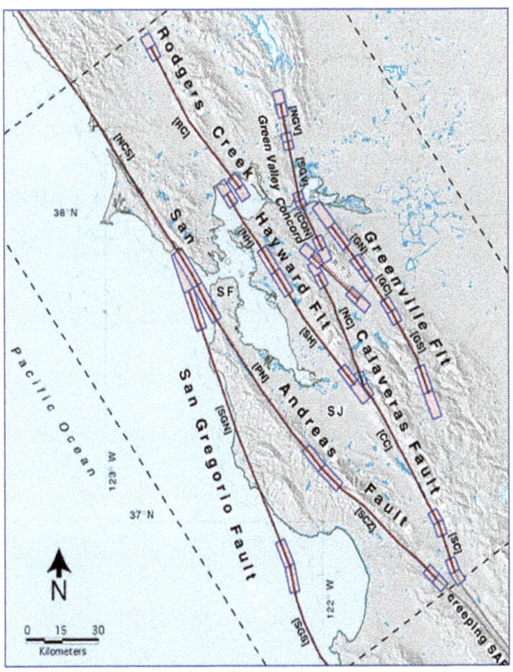

Figure 8.14 – Major Branches of the San Andreas Fault System

Figure 8.15 – Fault Movement on the San Andreas

Attenuation and Acceleration

Attenuation is the decrease in the amplitude of seismic waves as distance from the source increases. As a general rule, waves attenuate as they move away from the epicenter, but selected geologic conditions tend to cause local variations in amplitude, such that sites distant from the epicenter will show an amplification of ground motion.

Seismologists use sophisticated instruments to measure many different components of earthquake waves in an attempt to determine their effect on the local environment. An understanding of wave behavior with respect to its impact on the local geologic conditions is critical to establishing safe building codes, as the buildings must be able to sustain the forces exerted on the building from both the waves and the shaking sediment beneath them. During an earthquake, small particles on Earth's surface will be moved back and forth haphazardly. This movement can be described by its changing position/unit time (displacement), changing velocity/unit time (velocity), or changing acceleration/unit time (acceleration).

Acceleration is the rate of change in velocity, i.e., how much the velocity changes in a unit time. You experience acceleration every time you increase or decrease the speed of your car by applying the gas pedal or brake, respectively. When acceleration acts on a physical body, the body experiences the acceleration as a force. In the example with your car, your body experiences the force from the increasing velocity (depressing the gas pedal) by pushing you back into your seat, while it responds to the force of decreasing velocity (depressing the brake) by propelling you toward the windshield. Now picture the earthquake wave as the gas pedal/brake and the building as you in the car. The building will be pushed in multiple directions, just as you would be pushed back and forth if the car were abruptly changing speeds. The faster the change in velocity, the stronger the force (e.g., steadily braking versus slamming on the brakes). If the acceleration (change in velocity per unit time) was not constant, but fluctuated throughout the period of acceleration, the largest value of the acceleration would be the "peak" acceleration.

While displacement, velocity, and acceleration are all ways to describe the changing position of a material over time, acceleration is the most useful for our purposes, because the building codes prescribe how much horizontal force buildings should be able to withstand during an earthquake. Accelerographs, which measure acceleration of the ground, record data that are typically presented as a percentage of the force of gravity (g). Buildings are designed to withstand the vertical force of gravity (1 g, 981 cm/sec^2); hence the reason they are able to stand up under gravitational forces. They are not, however, necessarily designed to be moved sideways, and earthquakes generate some strong lateral forces against a building.

Figure 8.16 is a map showing horizontal acceleration from the Loma Prieta Earthquake at a number of sites in western California. Use this figure to answer the following questions:

3. A contour line has been drawn around the sites with acceleration rates of 20% g.

   a. What is the general compass orientation of the contour? ______________________

   b. Any other contour line would replicate this orientation. How does its orientation compare with the orientation of the faults that are shown in Figure 8.14? ______________________

   c. Had you drawn this line yourself (which I did, and it took me a while because all I had were numbers instead of the color coded dots I supplied you with), you would see there is a large number of irregularities that make drawing this contour line a challenge. What main geological factor contributes to difficulties in drawing the irregular 20% geoisoseismal line?

   ____

   d. What main engineering factor contributes to difficulties in drawing the irregular 20% geoisoseismal line? ______________________

   e. What happens to ground acceleration as you move progressively away from the epicenter?

   ____

   f. What does this suggest about the role of distance from the epicenter in attenuating earth-quake waves?______________________

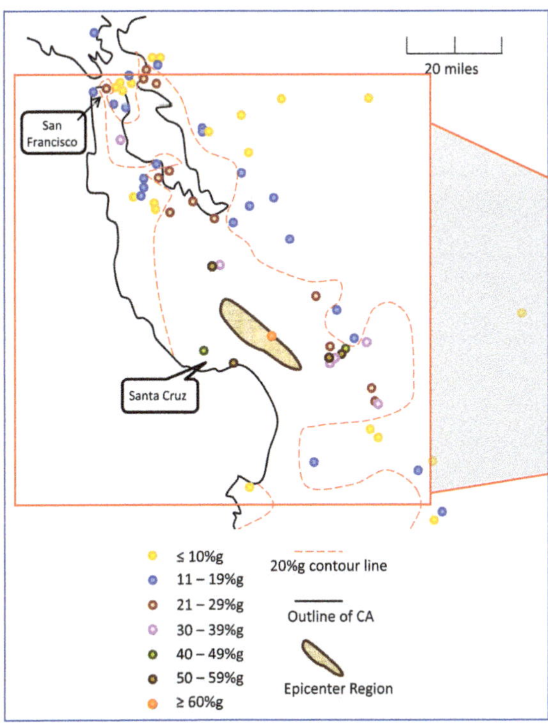

Figure 8.16 – Horizontal Acceleration from the Loma Prieta Earthquake Presented as % of Gravitational Force

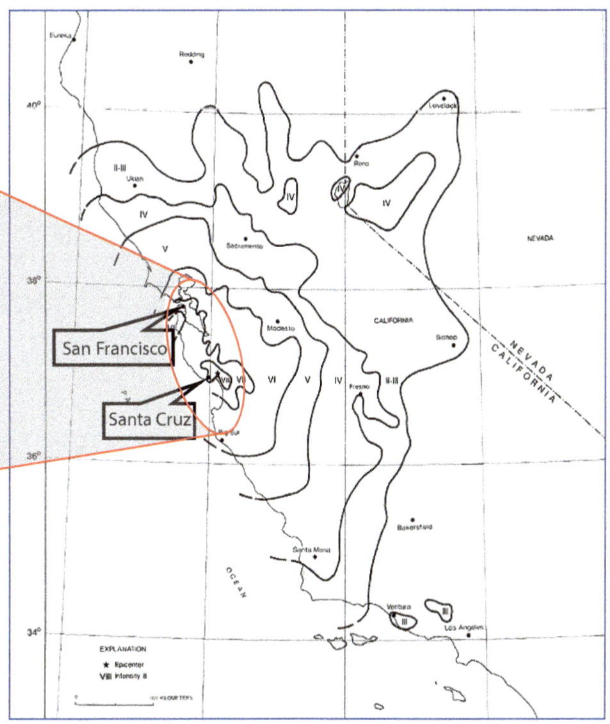

Figure 8.17 – Intensity Map of 1989, Loma Prieta Earthquake Region, USGS

4. Figure 8.17 is an isoseismal map showing areas that sustained similar structural damage as measured by the Modified Mercalli Intensity Scale. Compare this figure to Figure 8.16; describe the relationship between acceleration and structural damage.

____

Part D: Earthquakes and Plate Tectonics

As discussed previously, investigating the characteristics of a fault system with respect to the quantity, frequency, and magnitude of earthquakes along the fault, the length and width of the fault and the direction of movement, is pivotal to earthquake forecasting, land-use assessment, and building code regulations. While understanding how these factors affect the local environment is a necessity for the safety of the millions of people who live in these earthquake-prone regions, examining fault behavior also gives us a window into the larger-scale tectonic process, which not only permits scientists to gain further knowledge about earthquake production, but also helps to locate areas that are likely to develop earthquake-related phenomena in the future. The following exercises are designed to show you the correlation between earthquakes and tectonic activity, and demonstrate how that information can be used to forecast the short- and long-term effects of earthquakes in a given region.

Strike-Slip Faults

While primarily confined to spreading centers and harmless to civilization, strike-slip faults do on occasion span across continents and cause tremendous amounts of damage. Examples of such places include, but are not limited to, the Dead Sea Transform Fault in the Middle East, New Zealand's Alpine Fault, Pakistan's Chaman Fault, the San Andreas Fault in the United States, and Turkey's Anatolian Faults. Most of these faults are being carefully studied as they are in highly populated areas. Needless to say, the more scientists understand about the movement along these tectonically active regions, the better equipped they are to devise methods of keeping people safe.

Continental strike-slip faults result when a landmass spans across two tectonic plates that are grinding past each other in opposite directions. As such, the continent is essentially being sheared apart as the plates move and each claims its own piece of the continent. Because the motion involved is exclusively horizontal, the movement is classified as either "left lateral" (see Figure 8.18a) or "right lateral" (see Figure 8.18b).

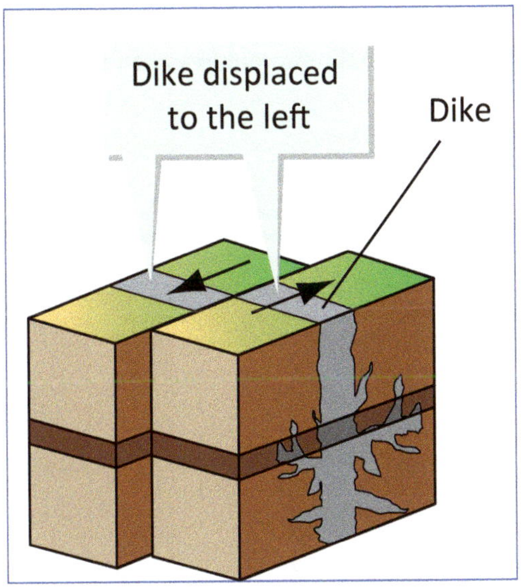

Figure 8.18a – Left Lateral Fault Movement

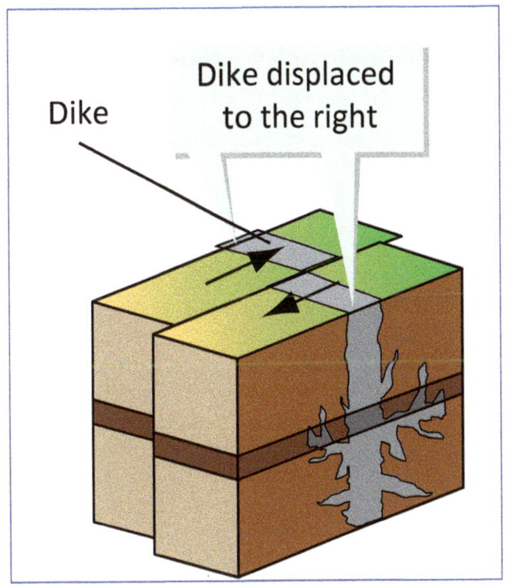

Figure 8.18b – Right Lateral Fault Movement

The laws of physics dictate that a moving mass experiencing friction will compress in the direction of movement, leaving the area behind the compressed part stretched or dilated. As such, the direction of plate movement is always away from dilation and toward compression.

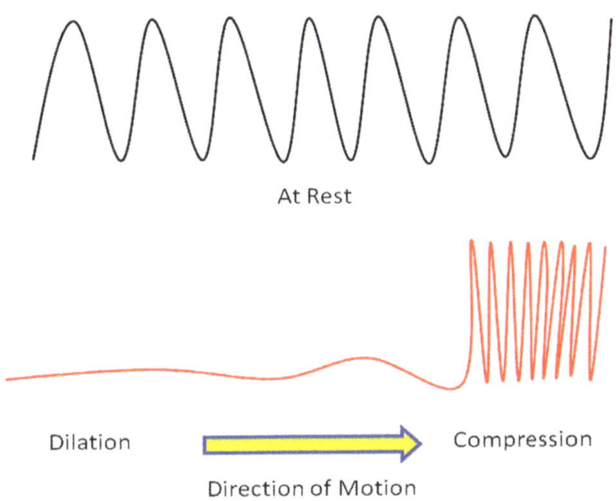

Figure 8.19 – Dilation and Compression—Key to Identifying the direction of Plate Motion

Scientists are able to evaluate which part of the plate is compressed and which part is dilated, by examining the seismograms from multiple locations along the fault after an earthquake. Essentially, when the P-wave enters the seismograph, if the area is one of compression, the first mark on the seismogram will be an up motion. If, however, when the P-wave enters the seismograph, the area is one of dilation, the first mark on the seismogram will be a down motion. When the areas of compression and dilation are plotted on each side of the fault, it becomes clear which direction each plate is moving—as each side of the fault will move away from dilation and toward compression— therefore allowing the region to be classified as having either right or left lateral motion. See Figure 8.20 for a visual representation of this concept.

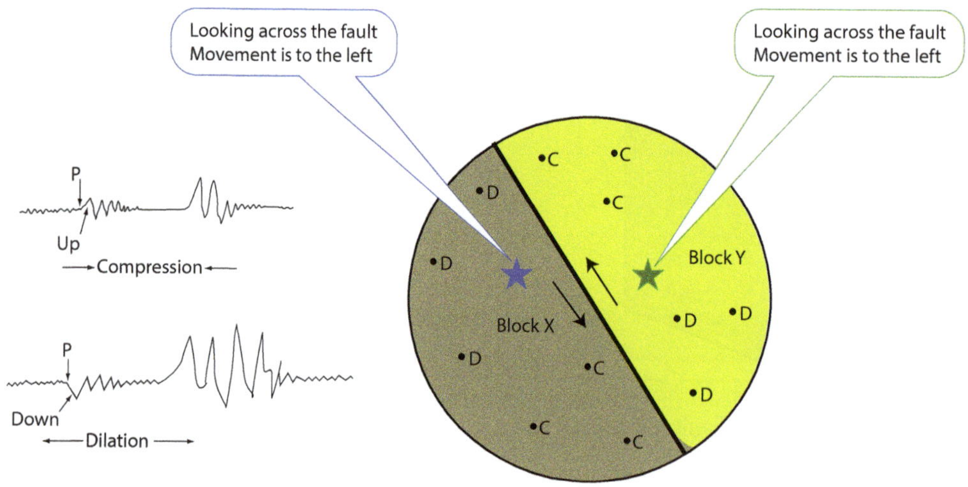

Figure 8.20 – Using Dilation and Compression to Determine Lateral Motion

Exercise 1: Determining Lateral Motion on Strike-Slip Faults

The North Anatolian Fault (NAF) (Turkish: Kuzey Anadolu Fay Hattı) is a major active strike-slip fault in northern Anatolia, running along the transform boundary between the Eurasian Plate and the Anatolian Plate. The fault extends westward from a junction with the East Anatolian Fault at the Karliova Triple Junction in eastern Turkey, across northern Turkey, and into the Aegean Sea. It runs about 20 km south of Istanbul.

Since the disastrous 1939 Erzincan earthquake, there have been seven earthquakes measuring over 7.0 on the Richter scale, each happening at a point progressively further west. Seismologists studying this pattern believe that earthquakes happen in "storms" over a number of decades, and that one earthquake triggers the next. By analyzing the stresses caused along the fault by each earthquake, they were able to forecast the shock that hit the town of İzmit with devastating effects in August 1999. It is thought that the chain is not complete, and that an earthquake will soon strike further west along the fault—perhaps in the heavily populated city of Istanbul. See Figure 8.21 and associated Table 8.C.

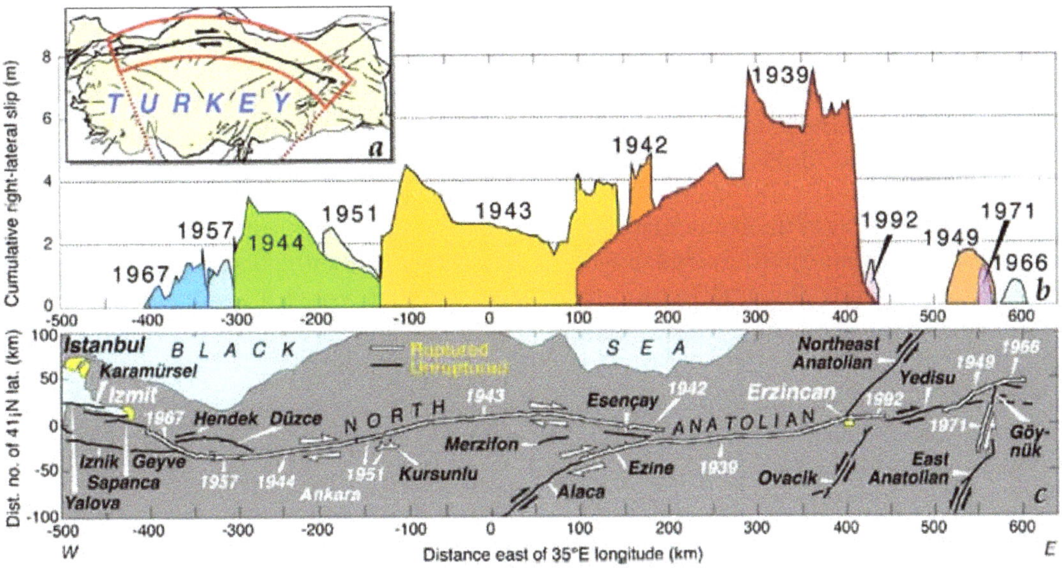

Figure 8.21 – The North Anatolian Fault and slip magnitudes of earthquakes in the 20th century

Event	Moment magnitude	Casualties
1939 Erzincan	7.9	32,962 dead
1942 Niksar-Erbaa	6.9	
1943 Tosya-Ladik	7.7	
1944 Bolu-Gerede	7.5	
1949 Karlıova	7.1	
1951 Kurşunlu	6.9	50 dead, 3,354 injured
1957 Abant	6.8	
1966 Varto	6.9	2,394 dead, 1,489 injured
1967 Mudurnu Valley	7.1	86 dead, 332 injured
1992 Erzincan	6.5	
1999 İsmit	7.4	17,480 dead and 43,959 injured

Table 8.C – Moment Magnitudes and Casualties of Select Earthquakes along the North Anatolian Fault in the 20th Century

1. Figure 8.22 represents 25 seismograms, each taken from a different seismograph surrounding the North and East Anatolian Faults in Turkey. Using those seismograms, determine whether the North and East Anatolian Faults are right or left lateral by following the procedure listed below.

   a. Examine each of the seismograms in Figure 8.22. Using the P-wave motion upon arrival into the seismograph, determine whether the seismogram came from a location of compression or dilation. Put a "C" next to the seismograms that represent compression and a "D" next to the ones that represent dilation.

   b. The location of each seismograph that produced the seismograms from Figure 8.22 is represented on Figure 8.23. Transfer the C's and D's from Figure 8.22 to their corresponding locations on Figure 8.23.

   c. Draw arrows on Figure 8.23 indicating which direction each side of the North AND East Anatolian Faults is moving (i.e., draw in two sets of arrows, one set for each fault), then answer the following questions.

Follow-up Questions:

2. Is the North Anatolia Fault right or left lateral? ____________________

3. Is the East Anatolia Fault right or left lateral? ____________________

4. Using a visible writing tool, outline on Figure 8.23 the part of the continent that is in the process of being sheared off from the mainland.

5. Using a visible writing tool, draw a large arrow on Figure 8.23 on the part of the continent that will be sheared off, indicating the direction it will be taken.

6. How many major plates are present and interacting in this area?____________________

7. What designates the plate boundaries—i.e., what was used to identify the boundaries of the plates? ____

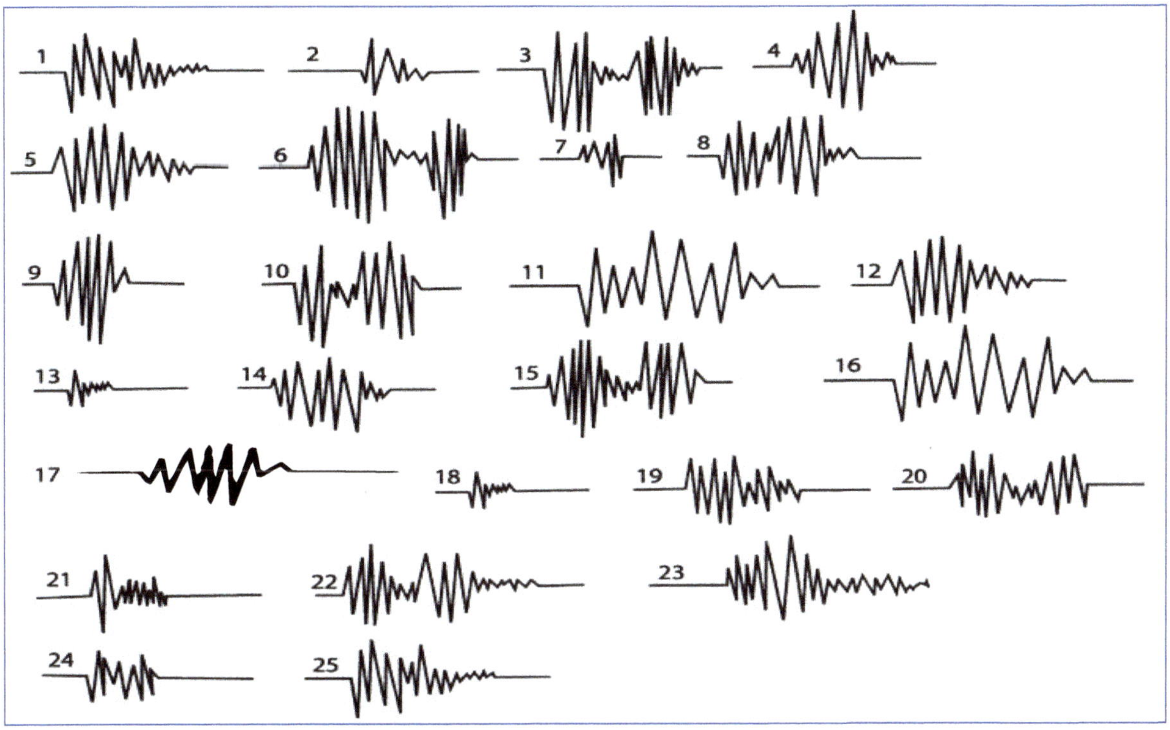

Figure 8.22 – North and East Anatolian Fault Seismograms

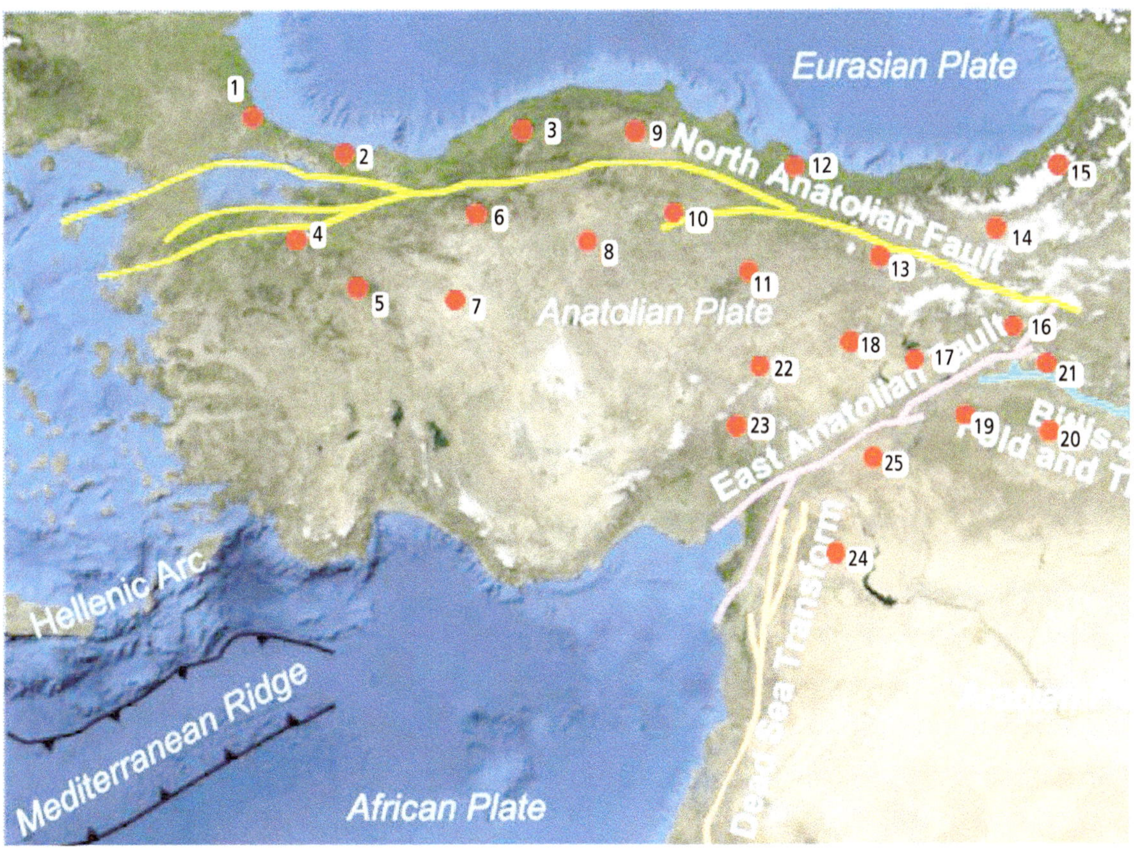

Figure 8.23 – Seismograph Locations along the Two Faults

Exercise 2: Motion of Reverse Faults

Unlike strike-slip faults, reverse faults involve vertical motion, for they develop as a result of a collision between two tectonic plates, as opposed to plates grinding horizontally past each other in opposite directions.

1. Using Figure 8.24, create an earthquake depth cross-section of the A–A' line by plotting the earthquake depths on the graph provided.

2. Figures 8.25 and 8.26 depict the historical seismicity and depth of earthquakes along two other convergent margins. Does the depth cross-section you created show the same pattern as that seen in Figures 8.25 and 8.26? Yes or No?

3. Notice that in all three figures, the earthquake hypocenters become progressively deeper in one direction away from the trench. Why are the earthquake depths changing in this fashion?

   ____

   ____

4. Examine Figure 8.27. This is a map showing M>2 aftershocks following the September 28, 2004, M6.0 earthquake in Parkfield, California. Below the map is a cross-section showing earthquake depth along the fault. In what two ways does the earthquake depth profile differ in this fault when compared to the depth profiles seen in Figures 8.24–8.26?

   - ____
   - ____

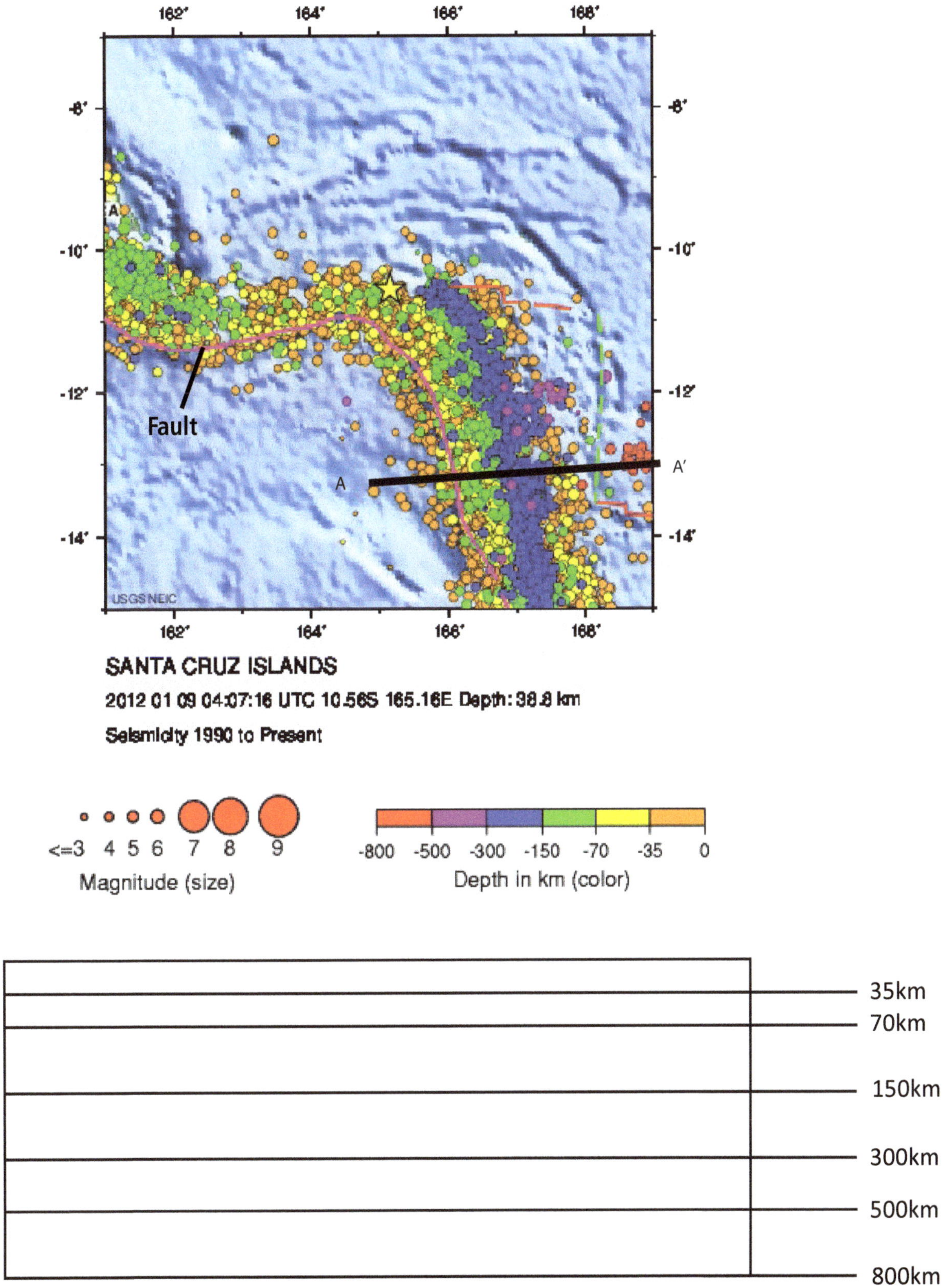

Figure 8.24 – Historic Seismicity, Santa Cruz Islands

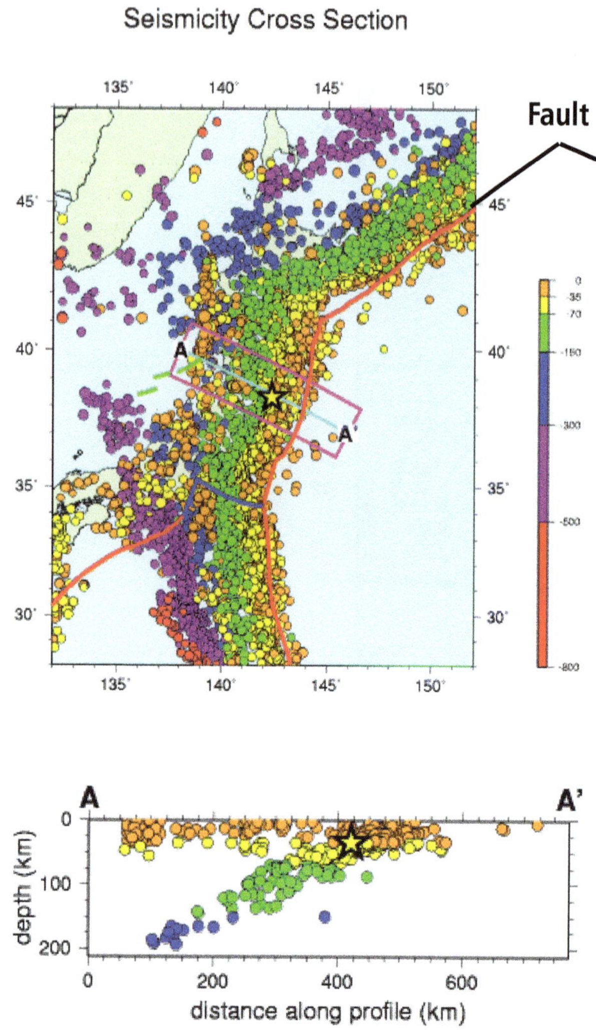

Figure 8.25 – Seismicity Cross-Section, Near East Coast of Honshu, Japan

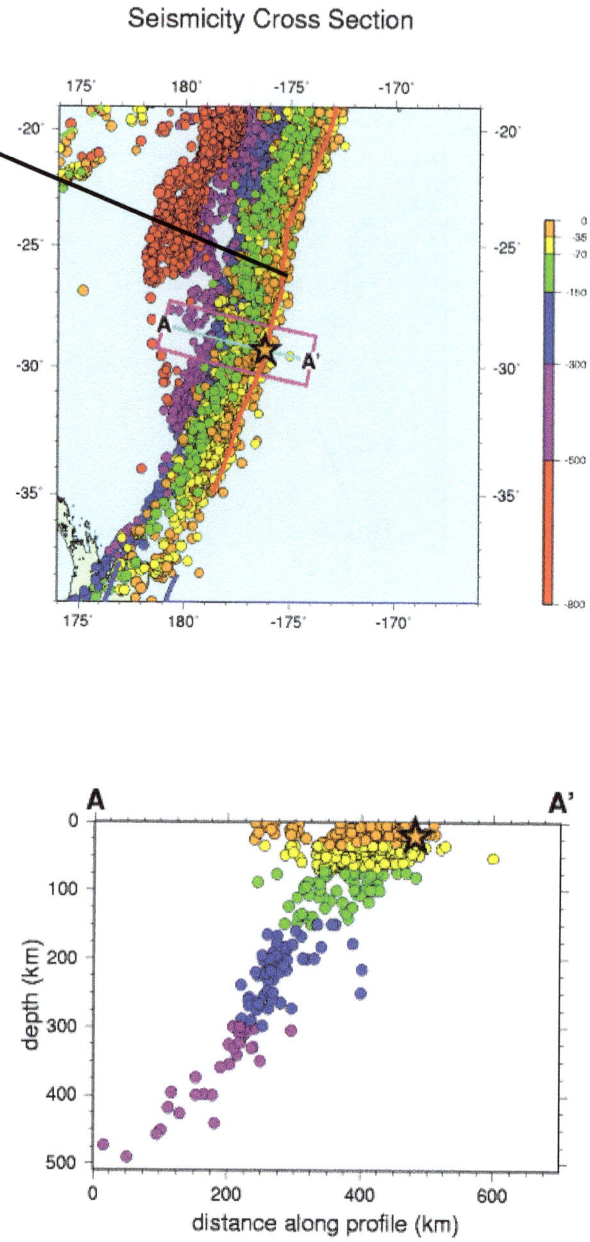

Figure 8.26 – Seismicity Cross-Section, Kermadec Islands Region

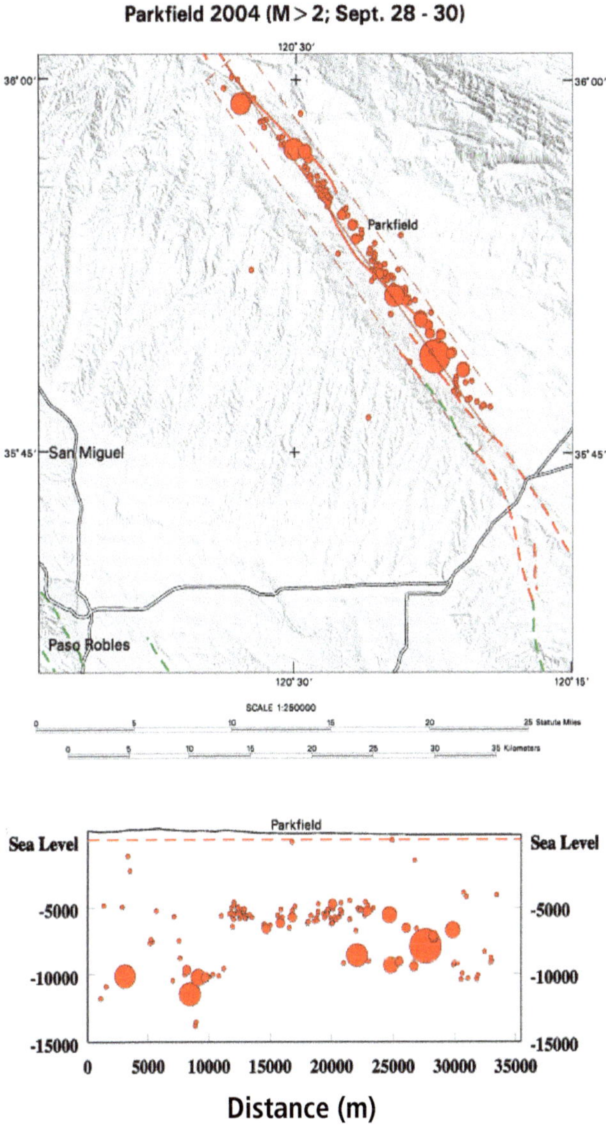

Figure 8.27 – Parkfield, California 2004 earthquake aftershocks.

5. Why are the convergent margins characterized by both deep and shallow earthquakes, whereas the transform fault margins have only shallow earthquakes?

____

____

____

6. Which are more dangerous, shallow or deep earthquakes? __________ Explain.

____

____

____

Based on the depth profiles in this exercise, do convergent and strike-slip boundaries put the local environment at high risk for earthquake damage? Yes or No? Explain.

____

____

____

____

____

Laboratory Exercise IX

Topography and Topographic Maps

Introduction

The Earth's surface is characterized by hills and valleys, plains and plateaus, riverbeds, mountain ranges, and many other geological features. Topography, aka "relief," describes differences in elevation on Earth's surface. Examining where the high and low ground are located throughout a region has many practical applications ranging from military strategic planning to locating acceptable places for water wells and landfills. As such, maps depicting Earth's relief is critical for multiple reasons. Topographic maps are one method by which the three-dimensional topography of Earth's exterior can be represented on a two-dimensional (i.e., flat) surface.

Objective

The objective of this lab is twofold: 1) introduce you to the concept of topography; 2) provide instruction on how to read, use, and create topographic maps.

The Beginning - Topography and Topographic Maps

1. **Latitude**: Global position relative to the equator. See figure 9.1.

2. **Longitude**: Global position relative to the prime meridian. See figure 9.1.

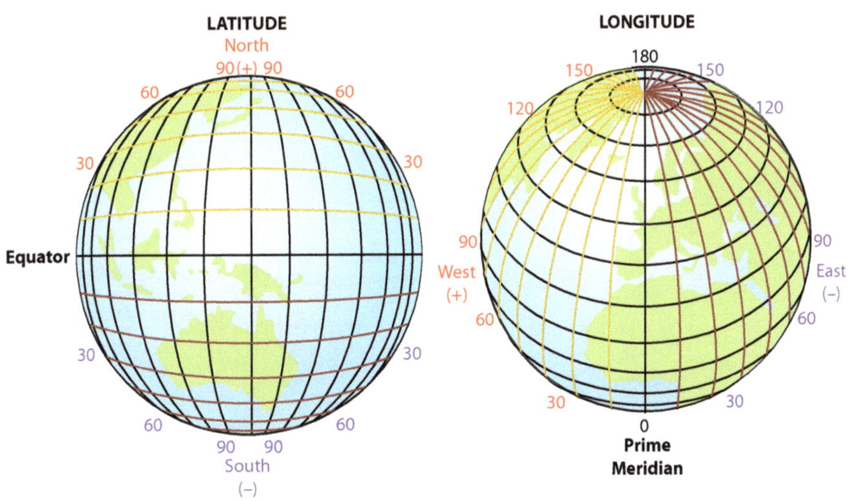

Figure 9.1 – Latitude and Logitude

3. **Datum:** A reference plane, or point of known location or value. For topographic maps, sea level is assigned an elevation of zero. All elevations are reported relative to mean sea level (msl) or zero.

4. **Map Scales:** A way of relating distances measured on the map to distances in the real world.

   a. Fractional scale

      - Distance of 1 unit on the map represents a distance of "X" units on the Earth's surface.
      - For example, 1/24,000 or 1:24,000 indicates that one unit on the map is equivalent to 24,000 of the same unit in the real world.
      - Can be any unit (inches, feet, meters, km, etc.), so long as the unit is the same for both numbers.

   b. Graphic/Bar scale (Figure 9.2)

      - Line symbol subdivided into several parts representing distances on the Earth's surface.
      - Caution: Make sure you know where the "zero" is located—it may not be at the far left end of the scale.

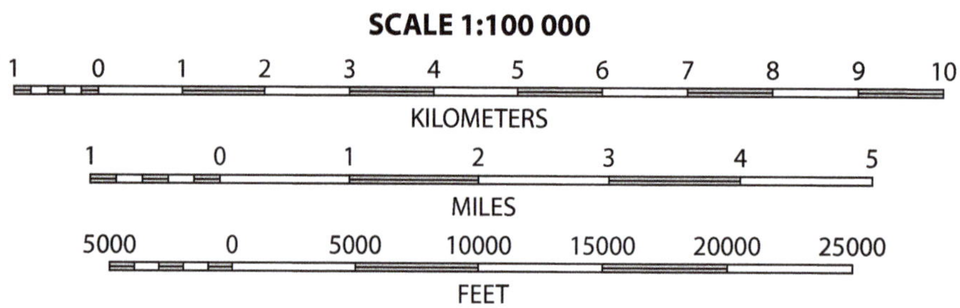

Figure 9.2 – Graphic/Bar Scale

5. **Contour lines** (Figure 9.3a and 9.3b): Lines drawn on a topography map that connect points of equal elevation.

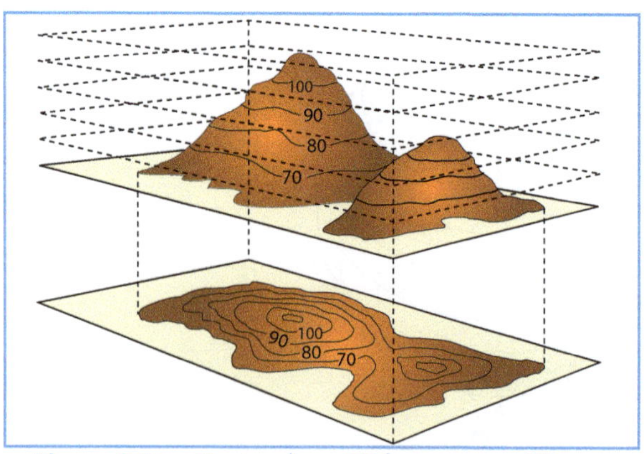

Figure 9.3a – Contour lines, 3D features on a 2D map.

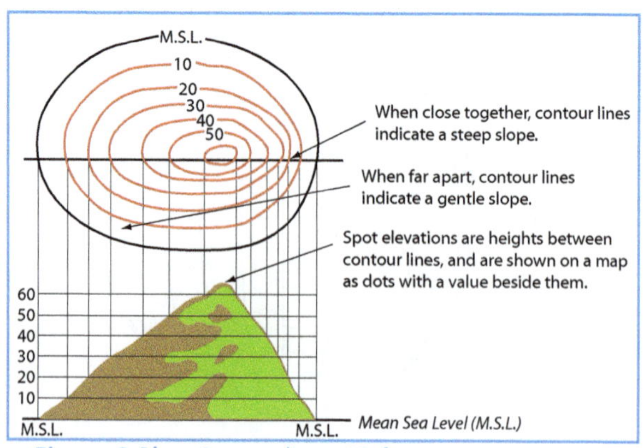

Figure 9.3b – Contour lines, 3D features on a 2D map.

6. **Contour interval** (Figure 9.4): Change in elevation between any two adjacent contour lines on the map. The contour interval is typically provided at the bottom of the map (see Figure 9.10). This interval is a fixed number for each map, but may differ from map to map. Map elevations are provided on *index* contours—typically every 5th contour line. For example, in Figure 9.4, index contour lines mark elevations of 0, 500, 1000, 1500, 2000, 2500, and 3000 meters. Elevations of all unmarked contours in between index contours can be deduced using the contour interval.

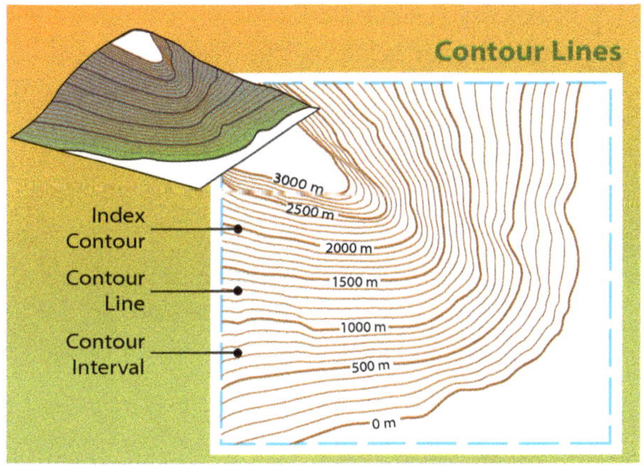

Figure 9.4 – Contour Interval: Reading between the Lines.

- Every point on a contour line is an area of equal elevation. As a result of this, every contour line must eventually close on itself to form an irregular circle. Contour lines on the edge of a map do not appear to close on themselves because they run off the map. This being the case, if the map you are working with and the map representing the adjacent region were sitting side by side, you could follow the contour throughout its complete, enclosed path (refer to Figure 9.5).

- Contour lines can never cross one another. Each line represents a separate elevation, and you can't have two different elevations at the same point. The only time two contour lines may merge is if there is a vertical cliff (refer to Figure 9.5).

- Moving from one contour line to another always indicates a change in elevation. To determine if it's an uphill or downhill change, you must look at the index contours on either side (refer to Figure 9.5).

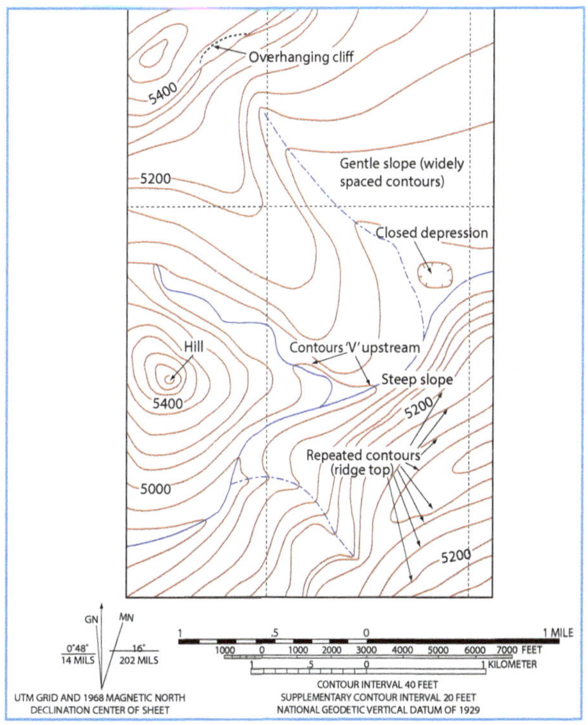

Figure 9.5 – Rules for Contour Lines.

- Contour lines crossing a stream valley will form a "V" shape pointing in the uphill (and upstream) direction (refer to Figure 9.5)

- On a hill with a consistent slope, there are always four contours between two adjacent index contour lines. If there are more than four intermediate contours, it means that there has been a change of slope and one or more contour lines have been duplicated. This is most common when going over the top of a hill or across a valley (refer to Figure 9.5 and 9.7).

- The closer contour lines are to one another, the steeper the slope is in the real world. By the same logic, the farther away the contours are from one another, the milder the slope (see Figure 9.6). Evenly spaced contours indicate a constant slope, while unequally distributed ones demonstrate a change of slope.

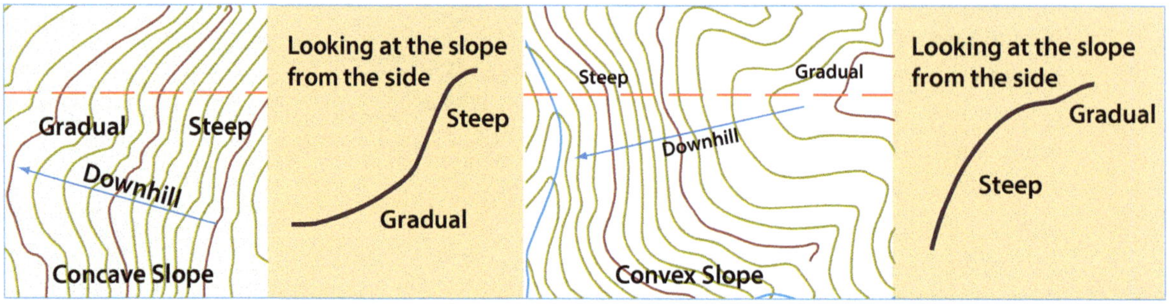

Figure 9.6 – Contour Lines and Slopes

- A series of closed contours (the contours make a circle) represents a hill. If the closed contours are hachured, it indicates a closed depression (refer to Figure 9.5). Note that the contour line repeats around hills, valleys, and when entering a depression (see Figure 9.7).

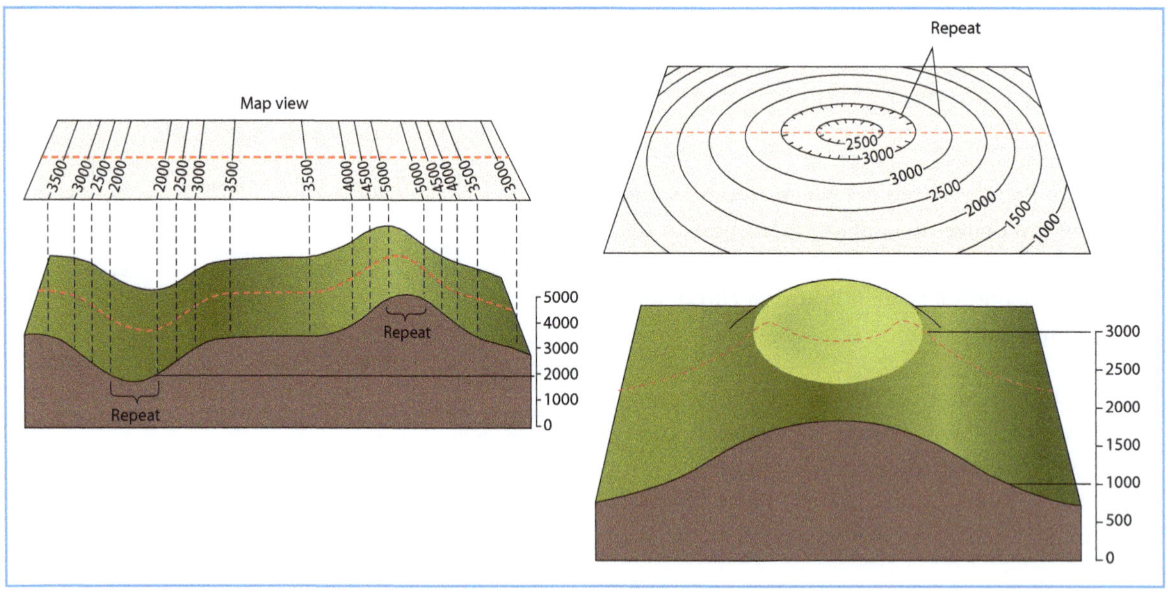

Figure 9.7 – Contour Lines representing Hills, Valleys, and Depressions

Figure 9.8 – Contour Line Overview

7. **Relief**: The difference in elevation between two points on a map.

8. **Gradient** = $\dfrac{\text{Change in elevation or relief (measured using contour lines)}}{\text{Change in horizontal distance (measured using map scale)}} = \dfrac{\text{rise}}{\text{run}}$

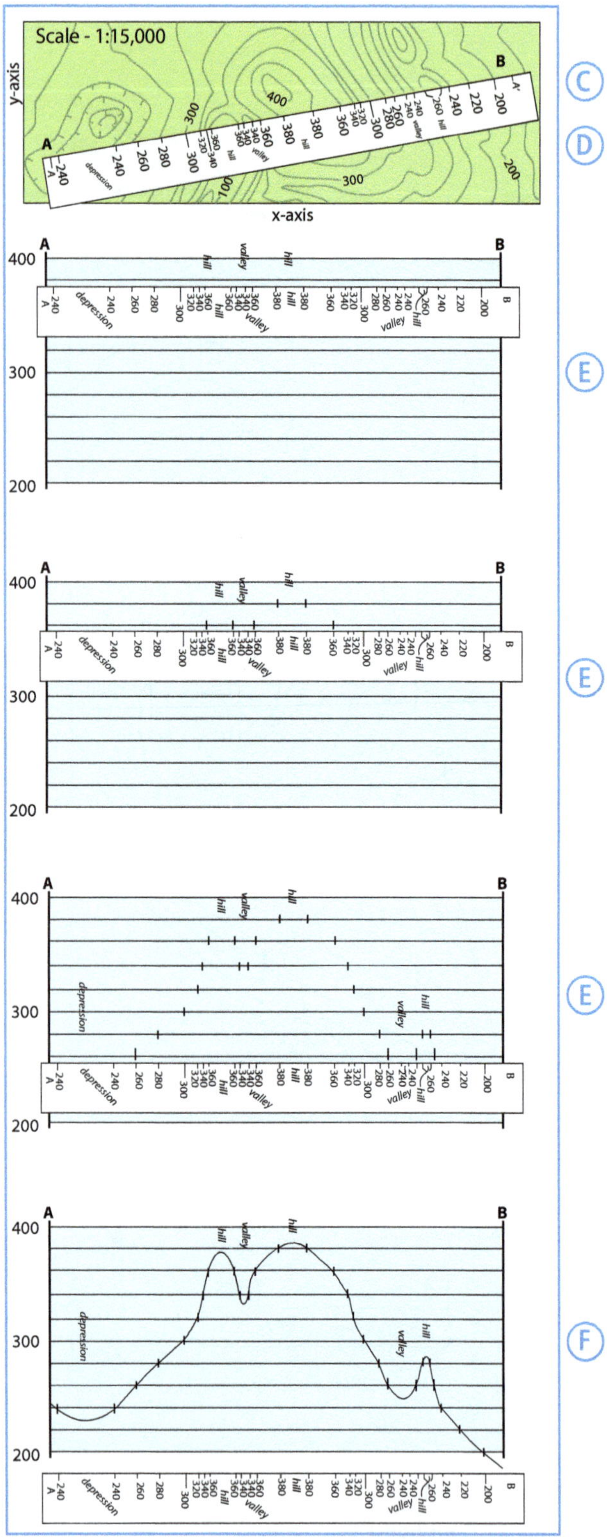

Figure 9.9 – How to draw a topographic profile

9. **Vertical Profile or Vertical Section**: A representation of a vertical slice through Earth's surface. This is done across a traverse line between two designated points on the map (ex: A–B). See Figure 9.9 for a schematic representation of constructing a vertical profile.

Creating a Vertical Profile

(A) Locate the traverse line, which will serve as the x-axis.

(B) Using the elevations on the topographic map, decide on an appropriate scale for the Y-axis of the profile graph, and label it accordingly. This can be done by putting the lowest and highest elevations on the graph, then marking even increments in between.

(C) Align a straight piece of paper along the traverse line.

(D) For every place the traverse line crosses a contour line, place a tick mark at the top of the paper and label the tick mark with the elevation the contour line represents.

(E) Once all the intersections have been marked, align the paper with the x-axis of the profile graph. Next, drop down vertically from each tick mark and put a point on the graph at the appropriate elevation represented by each tick mark.

(F) Connect the dots.

(G) Calculate the vertical exaggeration. See #10 for an explanation of this procedure.

10. **Vertical Exaggeration:** A representation of the amount of exaggeration along the vertical (Y) axis on a profile compared to the horizontal (X) axis. To determine the VE, simply divide the vertical scale by the horizontal scale.

$$\frac{\text{Vertical scale}}{\text{Horizontal scale}} = \text{Vertical exaggeration}$$

For example: If the vertical scale is 1:3500 and the horizontal scale is 1:35,000, the vertical exaggeration is:

$$\frac{1/3500}{1/35000} = \frac{35000}{3500} = 10\text{x}$$

This means that the profile plot is distorted … as it is stretched 10x more in the vertical direction than in the horizontal.

HINT: Watch YOUR UNITS!!!

★ See Figure 9.10 on the following page for a standardized layout of a topographic map.

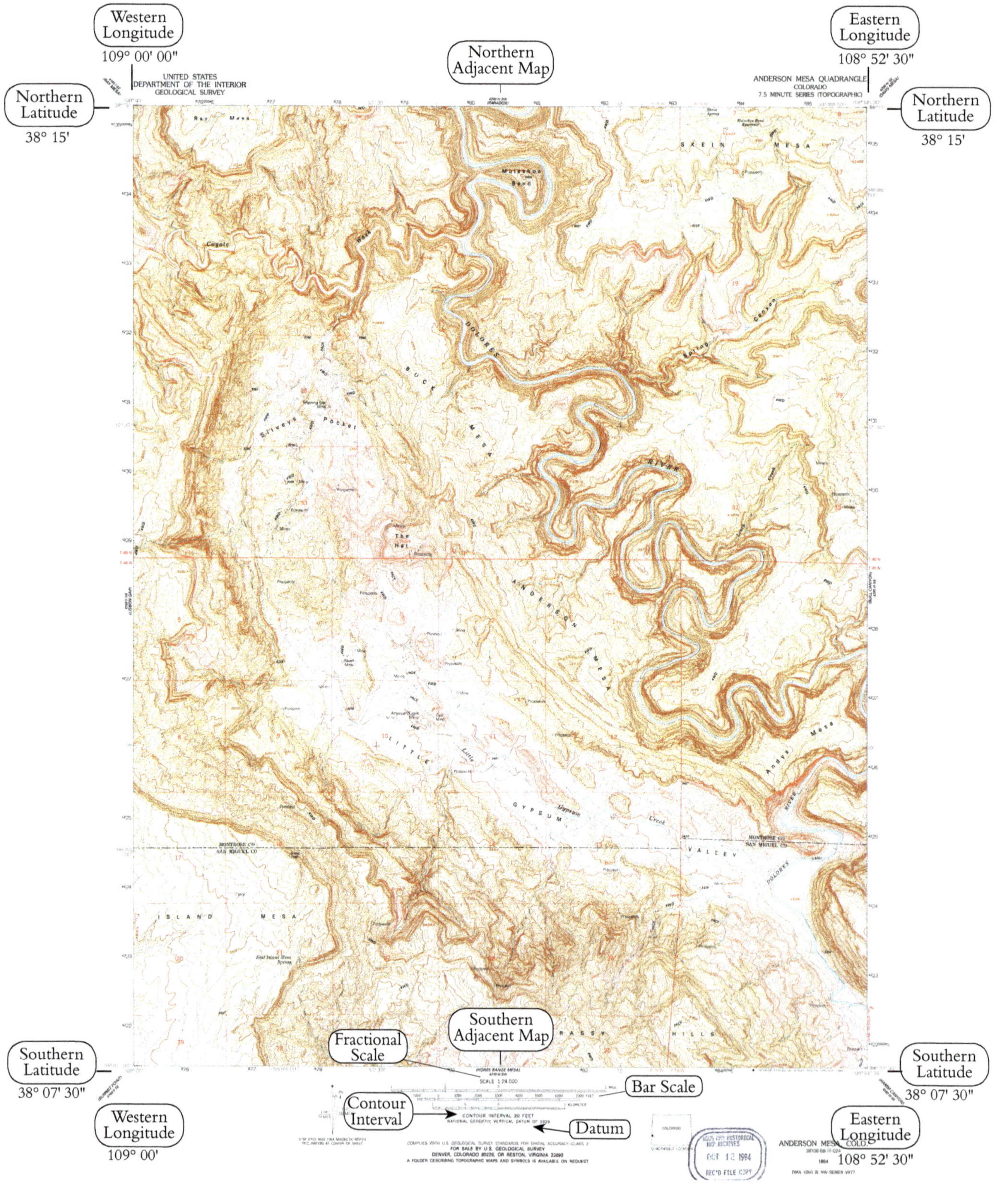

Figure 9.10 – Topographic map

Adjacent Maps

1	2	3
4		5
6	7	8

1. Ray Mesa
2. Paradox
3. Davis Mesa
4. Lisbon Gap
5. Bull Canyon
6. Summit Point
7. Horse Range Mesa
8. Hamm Canyon

PART A: LEARNING TO READ TOPOGRAPHIC MAPS

Exercise: Understanding Contour Lines

1.

a. In Figure 9.11, fill in the boxes on the contour lines with the correct elevation in feet above sea level. (*Note:* Contour interval is 20 ft.)

b. Using a red pencil, color in the area on the figure that represents the lowest elevation.

c. Label the closed depression by putting a * symbol in the middle.

Figure 9.11 – Learning to use contour lines

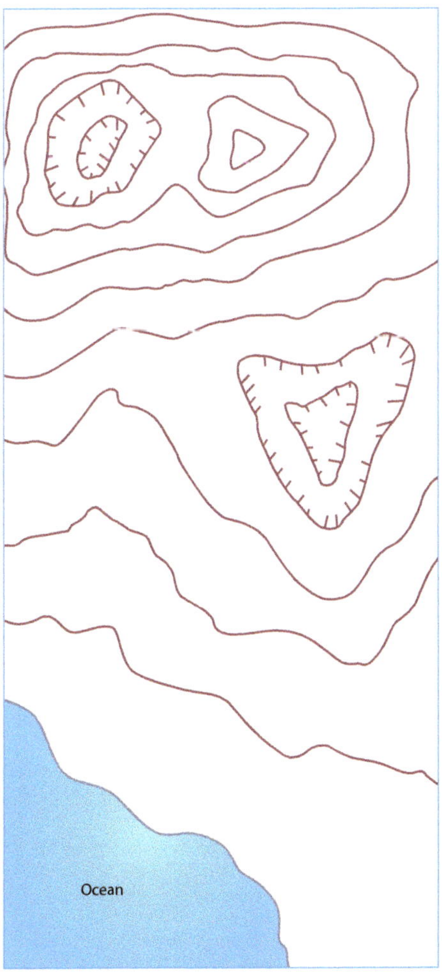

Figure 9.12 – Learning to use contour lines

2. Using a contour interval of 10m, label every line on Figure 9.12 with its exact elevation above sea level. Note: Before attempting this exercise, re-examine Figure 9.7 for a memory refresher on labeling closed depressions.

3. Use Figure 9.13 to answer the following questions:

   a. The contour lines on this map are labeled in feet. What is the index contour interval on this map? ________________ ft

   b. What is the contour interval on this map? ________________ ft

   c. What is the regional relief, i.e., total relief of the land represented on this map? ________________ ft

   d. What is the gradient from A to B? ________________ ft/mile *Show your work.*

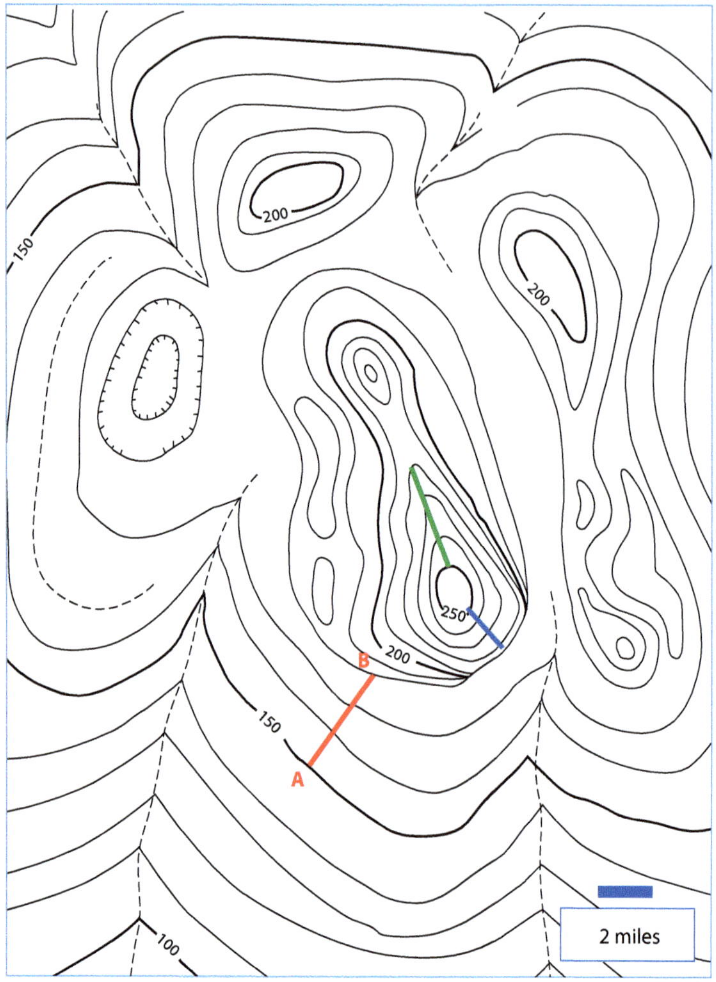

Figure 9.13 – Working with Topographic Maps

   e. Two potential routes to the top of a hill are marked on Figure 9.13. One is in blue and one is in green. Which of the two routes is more strenuous? The blue or the green? Why?

   ____

   ____

   ____

   f. How could you find the areas on this map that have a gradient of 20 ft/mile or greater? (*Hint:* Think about the contour interval and how many contour lines of map elevation must occur along one mile of map distance.) ____

4. Match each of the following contour sketches with the land features they represent by putting the letter of the appropriate land feature on the line provided.

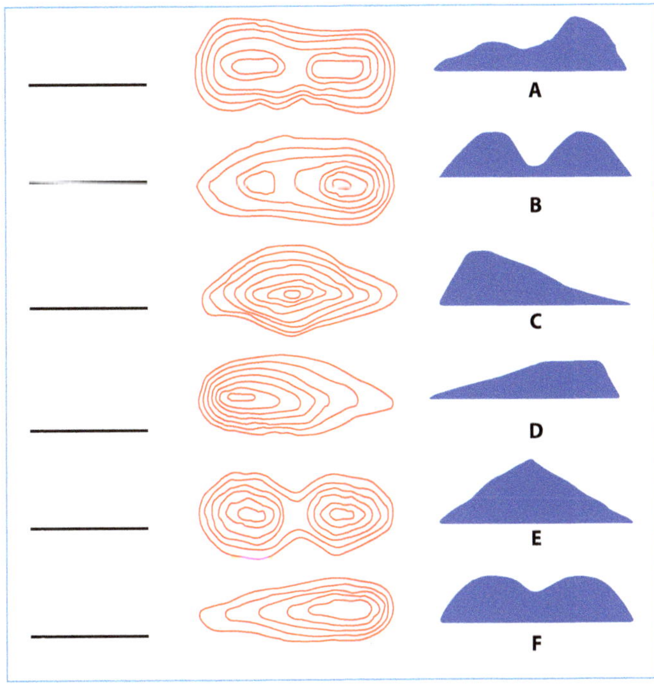

Figure 9.14 – Land features and contour lines

5. Draw a vertical profile of the A–B traverse line on Figure 9.15 *and* calculate the vertical exaggeration. Note: All elevations are in feet.

Show your work for the Vertical Exaggeration calculation:

VE = ____________________

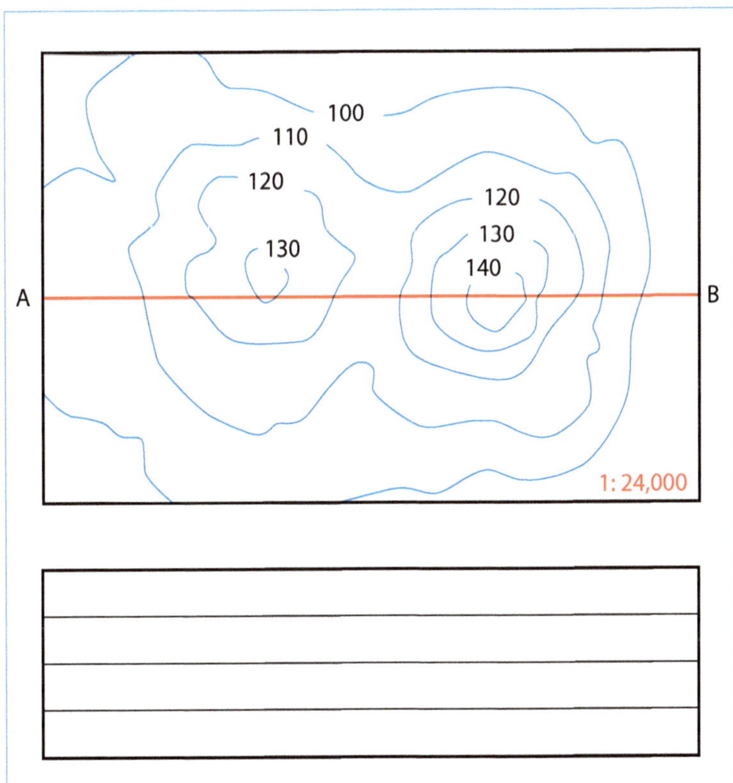

Figure 9.15 – Drawing Topographic Profiles

Part B: Working with Topography Maps

Exercise 1: Washington, D.C. Quadrangle

Use the Washington, D.C. quadrangle to answer the following questions.

6. If you wanted to see the topography of the area to the northwest of this map, which map would you need to view? ______________________

7. What is the contour interval? ____________ ft.

8. If the contour interval were twice what it actually is, would there be a **greater** or **lesser** number of contour lines? (Circle one)

9. Index contour lines appear every ____________ feet on this map.

10. What is the general flow direction of the Potomac River and Rock Creek? (Circle the correct answer.) S/SE, S/SW, N/NE, N/NW

11. Express the scale of this map in four ways:
    a. The fractional scale of this map is ____________.
    b. One inch on the map = exactly ____________ inches in real life.
    c. One inch on the map = exactly ____________ miles in real life.
    d. One inch on the map = exactly ____________ km in real life.

12. Name the parallels of latitude that serve as the northern and southern boundaries of the map
    North: ____________ South: ____________

13. Name the meridians of longitude that serve as the western and eastern boundaries of the map
    West: ____________ East: ____________

14. How many degrees of both latitude and longitude are covered on this map?
    Lat: ____________ Long: ____________

    Show your work.

Exercise 2: Front Royal, Virginia Quadrangle

Use the topographic map of Front Royal, Virginia, to answer the following questions. *Show your work.*

15. You're riding your bike in Shenandoah National Park on Skyline Drive. What gradient (in feet per mile) would you experience between the North Entrance, (near Front Royal), to the first stream crossing (Use your map to calculate the gradient of the area highlighted in figure 9.16)? *Show your work.* ______________________ ft/mile

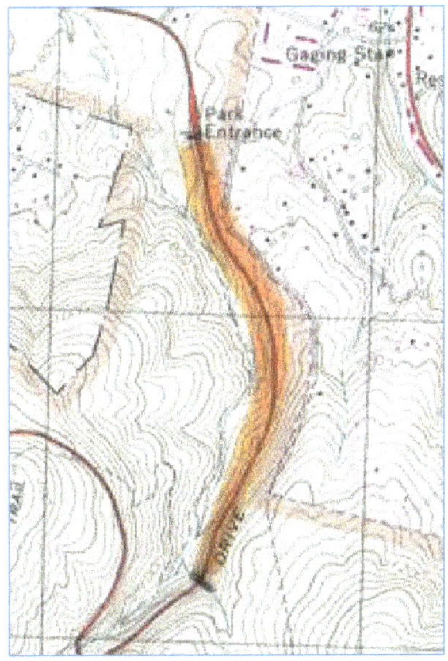

Figure 9.16 – Gradient.

16. Notice there is a series of streams surrounding the words "Shenandoah National Park" at the bottom, center of the map. See Figure 9.17a for a visual depiction of the area in question. In which direction are these streams flowing?

____

How do you know?

____

____

17.

Pluto and Goofy decide they want to go out and get some exercise and jointly agree that a moderate bike ride would be just the thing they needed. After looking at Front Royal topo map they concluded that a ride along the A–A′ line looked very promising (see Figure 9.17a).

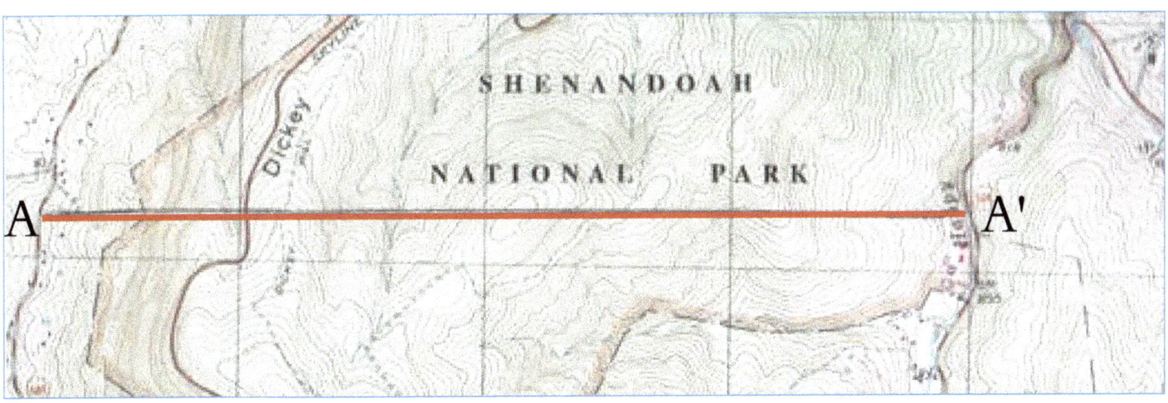

Figure 9.17a – Profile Line for Prospective Bike Path.

Just to make sure they weren't getting in over their heads, they quickly plotted out a profile of the transect so they could get an idea of just how steep the terrain is. Below (Figure 9.17b) is the profile they drew up using a vertical scale of 1 inch = 200 ft.

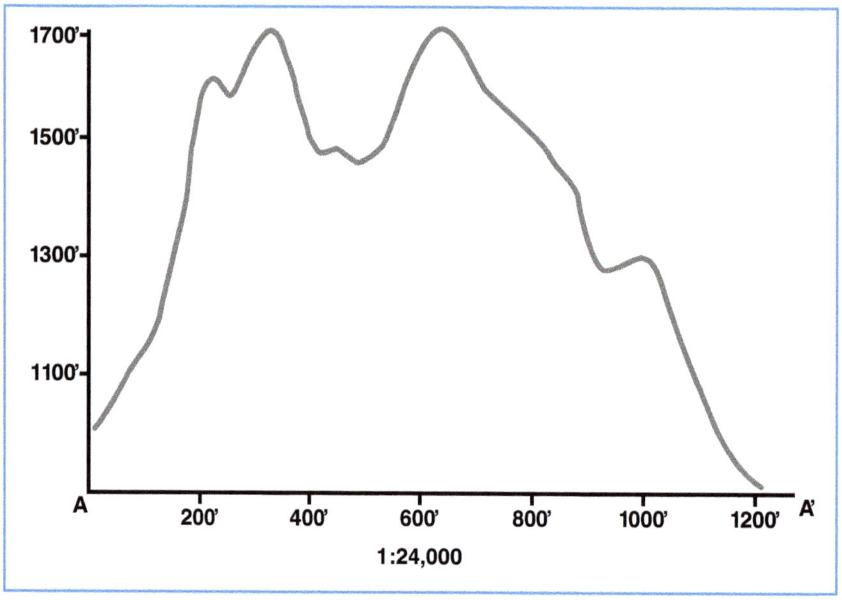

Figure 9.17b – Profile with vertical exaggeration.

a. Based on their profile, does this look like a ride two amateur bike riders should be taking? Yes or No? Explain.

____

b. Now calculate the vertical exaggeration of this profile plot. *Show work below.*

VE = __________

Now compare their profile to a profile drawn to scale (1 inch = 2,000 ft). See Figure 9.18.

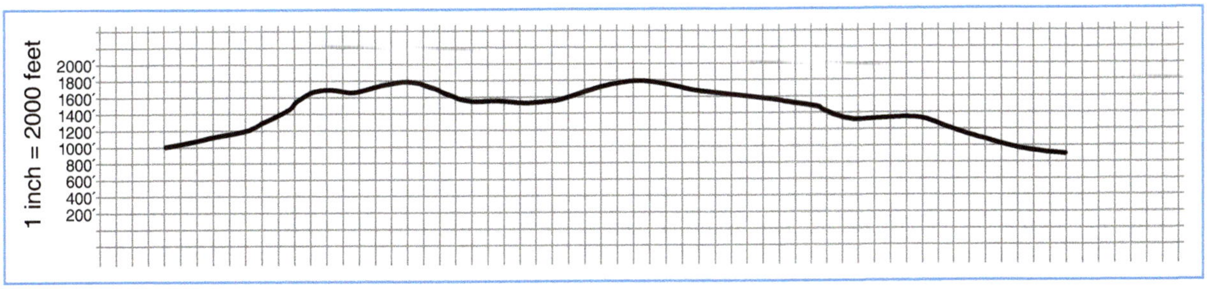

Figure 9.18 – Profile drawn to scale

1 inch = 2,000 ft

2,000 ft = 24,000 inches

Therefore, Vertical Scale is 1 inch = 24,000 inches

1/24,000 (Vertical Scale)
1/24,000 (Horizontal Scale)

VE = 1:1 ratio = no vertical exaggeration

c. What is your opinion now? ______________________________

____

18. If the South Fork of the Shenandoah River near Front Royal flooded 50 feet above normal due west of the Randolph Macon Military Academy, name four sites that would be flooded.

______________, ______________, ______________, ______________.

19. Assume that the elevation of the water surface in the river is ten feet below the adjacent land contour. Which specific sites pose a potential environmental threat in the event a flood occurs? ______________ and ______________.

Part C: Tying it all Together

On the center of each table is a plastic tub covered with a lid into which holes have been drilled at equal intervals. Inside each tub is a replicate 3D landscape full of mountains, valleys, stream beds, and other geographical features. You are going to create a topographic map of this 3D landscape by following the procedure below:

Exercise: Constructing Topographic Maps and Profiles

1. Examine figure 9.19. Figure 9.19 is a diagram that represents the lid of your tub. Each of the holes on the diagram corresponds to a hole in an identical location on the tub's lid. Starting at one of the corner holes on the tub's lid, use the stick that has been supplied for you to measure the vertical distance (elevation) between the top of the landform and the rim of the hole. You do this by inserting the stick into the hole and lowering it until it can go no further. Read the measurement from the ruler on the stick, then record the elevation (in cm) by the appropriate hole on figure 9.19. Repeat this procedure for all the holes in the lid.

2. Using a 2cm contour interval, connect all the holes of equal elevation. Don't forget to label your index contour lines. Your TA will also be around to help with this, as it can be a little confusing for beginners. If after you've given it a fair shot yourself and are still having problems, call over your TA and he/she will be happy to assist you.

Once the lines are all connected up, give yourself a sound pat on the back - as you have just created your first topographic map.

3. Using figure 9.20, create a vertical profile of the A – A' line on your self-made topography map.

4. Calculate the vertical exaggeration in the space provided below.

Vertical Exaggeration - ________x

5. NOW … TAKE THE TOP OFF OF THE TUB AND SEE IF YOUR TOPOGRAPHIC MAP AND PROFILE ACCURATELY DEPICT THE LAND FORMATIONS!

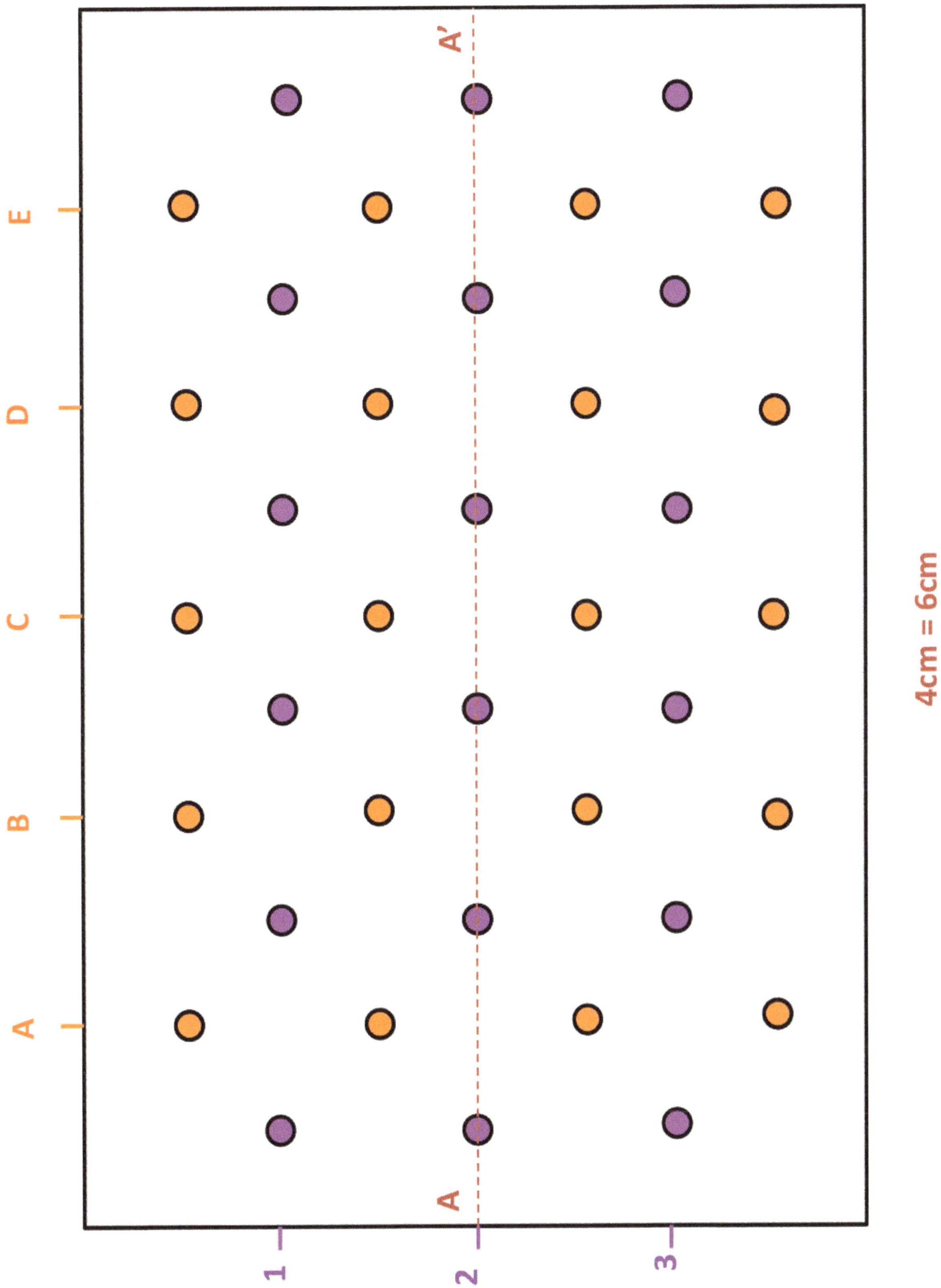

Figure 9.19

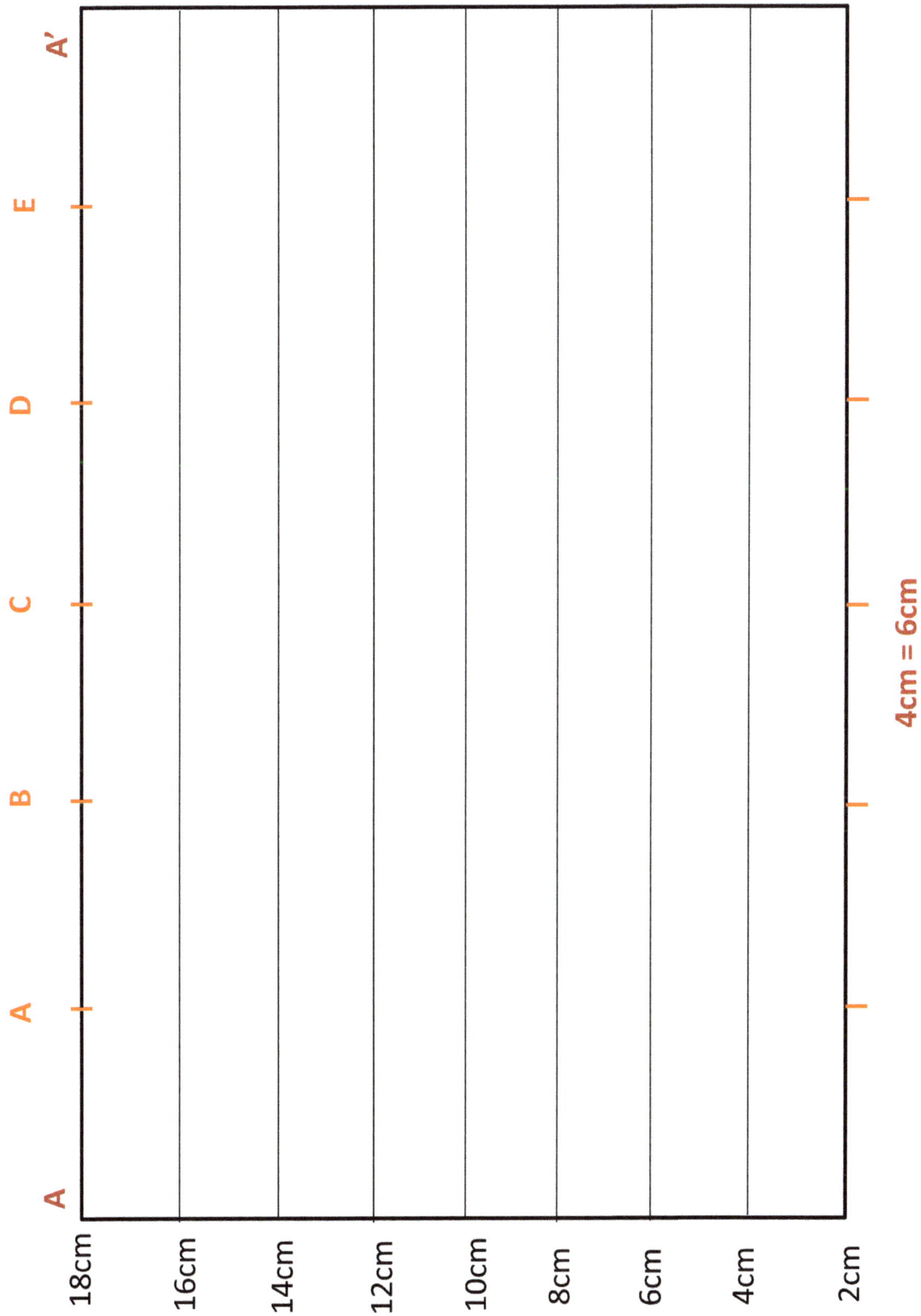

Figure 9.20

Laboratory Exercise X

Running Water

Introduction

When evaluating the countless rhythmic and episodic processes that jointly contribute to sculpting Earth's surface, it becomes clear that the work done by rivers is the most influential. To begin, rivers and streams erode, transport, and deposit billions of tons of sediment each year. If spread equally over the Earth's land surface, the volume of annual river sediment equates to a yearly elevation decrease of 7mm on each continent and would fill a cargo train that encircles the Earth twenty-four times at the equator (Smith and Punn, 2009). In addition to moving sediment, rivers also drain the land by confining surface water to distinct channels and funneling it to the coast—a task that requires downcutting into weaker substrate, winding around more resistant rock, and constantly changing size and shape in response to fluctuating water and sediment input. Due to the effects of its primary functions—water drainage and erosion, transportation, and deposition of sediment—river systems are instrumental in creating, maintaining, and altering multiple landforms throughout the surface landscape.

Objective

The objective of this lab is fourfold:

1) Understand what a drainage system is and how it's geographically delineated;

2) Examine how streams evolve due to their erosional and depositional processes;

3) Investigate why some streams behave differently than expected; and

4) Explore the human angle of stream-related processes.

Part A: Rivers—The Continent's Drainage System

River systems begin developing when the amount of precipitation on land exceeds the amount of water lost to evaporation and infiltration, resulting in excess water that pools on the surface. This "sheetwash"—a film of water less than a few mm thick—flows downslope under the pull of gravity and preferentially erodes substrate that is less protected by vegetation, less resistant to erosion, or where the water is flowing substantially faster. Because the erosion hollows out part of the ground, leaving it lower than the surrounding surface, sheetwash in neighboring areas flows into the trough, concentrating the flowing water into a confined area, and increasing the rate at which erosion takes place. This creates a positive feedback loop that continues to lengthen the trough via a process called "headward erosion," and subsequently attracts more water that further erodes the trough. In this way, a channel is born.

Multiple isolated channels eventually merge into a medium-size stream, resulting in the formation of a branch-like complex called a tributary. The tributaries from across the region merge into a major "trunk stream," which delivers the runoff to the coast. This interconnected system, consisting of the headwater channels, tributaries, and a trunk stream, is collectively referred to as a "drainage network" and the specific area the network drains is called a "drainage basin" (see Figure 10.1). The high ground that separates one drainage basin from another is dubbed a "drainage divide" and a "continental divide" forms the barrier between two drainage systems that empty into different oceans (see Figure 10.2).

Figure 10.1 – Organization of a Drainage Basin

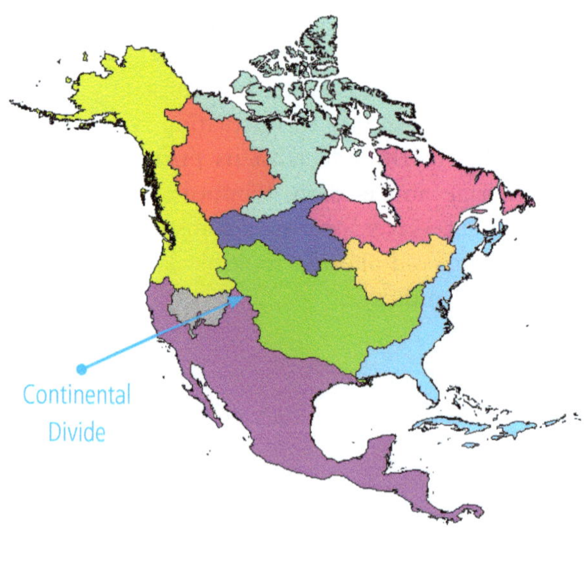

Figure 10.2 – Drainage Basins of North America

Exercise: Differences in Adjacent Drainage Basins

Map I depicts a part of the Mogollon Rim, an escarpment that defines the southwestern edge of the Colorado Plateau. It extends approximately 200 miles from northern Yavapai County eastward to near the border with New Mexico. It was formed by erosion and faulting, and is the home to numerous dramatic canyons.

Mongolian Rim at Promontory Butte

Examine Map I. Notice that there are two river systems in this locale, one flowing north of the rim and the other flowing south of the rim. Observe also that there is a substantial difference in the topography generated by the northern streams as compared to the topography generated by the southern streams. The following exercise is designed to help you investigate why the river systems to the north and south of the rim created such vastly different landscapes.

1. The drainage divide on Map I has been highlighted in black for you. How was the drainage divide identified?

   ____

2. Where is the Mogollon Rim located with respect to drainage divide—to the north, south, or relatively the same spot? (Circle one)

3. Just by examining the topography map, which area has the steeper gradient: north of the rim or south of the rim? (Circle one). On what did you base your conclusion?

   ____

   ____

4. Table 10.A below lists the gradients of six streams in the northern drainage basin and six streams in the southern drainage basin. Calculate the average gradient for the streams in both of the drainage networks.

Table 10.A Stream Gradients in the Northern and Southern Drainage Basins

Streams in the North	Gradient (ft/mi)	Streams in the South	Gradient (ft/mi)
Bear Canyon	125	See Canyon	318
Beaver Canyon	81	Spring Creek	568
Turkey Canyon	132	Doubtful Canyon	521
East Leonard Canyon	132	Horton Creek	333
Middle Leonard Canyon	167	Dick Williams Creek	639
West Leonard Canyon	96	Big Canyon Creek	405
Average		Average	

5. Do the numbers support or contradict your answer to #3? ______________________

6. How many times steeper are the stream gradients in the southern basin than in the northern one? ______________________ x

7. Figure 10.3 is a profile of the A–A' line on Map I. Examine this profile. Given that streams with steeper gradients downcut more efficiently, as time progresses, would you expect the divide between these two drainage basins to move toward the north, south, or remain in relatively the same place? ______________________

Explain: ______________________

____

Figure 10.3

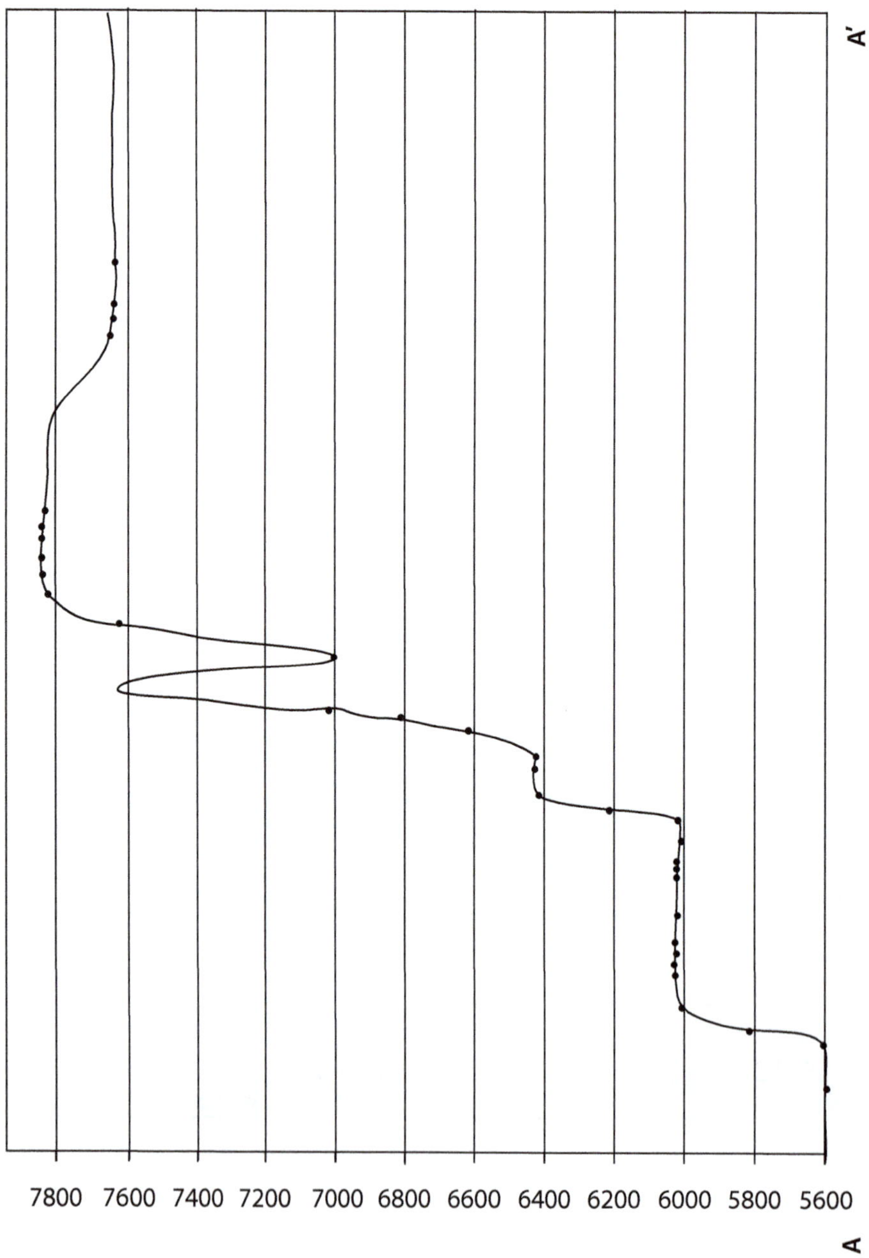

Map I – Promonatory Butte

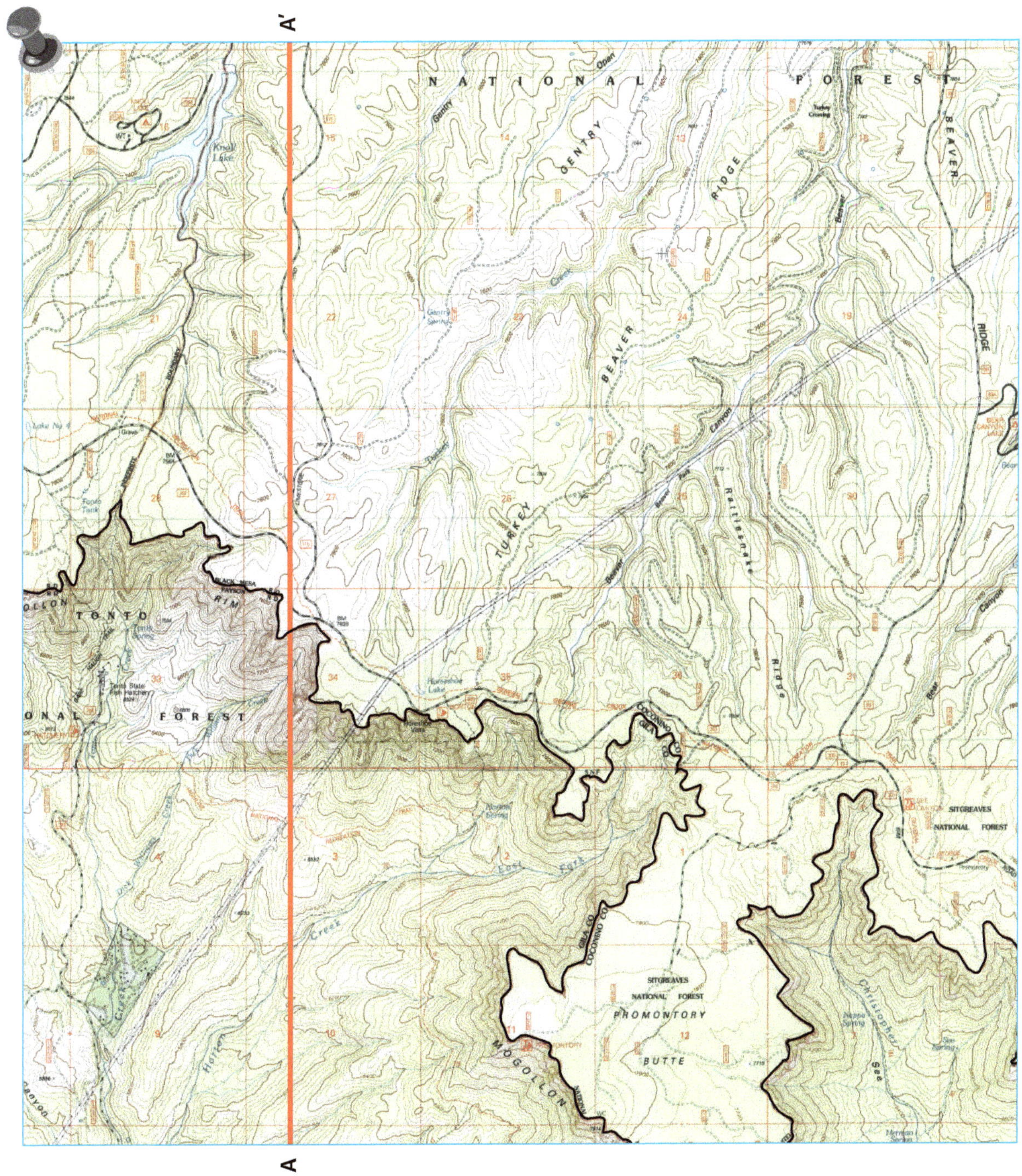

PART B: EVOLUTION OF A RIVER SYSTEM

As a river channel progresses from its birthplace in the highlands, through the plains and basins, and finally to the shores of an ocean or lake, changes in gradient, water quantity, substrate composition, and local geography require that the channel constantly adjust its size, shape, and path to accommodate the new set of conditions. Due to the regularly evolving nature of the channel over time, a river or river system will depict significant morphological differences during the course of its lifetime, in addition to differences along its length at any given time. Characteristics used to describe the channel consist of:

a. Cross-Sectional Area: Average width x depth. Used to characterize the size of the channel

b. Cross-Sectional Shape: Narrow and Deep vs Wide and Shallow

c. Sinuosity: A measure of the degree of curving. Obtained by dividing the length of the bends by the horizontal distance between its ends. Channels are considered "straight" when their sinuosity values are between 1 and 1.5 and "meandering" when their sinuosity values exceed 1.5. See Figure 10.4.

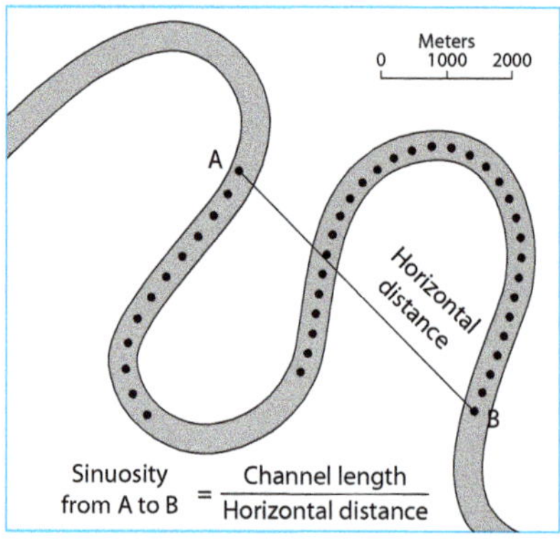

Figure 10.4 – Sinuosity

Differences in the above features both within a river system and between multiple river systems directly reflect variability in the erosional and depositional processes. Factors that affect a stream's ability to erode or deposit include but are not limited to:

a. Underlying Substrate: Structure and composition of the local geology will dictate a stream's path and ability to erode, respectively.

b. Gradient: Refers to the slope of a designated length of stream channel. Calculated by dividing vertical relief by horizontal distance and is expressed in ft/mile or m/km. Directly affects water velocity and a river's ability to downcut and erode laterally.

- Higher gradients promote downcutting
- Lower gradients promote lateral erosion

c. Discharge: Quantity of water passing a specific point per unit time. Expressed in ft^3/sec. The higher the discharge, the more channel space that is required to accommodate the flow.

- Discharge = Cross-Sectional Area x Water Velocity

d. Load: Quantity of sediment being transported by a stream. Sediment-laden streams have more erosive power.

e. Base level: The lowest possible level to which a stream can erode. Once base level is reached, the stream has no more erosional power and depositional processes prevail. Base level can and does change throughout the lifetime of a river due to tectonic uplift and ground subsidence. The ultimate base level of any stream is sea level, but lakes and resistant rock layers serve as local base levels throughout the length of the stream, intermittently interrupting the stream's ability to erode at isolated locations.

Exercise: A Quick Look at the Influence of Base Level

The waterfall known as Niagara Falls serves as a local base level between Lake Erie and Lake Ontario, due to the substrate beneath the falls being too resistant for efficient downcutting. The water instead falls over the resistant ledge, enabling the vertically flowing water to steadily erode the resistant layer horizontally in the direction of its source (see Figures 10.5a and 10.5b). As such, the falls will continue to retreat until they reach Lake Erie, at which point the stream will have reached grade and the great lakes will then drain out (see Figure 10.5c).

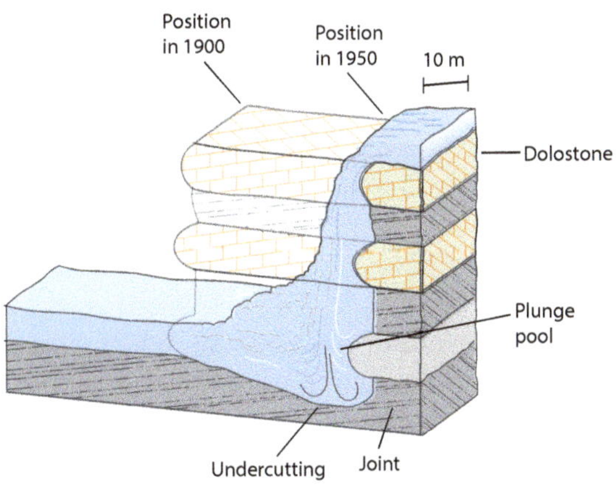

Figure 10.5a – Schematic Representation of Erosion by the Niagara Falls.

Figure 10.5b – Horseshoe Falls, the Canadian Side of Niagara Falls. City of Buffalo, NY in background.

Figure 10.5c – Satellite image of the Niagara River. Flowing from Lake Erie in the south (bottom of image) to Lake Ontario in the north, the river passes around Grand Island before going over Niagara Falls, after which it narrows in the Niagara Gorge. Two hydropower reservoirs are visible just before the river widens after exiting the gorge.

1. Niagara Falls has eroded 11.4 kilometers (7.1 miles) from Lake Ontario toward Lake Erie during the last 12,300 years. What was the rate of erosion during this time? __________ ft/yr

2. Recently, the erosion rate you calculated above has slowed to approximately 1 foot per year due to two primary factors:

   a. Currently, the falls are flowing over the exceptionally erosion-resistant cap rock. As the falls continue eroding southward, it will encounter the softer rocks of Navy Island and the erosion rate will increase.

   b. The development of hydroelectric-generating plants along the shoreline of the Niagara River has vastly reduced the rate of water flow due to the quantity of water being diverted from the falls.

3. Niagara Falls has approximately 20 miles to go before it reaches Lake Erie. Using an average future erosion rate of 2 ft/yr, in approximately how many years will Niagara Falls reach Lake Erie? ______________ years

As mentioned above, a river system's ability to consistently change the landscape results from its unparalleled power to erode, transport, and deposit sediment. Which of those processes dominates a locale, the degree to which it occurs, and whether or not the dominant process changes over time is in turn dependent on variability in the underlying substrate, gradient, discharge, load, and base level. The following exercises are designed to examine how these variables change over time and space and the result those changes have on a river's ability to alter the landscape.

SECTION I: EROSION—THE FIRST HALF OF THE STORY—WHEN STREAMS PICK UP THEIR LOAD

Rivers are able to effectively erode the landscape due to the joint use of downcutting, which cuts the substrate vertically, and lateral erosion, which widens the channel horizontally. Erosion is fastest and most powerful in upland environments, where the stream channels are characterized by steep gradients, high water velocities, and elevations substantially higher than base level.

Stream erosion generates a progressively evolving landscape that morphs through three distinct phases: youthful, mature, and old. During the youthful stage of stream development, gradients are steep, channels are straight, and intense downcutting results in V-shaped cross-sections. Aggressive downcutting as just described continues until base level is within reach. At this point, the stream is encountering significantly lower gradients, causing a gradual decrease in downcutting and simultaneous increase in lateral erosion, thereby substantially widening the channel. These features, in combination with more sinuous streams, define the mature phase of stream evolution. In an ideal scenario, as time continues, downcutting terminates and lateral erosion flattens out the landscape creating a vast, level surface called a "peneplain" that is occupied by highly sinuous stream channels—thereby characterizing the old stream phase. See Figure 10.6 for a visual depiction of this stream maturation process.

"Headwaters" is the term used to describe the streams generated in the highlands from the concentration and erosion of sheetwash. Because the headwaters within a drainage basin primarily reside in steeply sloped, topographically elevated locations, the waters are characterized by high flow velocities and intense downcutting. Due to the erosive nature of these streams, the headwaters will preferentially amass along the swiftest and least resilient routes to the ocean, thereby generating unique arrangements and trends that reflect the shape of the landscape and the composition of the substrate. These stream configurations are referred to as "drainage patterns," examples of which can be seen in Figure 10.7.

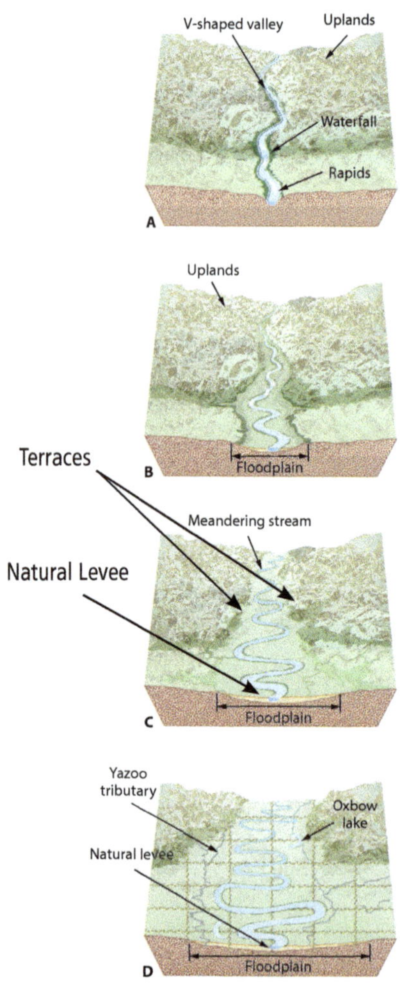

Figure 10.6 – Stream Maturation over Time

Figure 10.7 – Steam Drainage Patterns

Dendritic: Irregular pattern of channels that branch like a tree. Develops on flat lying or homogeneous rock.

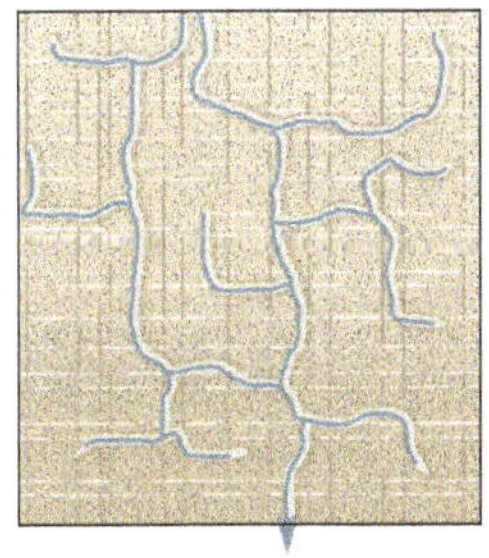

Rectangular: Channels have right-angle bends developed along perpendicular sets of rock fractures of joints.

Radial: Channels radiate outward like spokes of a wheel from a high point.

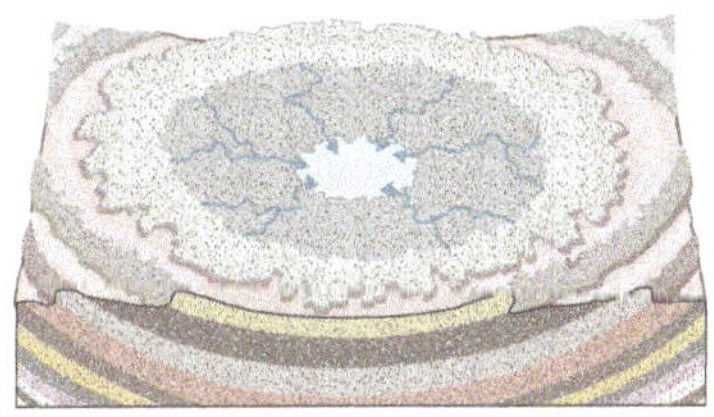

Centripetal: Channels converge on the lowest point in a closed basin from which water cannot drain.

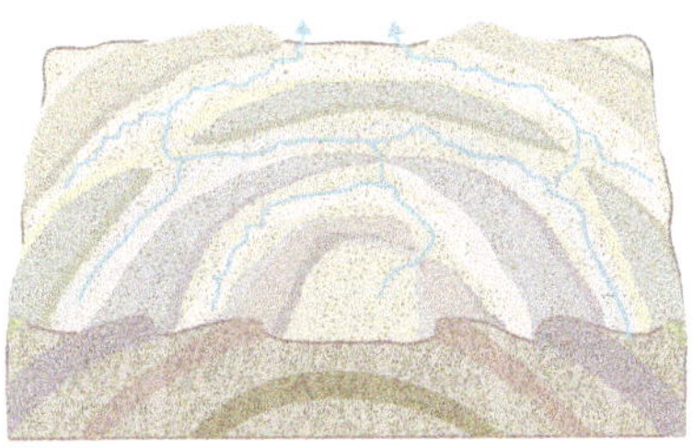

Annular: Long channels form a pattern of concentric circles connected by short radial channels. Develops on eroded domes or folds with resistant and nonresistant rock types.

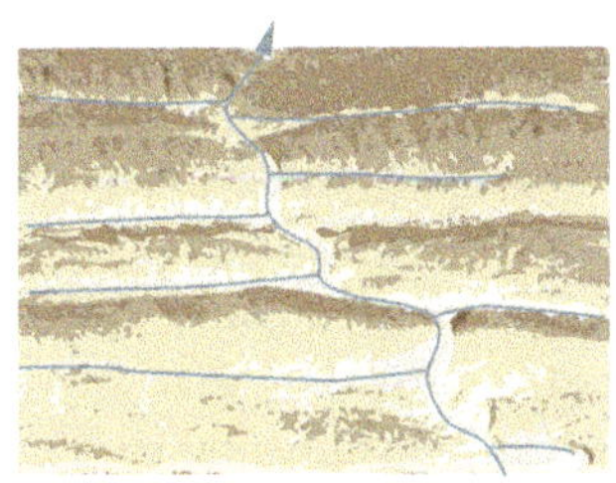

Trellis: A pattern of channels resembling a vine growing on a trellis. Develops where tilted layers of resistant and nonresistant rock form parallel ridges and valleys. The main stream channel cuts through the ridges, and the main tributaries flow along the valleys parallel to the ridges and at right angles to the main stream.

Exercise: Identifying Drainage Patterns

Examine Maps II–VII (located at the end of the chapter). For each map, state the drainage pattern present and the type of substrate the stream is likely flowing over.

a. Map II ______________________; Local geology ______________________

b. Map III ______________________; Local geology ______________________

c. Map IV ______________________; Local geology ______________________

d. Map V.a and V.b ______________; Local geology ______________________

e. Map VI ______________________; Local geology ______________________

f. Maps VII.a and VII.b___________; Local geology ______________________

SECTION II: DEPOSITION—THE SECOND HALF OF THE STORY—WHEN STREAMS DROP THEIR LOAD

As discussed earlier, a river system is designed to erode, transport, and deposit enormous quantities of sediment. Erosion dominates in the higher areas of the drainage basin, where the straight streams, steep gradients, and high water velocities incite aggressive erosion, while deposition primarily takes place in the lowland areas, due to a combination of high sediment loads and flatter terrain. Stream deposition generally takes place in four primary continental environments: Alluvial Fans, Meandering Streams and Floodplains, Braided Streams, and Deltas (see Figure 10.8).

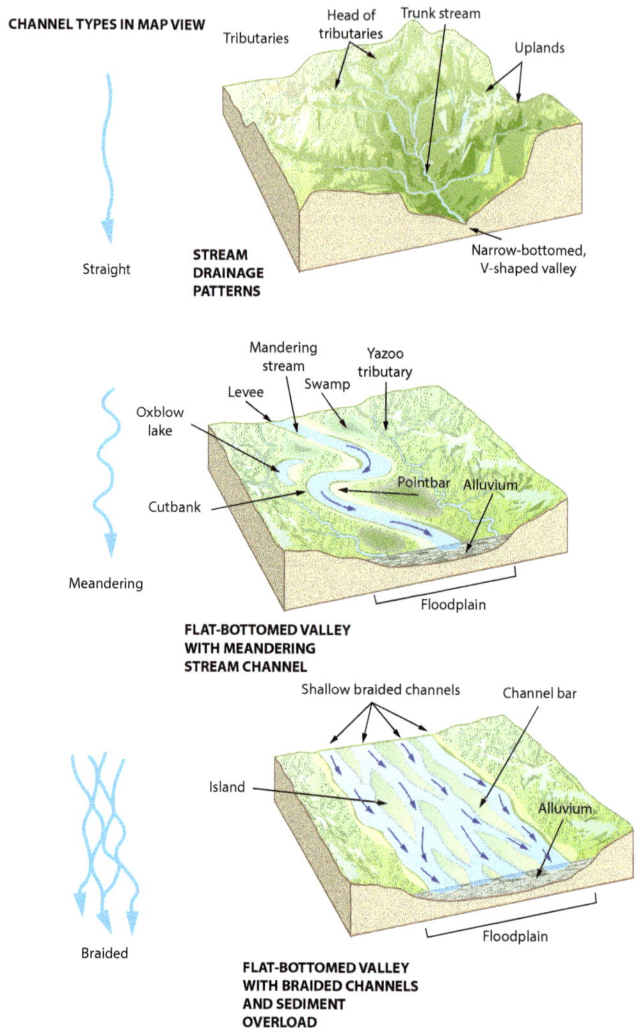

Figure 10.8 – Erosional vs. Depositional Environments

Alluvial Fans – See Figures 10.9a and 10.9b

Alluvial fans develop when a stream abruptly exits a canyon and plunges down onto a level plain. The sharp decrease in gradient results in an immediate velocity decrease and subsequently the coarse material is deposited on contact with the plain. The water will then break up into multiple "distributary channels" similar to the way cracks radiate outward from a bullet entrance. The various channels wind around the freshly deposited material while continuing to decrease in velocity as they travel down the mild slope from the point of contact. As the waters continue to slow, they will progressively deposit the sand and finally the mud, generating a fan-shaped deposit characterized by decreasing grain size from the inner fan to the outer fan.

Alluvial fans are most frequently associated with arid environments, the result of periodic flashfloods in the local hills. As such, they are a standard part of the landscape along the sedimentary margins of the North American Basin and Range province. While not as common, alluvial fans can also be found in temperate climates, as can be seen in the Himalayan Foothills.

Figure 10.9a – Mini Alluvial Fan

Figure 10.9b – Giant Alluvial Fan in China's XinJiang Province

Meandering Streams – See Figures 10.10 - 10.12

Meandering streams generally evolve in the lower parts of a stream's course, over flat, low-lying land adjacent to a seacoast. They consist of one large, highly sinuous channel with deep and wide cross-sections, low-flow velocities, and a sediment load consisting primarily of sand, mud, and clay. Meandering streams flow through a flat expanse of land called a "floodplain" (to be discussed in more detail later), which is generated by the constant migration of the stream channel as a consequence of local erosional and depositional processes. While this area does flood when the ability of the channel to accommodate its volume of water is exceeded, the floodplain is not created by flooding, but by the intense lateral erosion of the channel, and is replenished with sediment by flooding episodes.

Figure 10.10 – Meandering Stream in the Innoko National Wildlife Refuge

Streams begin with straight channels (see Figure 10.11a), but variations in the water velocity and substrate cause the flowing current to swing back and forth. Because erosion takes place more effectively in areas where the water flows faster, the river is able to carve out arcs as it oscillates and subsequently cuts into the walls of the channel at intermittent spots (see Figure 10.11b). The brisker part of the flow will continue to cut away on the outer bank of the bends, now known as the "cutbank," causing each curve to migrate laterally and enlarge into a full-fledged meander. While lateral erosion continues to eat away at the cutbank, the slower-flowing water on the inside of the bend deposits sediment, creating a wedge-shaped accumulation called a "point bar" - see Figure 10.11c.

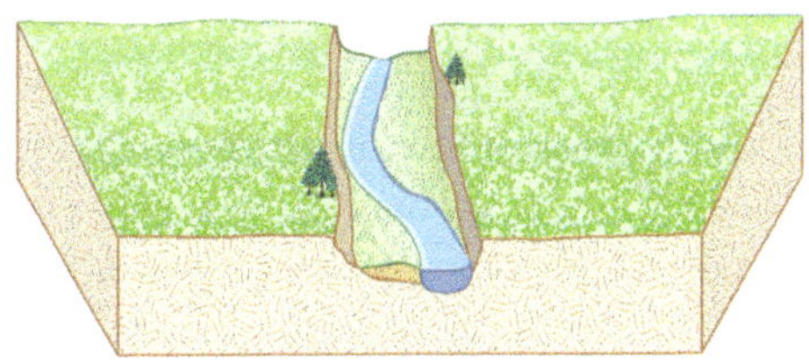

Figure 10.11a

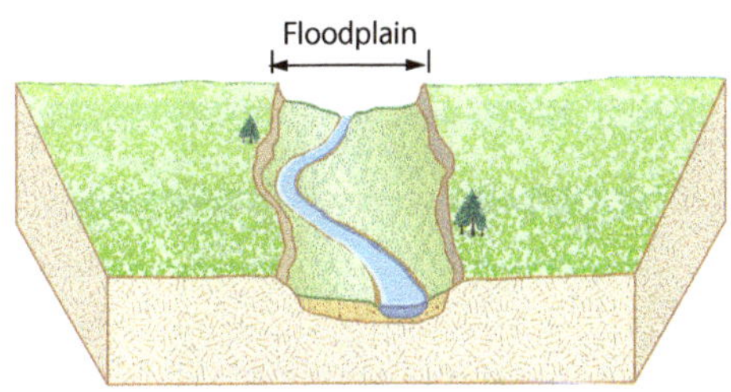

Figure 10.11b

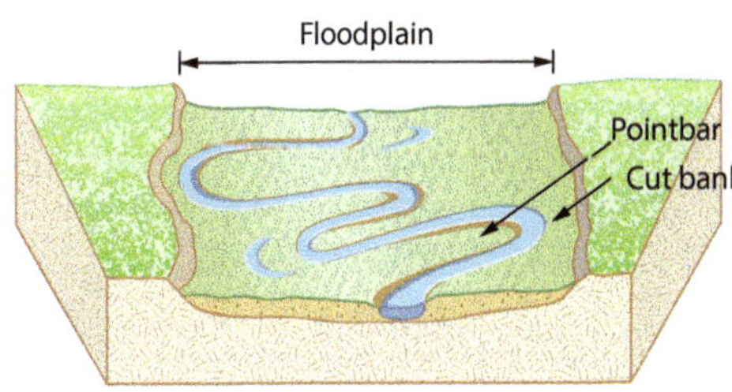

Figure 10.11c

The continued erosion on the cutbanks progressively increases the curvature of the meanders. Eventually, the meander may curve to such a degree that the ends of the meander almost meet, creating a "meander neck." When the stream erodes through a meander neck in an attempt to find a shorter route to the mouth of the river, a "cutoff" develops. The meander that has been cut off is called an "oxbow lake" if it remains filled with water, or an "abandoned meander" if it dries out (see Figure 10.12). Cutoffs straighten the stream in the areas they occur, effectively re-centering the channel such that new meanders can develop and lateral migration can start anew. It is the simultaneous erosion of the cutbank and deposition on the point bar that generates the lateral channel migration, and in turn carves out the floodplain through which the channel flows.

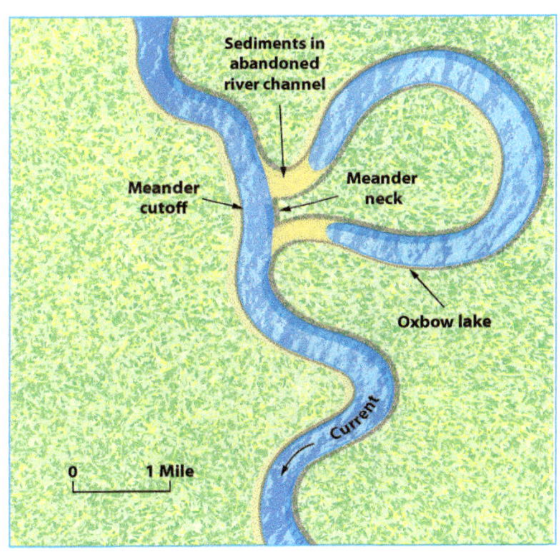

Figure 10.12 – Formation of a Cutoff

Rivers carve their channels such that they are the appropriate size and shape to accommodate the average amount of water and sediment they are responsible for transporting throughout the year. However, due to natural factors, such as spring melts or excessive precipitation, for example, there will be times when the ability of the stream to contain the water and sediment is exceeded and the water will overflow its banks and flood. Natural flooding of the type just described takes place on the floodplain.

The floodplains are characterized by ridges of sand, called natural levees, that border the banks of the river and gradually bleed into silt and mud deposits—the latter of which extends outward into a wide, low-lying, swampy plain. This sediment distribution occurs because the water decreases velocity as it moves away from channel, thereby dropping coarser material first, followed by progressively finer sediment. In older river systems, the natural levees sometimes become too high for tributaries to breach, and in such cases, the tributaries will run parallel to the trunk stream until they find an open entrance. Such a stream is called a *Yazoo stream* and is common in highly evolved river systems. While extremely lush and nutrient-rich from the new supply of freshly weathered sediment, the floodplains are flooded annually and are not designed for any kind of permanent structures or indiscriminant use. It's important to remember that floodplain formation and flooding are a natural part of river evolution and become a hazard only when we fail to accept the limitations involved in safely taking advantage of the resources the floodplains have to offer.

Figure 10.13 – Alaska, Chichagof Island, Deep Bay Braided Stream

Braided Streams - See Figure 10.13

Braided streams are composed of many channels that intertwine in a braid-like network. They are generated when the discharge is insufficient to transport its load. As a result sediment gets deposited on the floor of the channel creating "channel bars," which forces the stream to divide up into multiple strands and weave around the bars of sediment.

With some exceptions, braided channels are primarily located in middle- to low-relief landscapes, where the decreasing water velocity and increasing sediment load overwhelm the capacity of the river to transport the sediment and subsequently, the sediment is deposited. Braided channels are common in the valleys below glaciated landscapes where streams are fed by debris-choked meltwater. Other environments characterized by low relief, low-velocity flows, and very high sediment loads include, but are not limited to, river deltas, alluvial fans, and peneplains.

Figure 10.14 – Indus River Delta

Deltas - See Figure 10.14

Deltas develop where the running water of a stream enters standing water. It is an identical effect as described for an alluvial fan, where the stream velocity abruptly drops to zero and all remaining sediment is dropped and spread. Geologists refer to any wedge shape of sediment formed at a river mouth as a delta, even though relatively few have the triangular shape of a delta.

Exercise: Understanding Channel Shapes

Map VIII, located in the back of the chapter, depicts the geology of Ennis, Montana. The east half of the map shows a high, mountainous region modified by alpine glaciation. To the west lies a semi-arid, lowland basin. The Cedar Creek Alluvial Fan emanates from the mountains westward into the basin. Further west, the Madison River empties into Ennis Lake, which is drained through a deeply cut canyon to the north. Use Map VIII to answer the following questions:

1. What mountain range is the sediment source for the alluvial fan? (Hint: consider the location of the landscape relative to a major mountain range in this area.) ________________

2. What is the drainage pattern in the alluvial fan? ________________________

3. Segments of each of the following rivers have been highlighted in yellow on Map VIII. Figures 10.15a, 10.15b, and 10.15c are enlargements of the areas in which each segment is located. Using the enlargements, determine the relief along the segment of slope highlighted by the yellow line, then return to Map VIII to measure the distance of the highlighted segment (this is necessary as the enlargements will not match the map scale). Finally, calculate the stream gradient of:

   a. The stream that flows into the alluvial fan _______ ft/mi

   b. A stream within the Cedar Creek Alluvial Fan __________ ft/mi

   c. The Madison River ______ ft/mi

4. What type of stream (straight, meandering, or braided) is:

   a. The stream that flows into the alluvial fan? _____________

   b. Most the streams in the Cedar Creek Alluvial Fan? __________

   c. The Madison River? ________________

   d. Bear Creek in the SW corner of the map? ___________

5. Do the gradients you calculated for the first three rivers above make sense based on the type of stream channel these streams display? Yes or No? Explain

   ____

   ____

6. Were you to calculate the stream gradient for Bear Creek yourself (located in the SW corner of Map VIII), you would find that Bear Creek and the Madison River have very close gradients. However, as you can see, they depict two different channel types. Explain why this is the case.

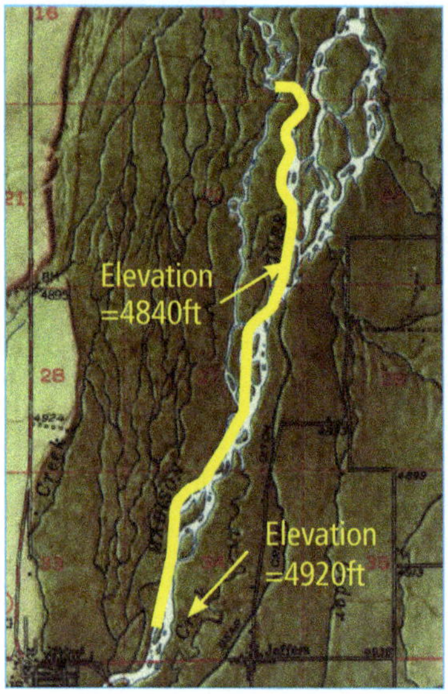

Figure 10.15a

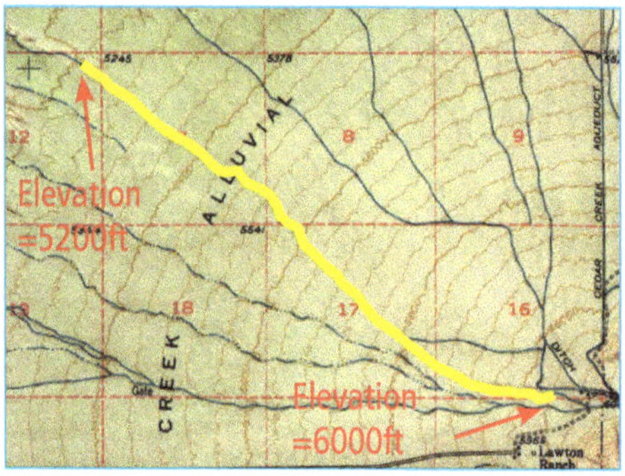

Figure 10.15b

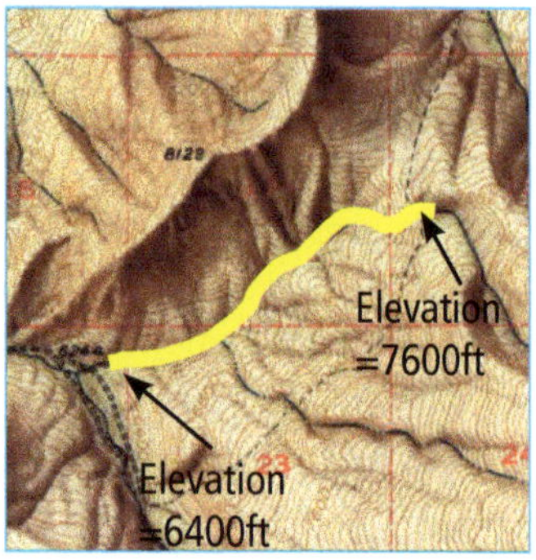

Figure 10.15c

Section III: Putting the Pieces Together

Exercise 1: Evolution Along the Length of a River

1. As a river evolves over time it simultaneously cuts down into the highlands and extends its length into the lowlands. Based on what you know about the evolution of a river system, state whether you think each of the criteria below would increase or decrease as you move from the head to the mouth of a stream channel.

Cross-sectional area:	Increase or Decrease
Sinuosity:	Increase or Decrease
Gradient:	Increase or Decrease
Discharge:	Increase or Decrease
Sediment Load:	Increase or Decrease

2. Examine maps IX and X at the end of the chapter. Map IX depicts the Missouri River in Portage, Montana, in the midst of the Rocky Mountains. Map X shows the Missouri River in Leavenworth, Kansas, well on its way to the confluence with the Mississippi River. Compare these two segments of the river. Document the differences between the channel segment in Kansas and the segment in Montana with respect to:

	Montana	Kansas
Gradient (steeper vs shallower)		
Cross-Sectional Area (larger vs smaller)		
Cross-Sectional Shape (v-shape vs u-shape)		
Sinuosity (straighter or meandering)		
Presence of Depositional Features (natural levees, floodplains, cutoffs, oxbow lakes, etc.)		

3. Explain why the channel shape changed as it progressed from Montana into Kansas. ____________
____
____
____

Exercise 2: Identifying the Stages of River Evolution

1. Maps XI, XII and XIII, located at the end of the chapter, illustrate segments of three U.S. rivers: Tower Creek in Yellowstone, the Red River in Louisiana, and Marias River in Montana. Each of these three rivers shows very different characteristics. To begin examining their differences, draw a topographic profile (A–A' line) for each of the three maps. Please do this on the three pieces of graph paper provided at the end of this laboratory. **USE A VERTICAL SCALE AS FOLLOWS:**

Tower Creek MAP – 1 inch = 100 ft.

Red River MAP – 1 inch = 20 ft.

Marias MAP – 1 inch = 100 ft.

A profile line has been placed on each map with tick marks indicating the contour line you should use to draw your profile.

2. Examine all three maps and **THOROUGHLY** fill in the table below

Stream characteristics	Tower Creek	Marias River	Red River
Presence of:			
Floodplain			
Meanders			
Natural levees			
Back swamps			
Yazoo streams			
Oxbow lakes			
Cutoffs			
Slope of valley wall and surrounding topography in ft/mile (from east side of the river, to the highest point on the east part of the map along the A–A' line)			
Sinuosity from X–Y			
Shape of the river valley in terms of width and depth (check your profiles)	V-shaped U-shaped Wide & Flat	V-shaped U-shaped Wide & Flat	V-shaped U-shaped Wide & Flat
Probable importance of : Vertical erosion	High, med, or low	High, med, or low	High, med, or low
Lateral erosion	High, med, or low	High, med, or low	High, med, or low

3. The three rivers in Maps XI, XII, and XIII are examples of streams with characteristics of early/youthful, middle/mature, and late/old stages. Based on the information you have recorded in the above table, which is which? JUSTIFY YOUR ANSWER

Tower Creek is: ______________________________ stage

Because: ____

____

Red River is: ______________________________ stage

Because: ____

____

Marias River is: ______________________________ stage

Because: ____

____

PART C—WHY DID IT DO THAT?

So far, we have discussed a number of ideal scenarios concerning how river systems evolve if all things are equal and all the variables support change in a unilateral direction throughout the lifetime of the river system. However, in reality, the geologic factors that dictate a stream's behavior are significantly more complex, and there are many cases where streams appear not to follow the rules. Closer examination of a few such examples will demonstrate that these river systems are in fact doing exactly what's expected, but untangling their story requires a more in-depth understanding of the fluctuating conditions streams must adjust to. Consider the following scenarios.

Exercise 1: Anderson Mesa, Colorado

1. Examine Map XIV of Anderson Mesa, Colorado at the end of the chapter. Is this river straight, meandering, or braided? ______________________

2. Based on your answer to the question above, would you expect this river to have a large cross-section with wide floodplains, or a v-shaped cross-section with virtually no floodplain? (Underline your answer)

3. Looking at the map, which of the choices in the question above does the river depict: cross-section with wide floodplains, or a v-shaped cross-section with virtually no floodplain? (Underline your answer)

4. While this landscape appears not to follow the rules, keep in mind that the examples we have examined thus far demonstrate ideal river evolution. In reality, the progression described can be interrupted by large scale tectonic processes that uplift the land, steepen the gradient, and "rejuvenate" a stream, causing it to morph from extensive deposition to aggressively downcutting into its own alluvium-filled floodplain. Now that you are aware of the stream-rejuvenation process, explain why this highly sinuous channel has steep canyon walls. (Hint: Remember that canyons form in strong rock, not soft sediment.)

____

____

____

Exercise 2: Lake Scott, Kansas and Voltaire, North Dakota

Map XV is a topographic map of Lake Scott, Kansas and Map XVI is topographic map ofVoltaire, North Dakota. Use these map, located at the end of the chapter, to answer the following questions.

1. Using the chart below, describe the Voltaire region with respect to cross-sectional shape, sinuosity, and importance of lateral vs. vertical erosion.

2. Using the chart below, describe the Lake Scott region with respect to cross-sectional shape, sinuosity, and importance of lateral vs. vertical erosion.

	Voltaire Region	Lake Scott Region
Cross Sectional Shape (V-shaped, U-shaped or flat and wide)		
Sinuosity (high or low)		
Primary Erosion Method (Lateral or Vertical)		

3. Geologic evidence has demonstrated that streams have been shaping both landscapes for approximately 11,000–12,000 years. Why are there such differences in channel shapes, cross-sectional areas, and erosion and deposition patterns between the Voltaire region and the Lake Scott region—even though they are the same age?

Part D—Human Angle

Because of the extensive impact rivers have on the landscape that humans inhabit and the plethora of resources they provide for the sustenance of mankind, it's no surprise that humans and rivers have regularly crossed paths throughout history, not always with positive results. Below are some examples of situations that depict the complex relationship between natural river processes and anthropogenic use of its resources.

Exercise 1: Rivers and Changing Borders

Throughout history, many geographic boundaries, both national and international, have been drawn using rivers to define the borders. Examples include but are not limited to, the Rhine River of Europe, which defines the border between six countries; The Red River, Missouri River, and Mississippi Rivers in the US, each of which serve as borders between various states; and the Rio Grande, which delineates the border between the United States and Mexico.

As has been demonstrated, rivers are prone to constant migration, and therefore are not permanent borders. This being the case, many places where geographic boundaries are defined by rivers, have ended up in boundary disputes with their neighbors. The exercise below is designed to help you understand the nature of these disputes, such that you can make an educated assessment of whether rivers make sustainable boundaries.

Background

The **Chamizal dispute** was a border conflict over about 600 acres on the U.S.-Mexico border between El Paso, Texas, and Ciudad Juárez, Chihuahua. The U.S.–Mexico border was defined based on the location of the Rio Grande in 1852. Within a few decades, the river path had substantially altered resulting in a dispute between the United States and Mexico that was not officially resolved until 1963.

The Treaty of Guadalupe Hidalgo (which officially ended the Mexican–American War) and the Treaty of 1884 were the agreements originally responsible for the settlement of the international border, both of which specified that the middle of Rio Grande was the border– irrespective of any natural, gradual alterations in the channels or banks. This provision followed the long–established doctrine of international law that when changes in the course of a boundary river are caused by *natural, gradual* changes, the boundary changes with the river, but when changes are due to sudden, catastrophic circumstances, the old channel remains the boundary.

The river continually shifted south between 1852 and 1868, with the most radical shift in the river occurring after a flood in 1864 (see figure 10.16).

By 1873 the river had shifted approximately 600 acres from its earlier position. The newly exposed land came to be known as *El Chamizal*. Both Mexico and the United States claimed the land.

1. Before you examine the details surrounding this conflict, consider the flood of 1864. Based on the international law, should the border change with the new path, remain with the old channel boundary, or is there a case to be made for both sides? Explain

____

____

____

____

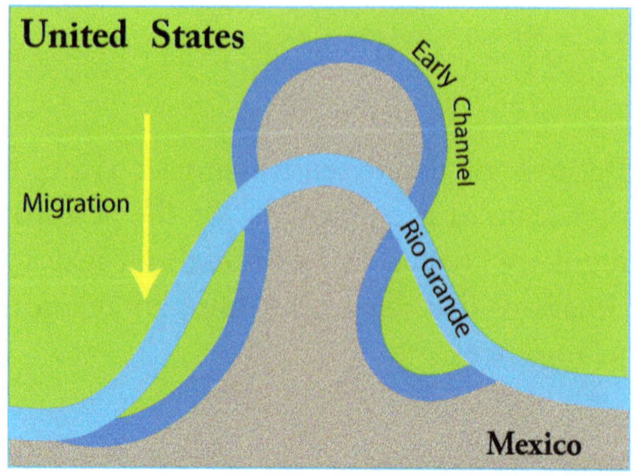

Figure 10.16 – Southern migration of river channel over time

2. Figure 10.17 is a map showing the migration of the river from 1852 to 1963 when a settlement was finally reached. Using a red pen, trace the boundary of *El Chamizal*—the area that prior to 1852 belonged to one country and after decades of river migration, was then considered part of the other country.

   a. Based exclusively on the river channel location, which country did *El Chamizal* belong to prior to 1852? ____________________

   b. Based exclusively on the river channel location, which country did *El Chamizal* belong to following 1873? ____________________

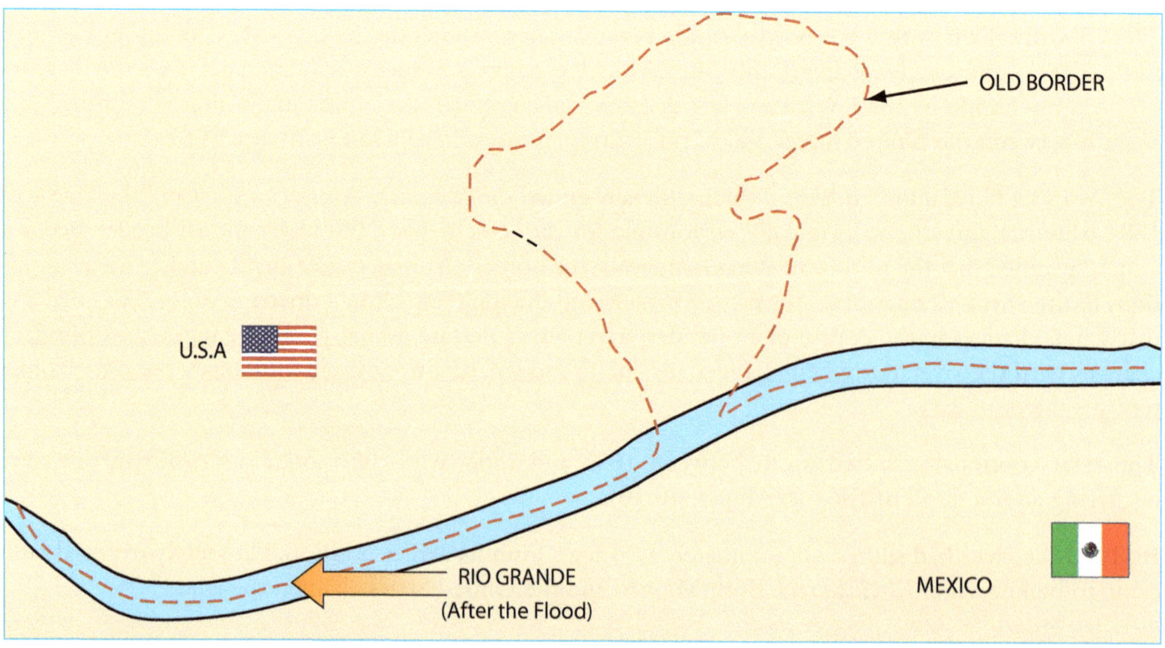

Figure 10.17 – Changes in river path from 1852–1963

3. Based on your answer to the questions above, do you think Mexico claimed the flood was catastrophic or natural? ____________ How about the US? ____________

Explain. ____________

In 1895, Mexican citizens filed suit in the Juárez Primary Court of Claims to reclaim the land. The dispute was formally settled on 14 January 1963. The agreement awarded to Mexico 366 acres of the Chamizal area and 71 acres east of the adjacent Cordova Island. Although no payments were made between the two governments, the United States received compensation from a private Mexican bank for 382 structures included in the transfer. The United States also received 193 acres of Cordova Island from Mexico (see Figures 10.18a and 10.18b). On 17 September 1963, the U.S. Congress introduced the American–Mexican Chamizal Convention Act of 1964, which finally settled the matter. In October 1967, President Johnson met with President Gustavo Díaz Ordaz on the border and formally proclaimed the settlement.

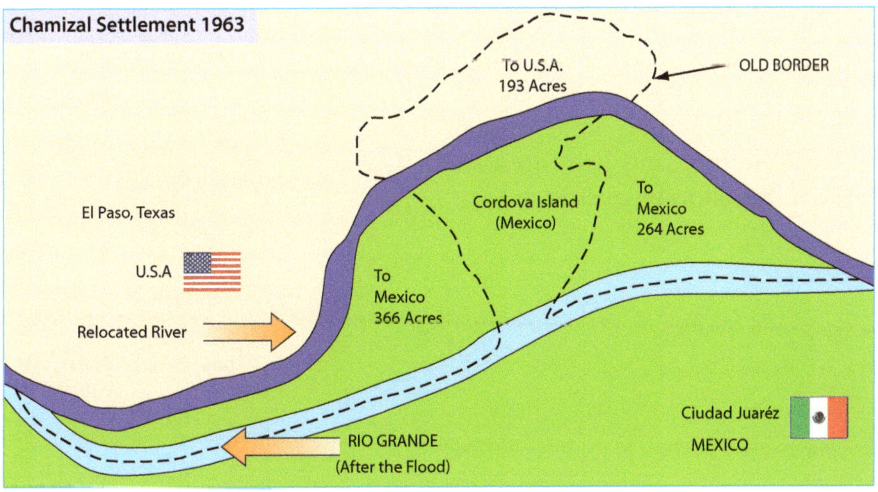

Figure 10.18a – Dispute Resolution

Figure 10.18b – Arial view of dispute resolution

4. Examine figure 10.18a. Note that the river has been "relocated." This part of the channel is man-made and fortified with concrete. The two governments shared the cost of stabilizing this part of the channel. Why would the governments have cooperated to do this?

____

____

Exercise 2: Damming up River Systems

Damming up a river system causes a plethora of unfavorable side effects, including but not limited to: blocking fish migratory routes, depleting the channel of water on one side of the dam and flooding it on the other, and interrupting the flow of natural sediment transport processes. Furthermore, dams serve as a man-made base-level, resulting in devastating displacement of erosional and depositional processes (see Figure 10.19). Map XVII, located at the end of the laboratory exercise, shows a portion of the Missouri River, in which three hydroelectric dams have been built. Use map XVII to answer the following questions.

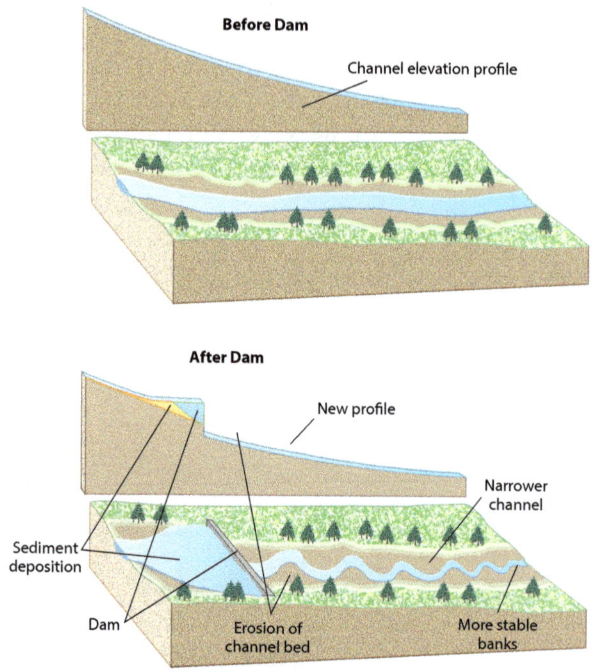

Figure 10.19 – Dams and their effect on a river's profile

1. What direction is the river flowing?

   ____

   How do you know? ______________________________

   ____

   ____

2. How many unnatural base levels have been created due to the installation of the dams? ______

3.

   a. Based on what you know about the direction of flow, would you expect the area directly to the west of Dam #2 to be one of erosion or deposition? (Circle one)

   b. Based on what you know about the direction of flow, would you expect the area directly to the east of Dam #2 to be one of erosion or deposition? (Circle one)

   c. This pattern is found at each of the three dams along the river. This being the case, looking exclusively at the segment between Dam #1 and Dam #2, is the river progressing from erosion to sedimentation or sedimentation to erosion? (Circle one)

d. Based on your answer to the question above, what problems do you foresee developing with respect to the rivers ability to continue flowing? Explain.

____

____

e. What maintenance would have to be regularly done in order to ensure the river continues to flow in its natural direction? ______________________________

____

4. The gradient of the river segment between Dam #1 and Dam #2 is approximately 35ft/mile, whereas the average gradient over the whole 2,500-mile stretch of the Missouri River is approximately 5.4 ft/mile.

a. How many times larger is the gradient in the dammed-up area than the rest of the river?

__________x

b. What is causing this effect?______________________________

5. Describe three major environmental problems that interfering with the natural flow causes for wildlife:______________________________

- ______________________________
- ______________________________
- ______________________________

Supplemental Learning: Rivers and Flooding

Part E: Flood Frequency and Major Floods

Where flood records are available, computations of flood frequency are based on peak annual floods (the maximum discharge for the year as measured at a specific station). Flood frequency is expressed as a recurrence interval (or return period), which is the likelihood that a flood of a given discharge can be expected to occur X times within a fixed time interval. The recurrence interval (T, in years) for a flood of a given discharge is determined by the formula below:

$$T = (n + 1)/m, \text{ where}$$

n equals the number of years of record and
m is the rank or order (1 = greatest flood and so on) of the annual peak flood discharge.

Exercise 1: Flood Frequency of the fictitious "Risky River"

The two tables below contain data sets of Peak Flood discharge for "Risky River." Each data set spans 15 years of recording. Using one data set at a time, estimate the likely discharge for Risky River by following the steps below:

1. Rank the peak flood discharge for data set #1 in order of magnitude, starting with 1 for the largest and ending with 15 for the smallest. Write these results in the "rank" column.

2. Use the formula T = (n + 1)/m and determine the recurrence interval for each of the 15 floods in data set #1. Write the results for each year in the "recurrence interval" column.

3. Repeat this procedure for data set #2.

Risky River Data Set 1

Year	Peak Flood Discharge (cfs)	Rank	Recurrence Interval
1949	165		
1950	235		
1951	198		
1952	183		
1953	170		
1954	152		
1955	139		
1956	220		
1957	169		
1958	162		
1959	257		
1960	178		
1961	189		
1962	200		
1963	173		

Risky River Data Set 2

Year	Peak Flood Discharge (cfs)	Rank	Recurrence Interval
1985	499		
1986	387		
1987	635		
1988	585		
1989	458		
1990	312		
1991	817		
1992	525		
1993	502		
1994	463		
1995	208		
1996	315		
1997	560		
1998	430		
1999	708		

The graphs in Figure 10.20 depict the peak flood discharge (given on the table) plotted against the recurrence interval you calculated for each of the data sets on the previous page. The graphs on the left show the data from the two time periods as a logarithmic curve, which is the way it plots on a normal grid, while the graphs on the right depict the two data sets on a logarithmic grid so that the curve is straightened out allowing you to view the data more easily. As such, the top two graphs depict the same data and the bottom two graphs depict the same data, just presented two different ways. A best-fit curve has been applied to the data, and the equation for each curve is displayed on the graph.

4. Based on the graphical representation of data set 1, what is the predicted discharge for a 100-yr flood? ________________ cfs (*Hint:* ln(100) = 4.60517)

5. Based on the graphical representation of data set 2, what is the predicted discharge for a 100-yr flood? ________________ cfs (*Hint:* ln(100) = 4.60517)

6. How much difference is there between the prediction from the first data set and the prediction from the second data set? ____________ cfs

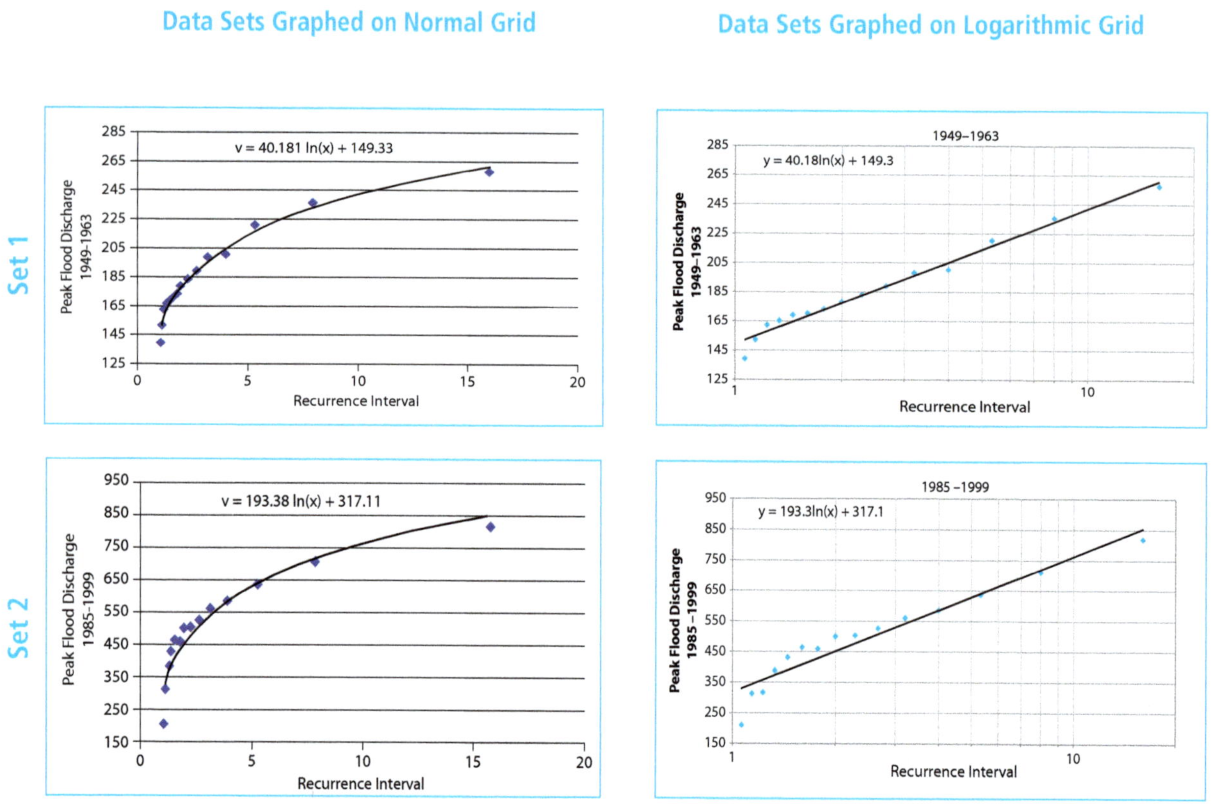

Figure 10.20

7. List some likely anthropogenic reasons for the increased discharge and therefore higher flood prediction in data set 2.

- ______________________________
- ______________________________
- ______________________________

8. Based on your answers above, would you say using data sets to extrapolate far into the future is an accurate and useful method for land use planners? **Yes** or **No**?

   Explain: ______________________________

9. How might your answer to the previous question affect your decision to buy a home adjacent to but just outside the 100-year flood line?

Part F: Case Study
Cumberland River Flooding, Nashville, TN, 2010

Rivers throughout middle Tennessee crested at record high levels on May 2 and 3, 2010. They exceeded previous highs at many stream gages by as much as 14 feet, according to preliminary estimates released by the US Geological Survey (USGS). The highest flood levels were recorded on May 2 and 3, from Nashville west toward Jackson, extending about 40 miles north and south of Interstate 40, and affecting major tributaries to the Cumberland and Tennessee rivers.

The flood peak on the Cumberland River in downtown Nashville ranks as only the tenth highest in more than 200 years of records at that site. This peak was, however, the highest observed during the past 73 years in which much of the basin upstream of Nashville was regulated by several large flood-control reservoirs. During this flood, at least four major tributaries to the lower Cumberland River met or exceeded warning levels established by the National Weather Service (NWS) for major flooding conditions.

"Most tributaries to the lower Cumberland River had flows with only a 1 in 500 chance in any given year, causing the lower Cumberland to flood with a severity that was almost entirely unexpected," according to the USGS Tennessee Water Science Center. "That a regulated river like the Cumberland could have such high flooding is unusual and is a testament to the severity of this event. The extreme rainfall in tributaries that enter the Cumberland River downstream from the flood storage area made this a very difficult event to regulate."

Notable Floods on the Cumberland River, Nashville, TN

Year	Discharge	Gage Height (MSL)
1793	193,067 (estimated)	426.85
1808	174,531 (estimated)	422.17
1847	178,238 (estimated)	423.07
1882	179,886 (estimated)	423.47
1927	203,000	424.37
1937	186,000	422.07

A View of Some Notable Past Floods

The Great Ohio River Flood of January–February 1937 surpassed all prior floods during the previous 175 years of modern occupancy of the Ohio River Valley and geological evidence suggests the 1937 flood outdid *any* previous flood. NOAA.

Tennessee River Flood, Chattanooga, Tennessee, 1917. RG 82, Department of Conservation Photograph Collection. Tennessee State Library and Archives.

The following data were acquired from a stream gaging station located on the Cumberland River in Davidson County, TN, at an elevation of 368.17 feet above mean sea level (msl). The peak flood stage (recorded number of feet above local baseflow height) of the 2010 flood at this gage was 51.86 ft, which means that the elevation of the flood at this point was 420.03 feet above sea level. Any structure in the immediate area of the gaging station and lower than this elevation would have suffered flood damage. For your reference, at this gaging station, the river is considered to be in "flood stage" at 40 ft above the gaging station or 408.17 ft above sea level.

In this exercise, we examine the extent, frequency, and causes of some river floods, and ways to reduce loss of life and property.

1. Using the data in Table 10.B plot the recurrence interval against elevation on the graph provided and draw a best-fit line for frequency of floods at Nashville, TN.

2. What are the expected elevations of the following floods at the Nashville gaging station as determined from the graph you just plotted?

   20-year flood: ____________________

   75-year flood: ____________________

3. Figure 10.21 is a graph of floodwater elevation vs. discharge for the Cumberland River at Nashville, TN between the years 1815–2010. What trend do you see between stage level and discharge?

   ____

4. One main anthropogenic factor contributing to this trend is increased regulation by the Army Corps of Engineers. What's the other? ______________________________

5. Discharge can be obtained for a flood of a given frequency by using information on the flood frequency-elevation plot (obtained from the data in Table 10.B), and the Stage-level–Discharge plot (Figure 10.21). What is the expected discharge at Nashville for:

   a. A 20-year flood? ______________________________

   b. A 75-year flood? ______________________________

   c. A 125-year flood? ______________________________

Table 10.B – Recurrence Interval, Stage, and Discharge at Nashville, TN

Recurrence Interval at gaging station (yr)	Flood Stage	Elevation above mean Sea Level Stage (Ft)	Discharge (cfs)
2	38.6	406.77	110,000
5	44.8	412.97	137,500
10	48.4	416.57	154,500
30	52.7	420.87	175,100
50	55.6	423.77	189,900
100	58.4	426.57	204,300
150	61.1	429.27	218,500
200	64.6	432.77	237,000

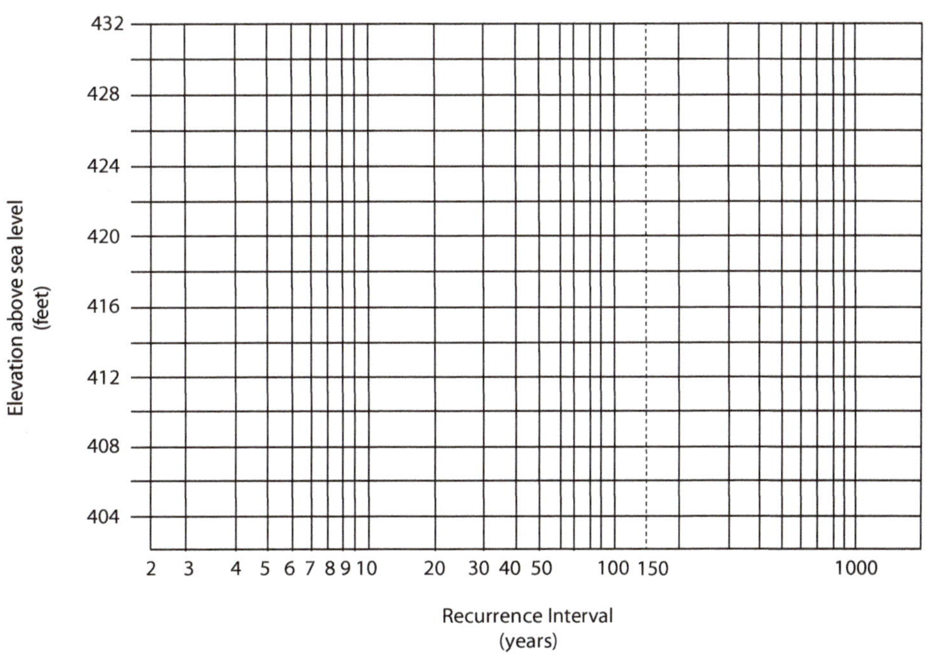

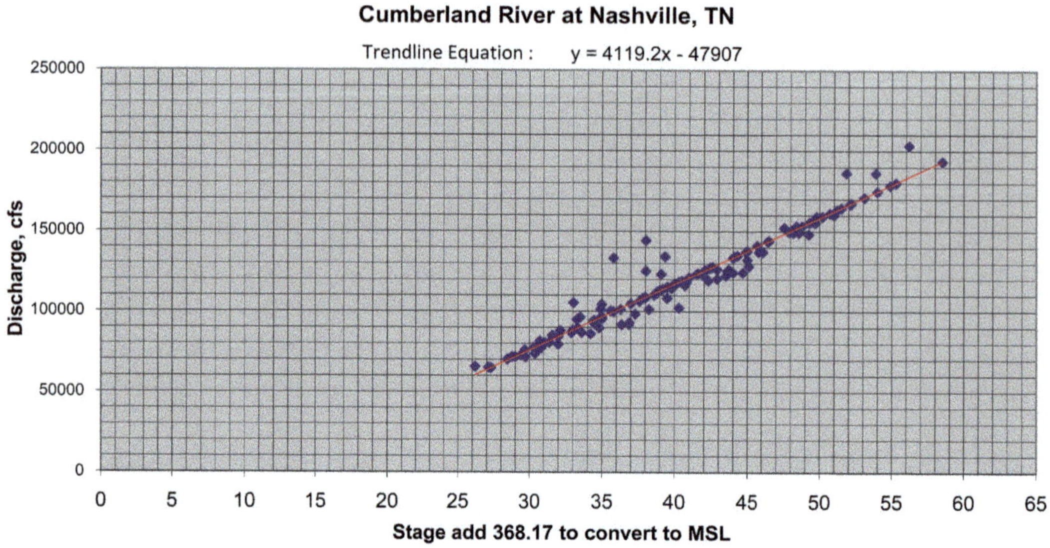

Figure 10.21 – Stage vs. discharge for the Cumberland River at Nashville, TN from 1815 - 2010

6. The May 2–3, 2010 event had a peak flood elevation of 51.86 ft and a peak discharge of 186,000 cfs. Using the data from this gaging station, calculate how often a flood of this magnitude can be expected based on:

   a. The flood elevation ______________________________

   b. The discharge ______________________________

   c. Name a few reasons why the recurrence intervals you calculated in a and b above are different from each other, even though the data came from the same flood?

   ____

   ____

   ____

   d. Is flood data useful for "predicting" or "forecasting" (circle one)? Explain:

   ____

   ____

   ____

7. If you visit a flood area after the water recedes, what indications might there be on buildings, trees, and land surfaces to indicate the level of the flood maximum?

- ______________________________
- ______________________________
- ______________________________

Map II – Drainage Patterns, Green River

Map III – Drainage Patterns, Mount Shasta

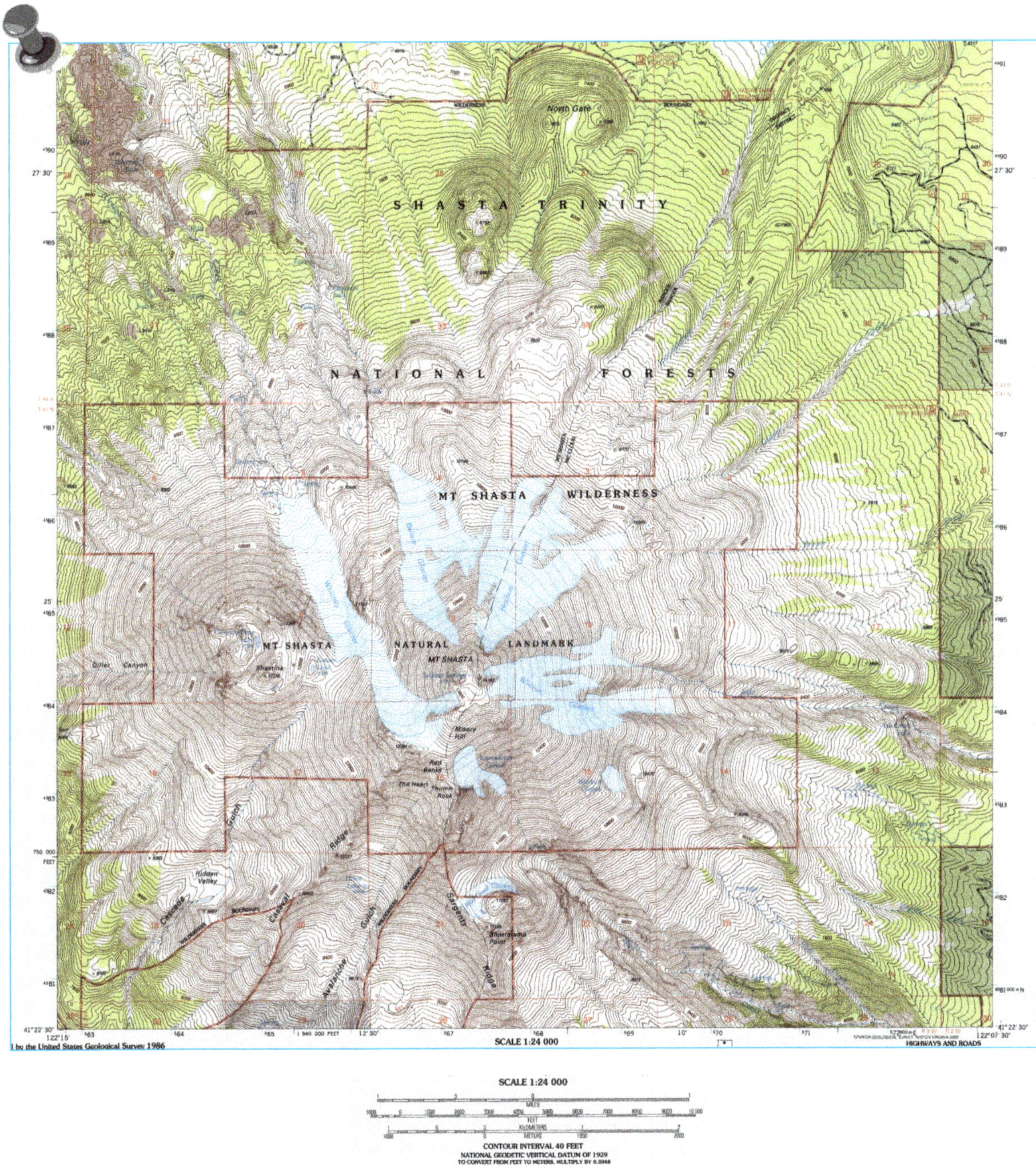

Map IV – Drainage Patterns

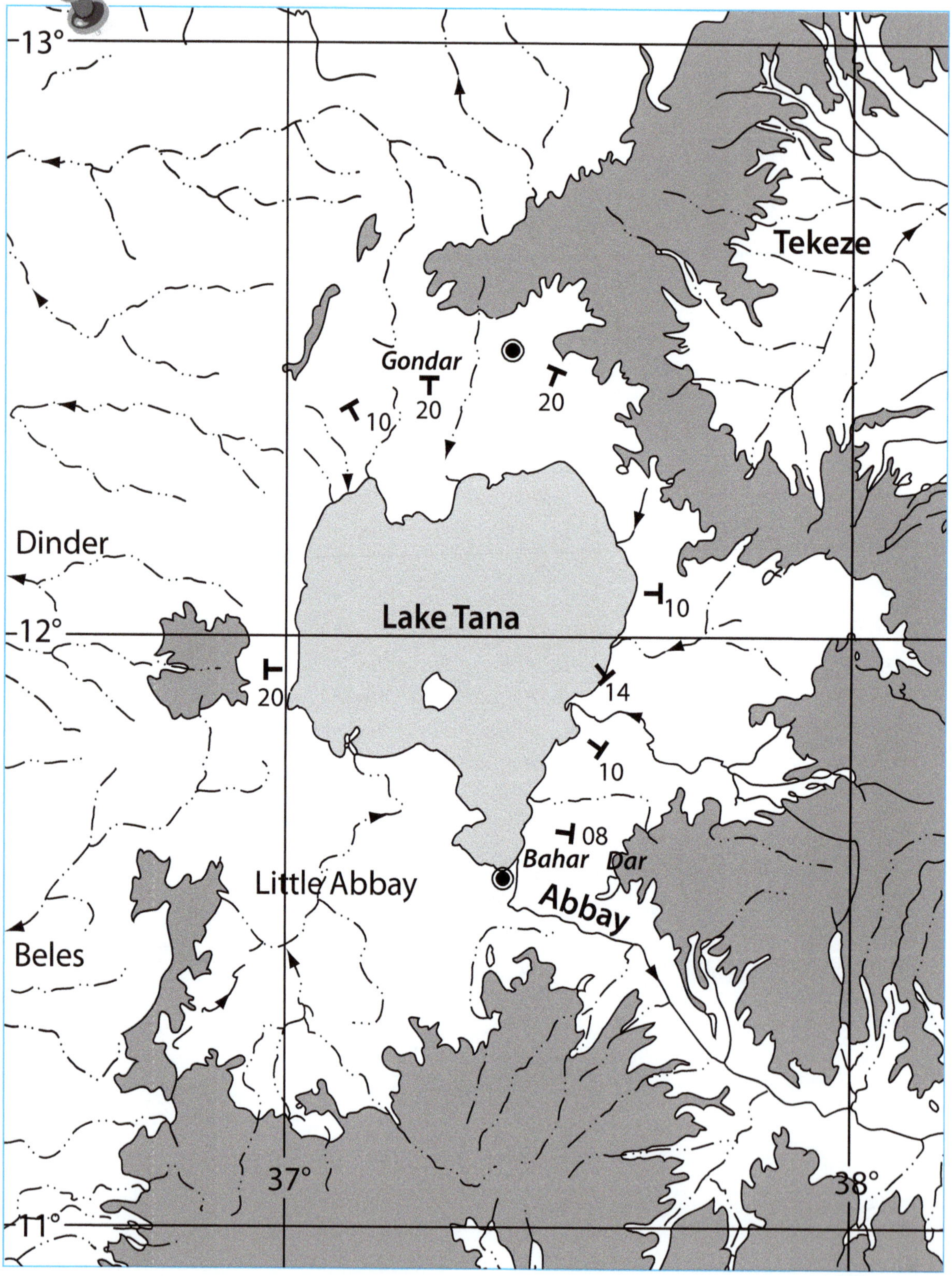

Map V.a – Drainage Patterns

Map V.b - Drainage Patterns

Map VI – Drainage Patterns

Map VII.a – Drainage Patterns

Map VII.b – Drainage Patterns

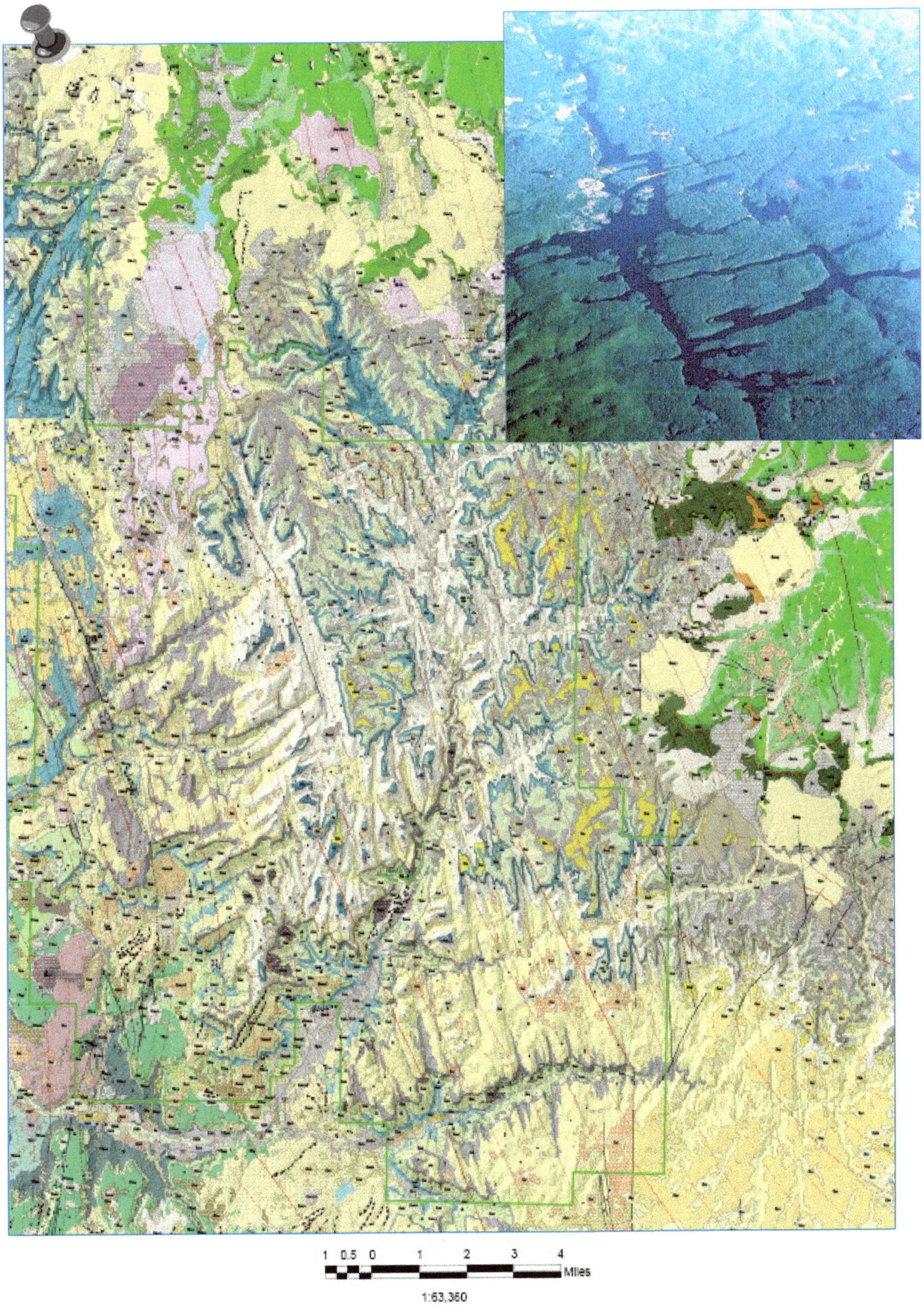

Map VIII – Ennis

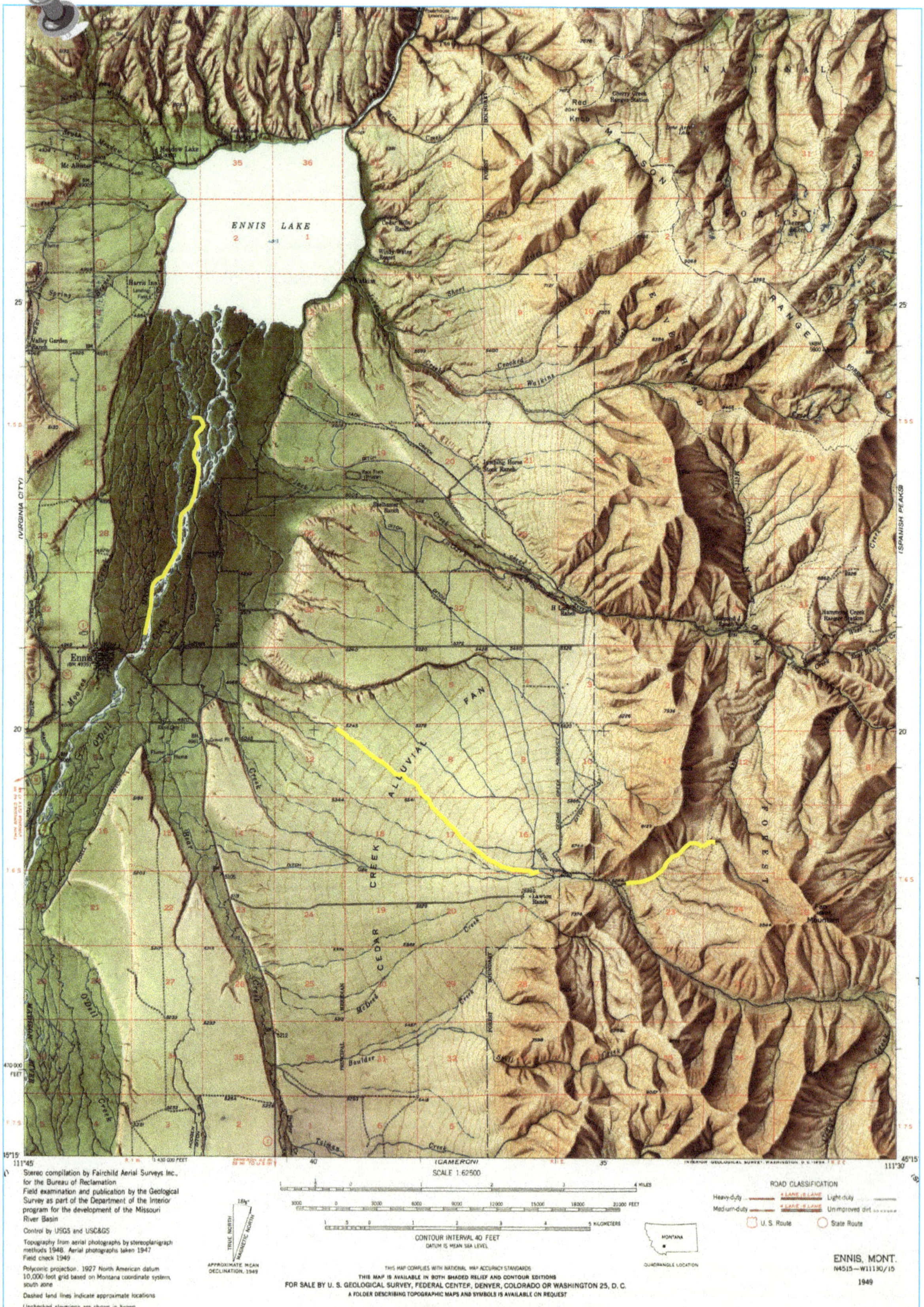

Map IX – Missouri River

Map X – Missouri River

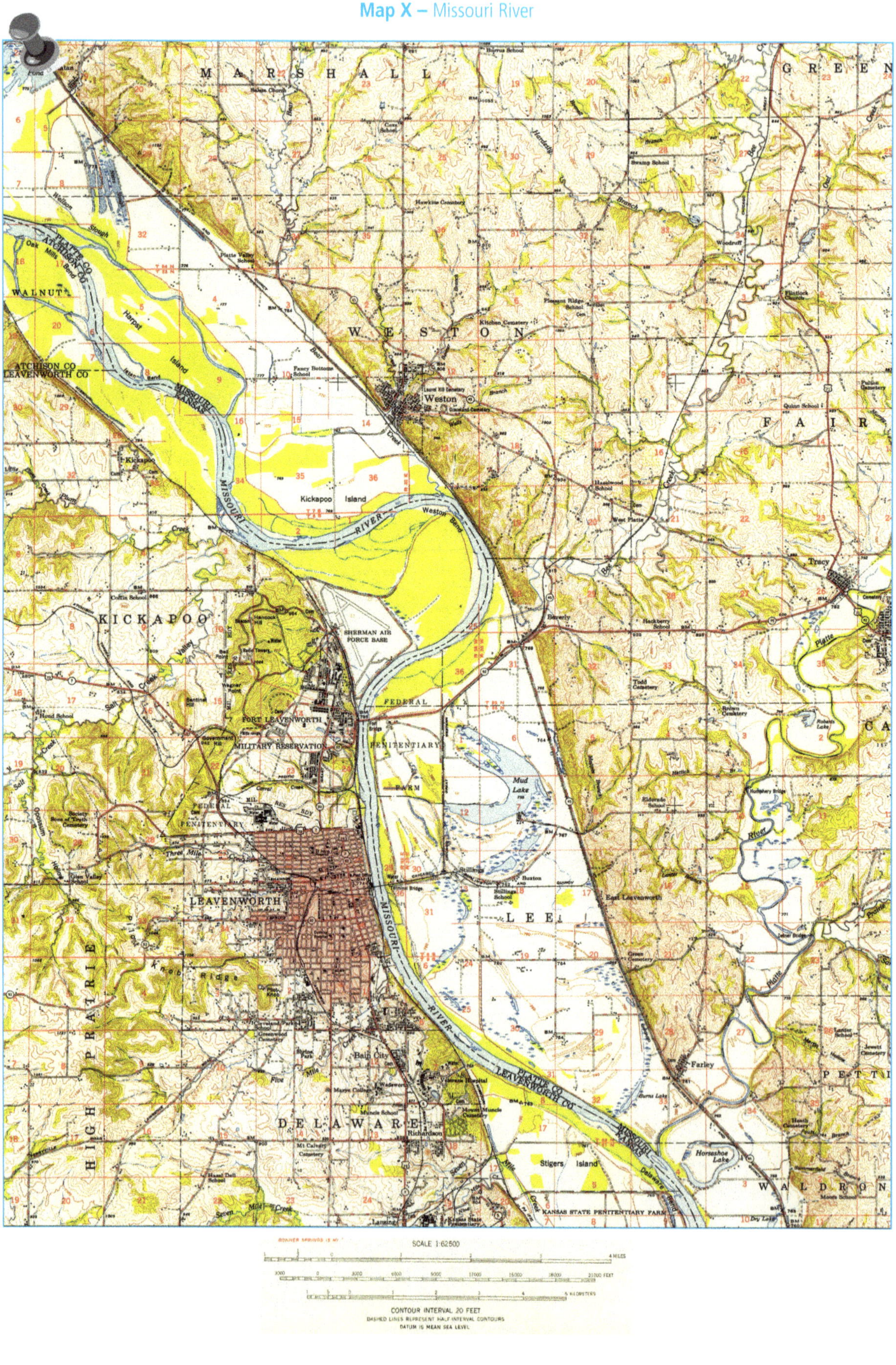

Map XI – Tower Creek

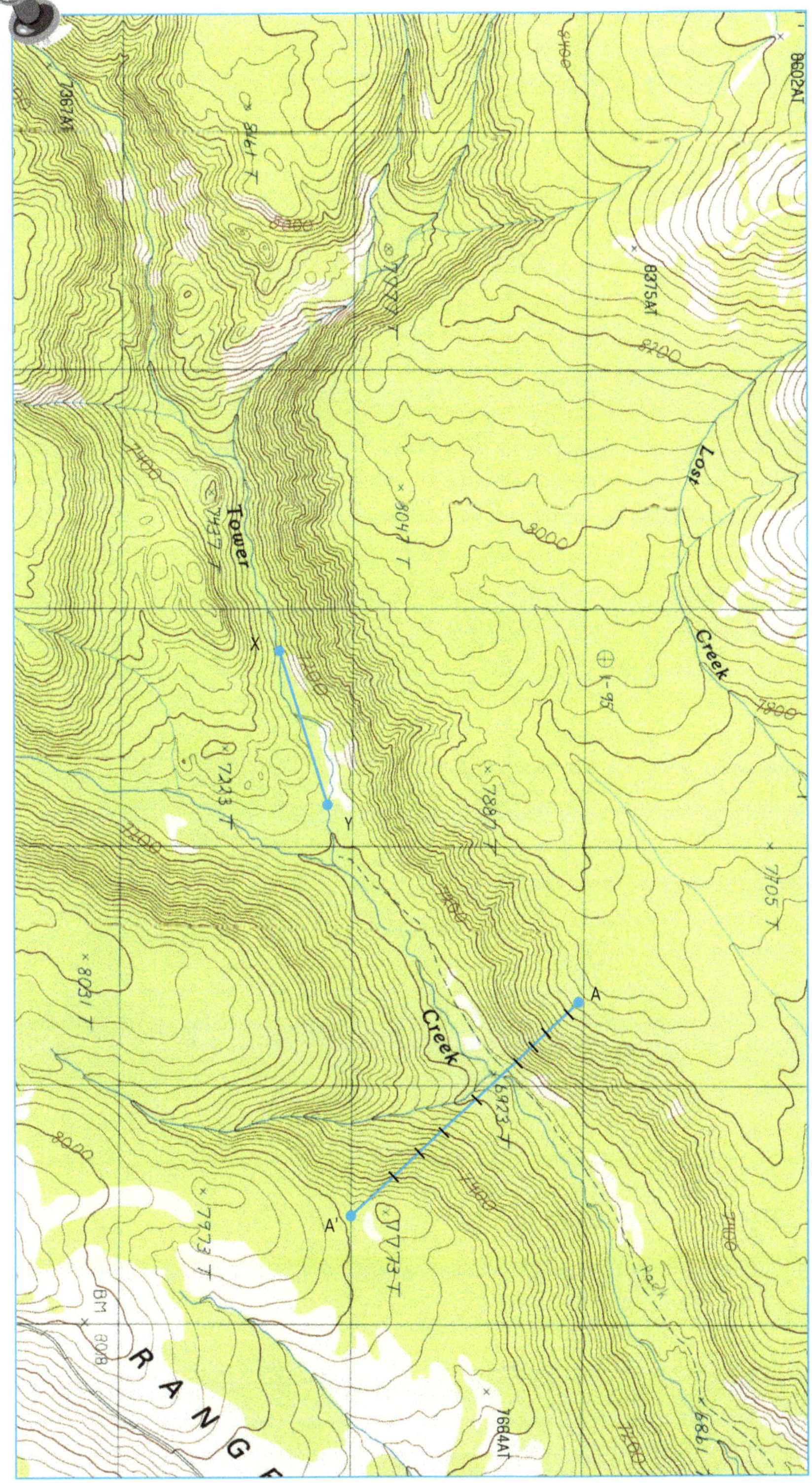

SCALE 1:24 000

1 .5 0 1
MILES

1000 0 1000 2000 3000 4000 5000 6000 7000 8000 9000 10 000
FEET

1 .5 0 1 2
KILOMETERS

1000 0 1000 2000
METERS

Map XII – Hanna

SCALE 1:24 000

INTERIOR—GEOLOGICAL SURVEY

KILOMETERS 1 5 0 1 2

METERS 1000 0 1000 2000

MILES 1 5 0 1

FEET 1000 0 1000 2000 3000 4000 5000 6000 7000 8000 9000 10 000

CONTOUR INTERVAL 10 FEET
SUPPLEMENTARY CONTOUR INTERVAL 5 FEET

*Due to the light coloring on this map elevations on the profile line have been supplied for you.

Map XIII – Loma West

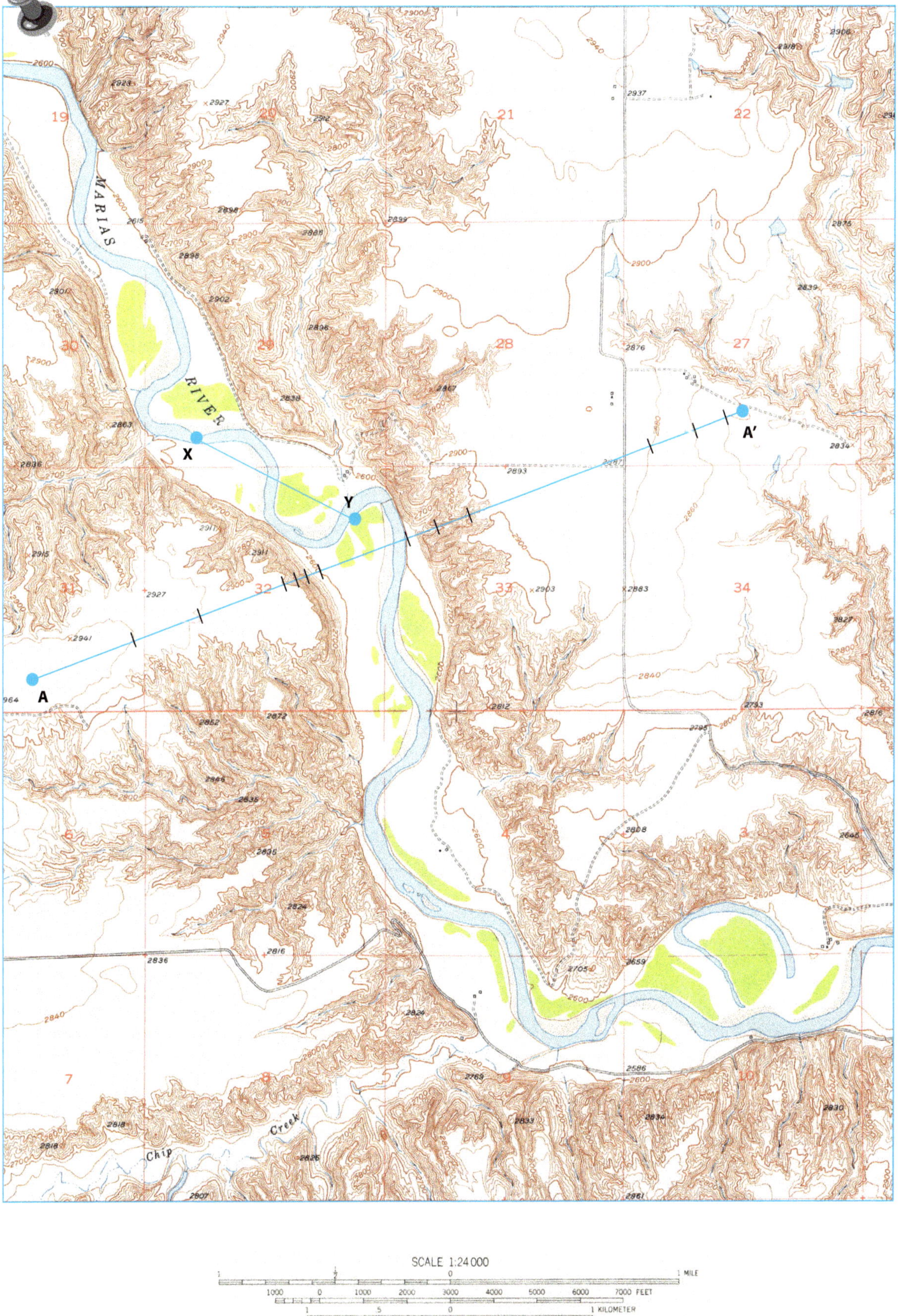

Map XIV – Anderson, CO

Map XV – Lake Scott

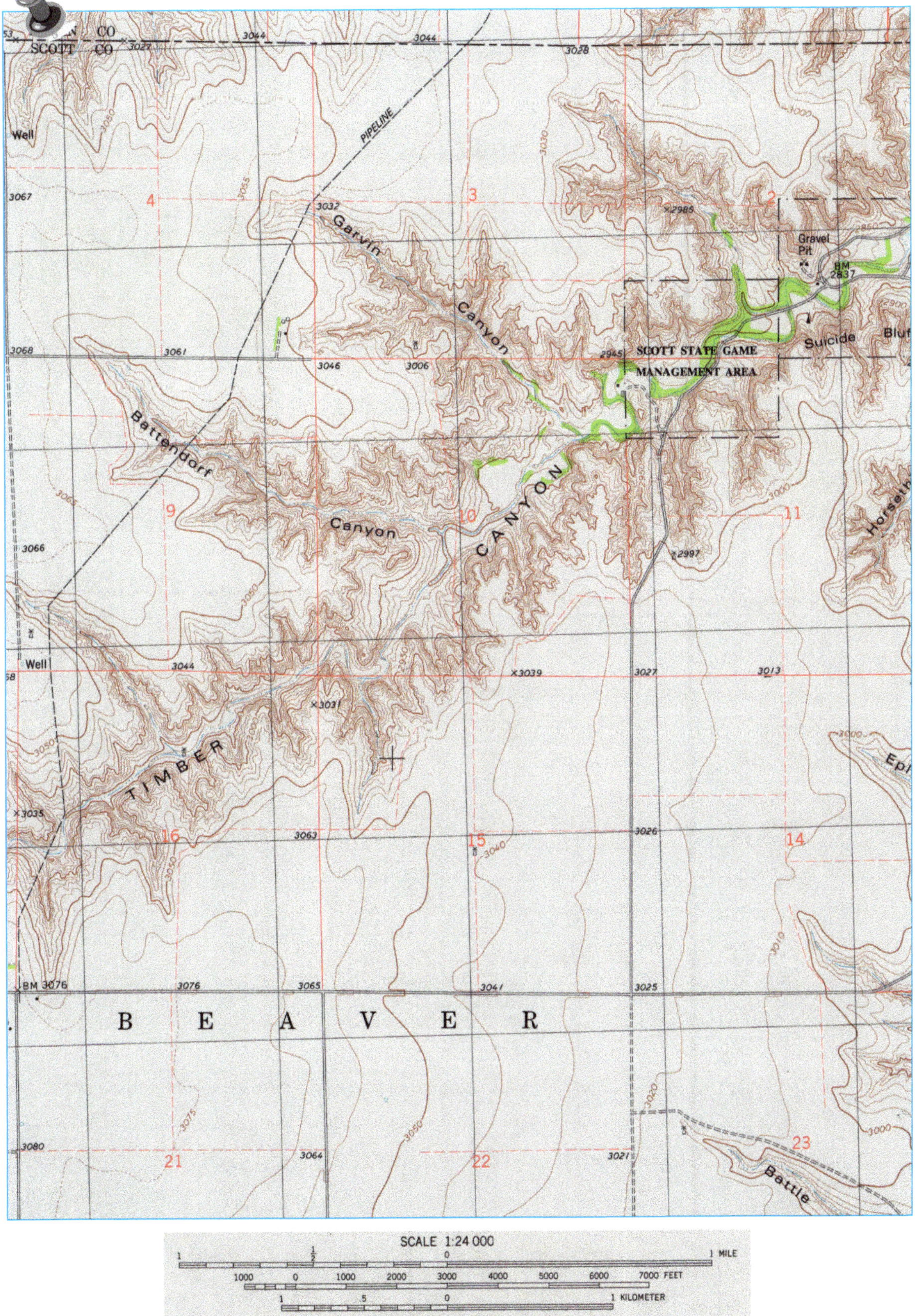

Map XVI – Voltaire

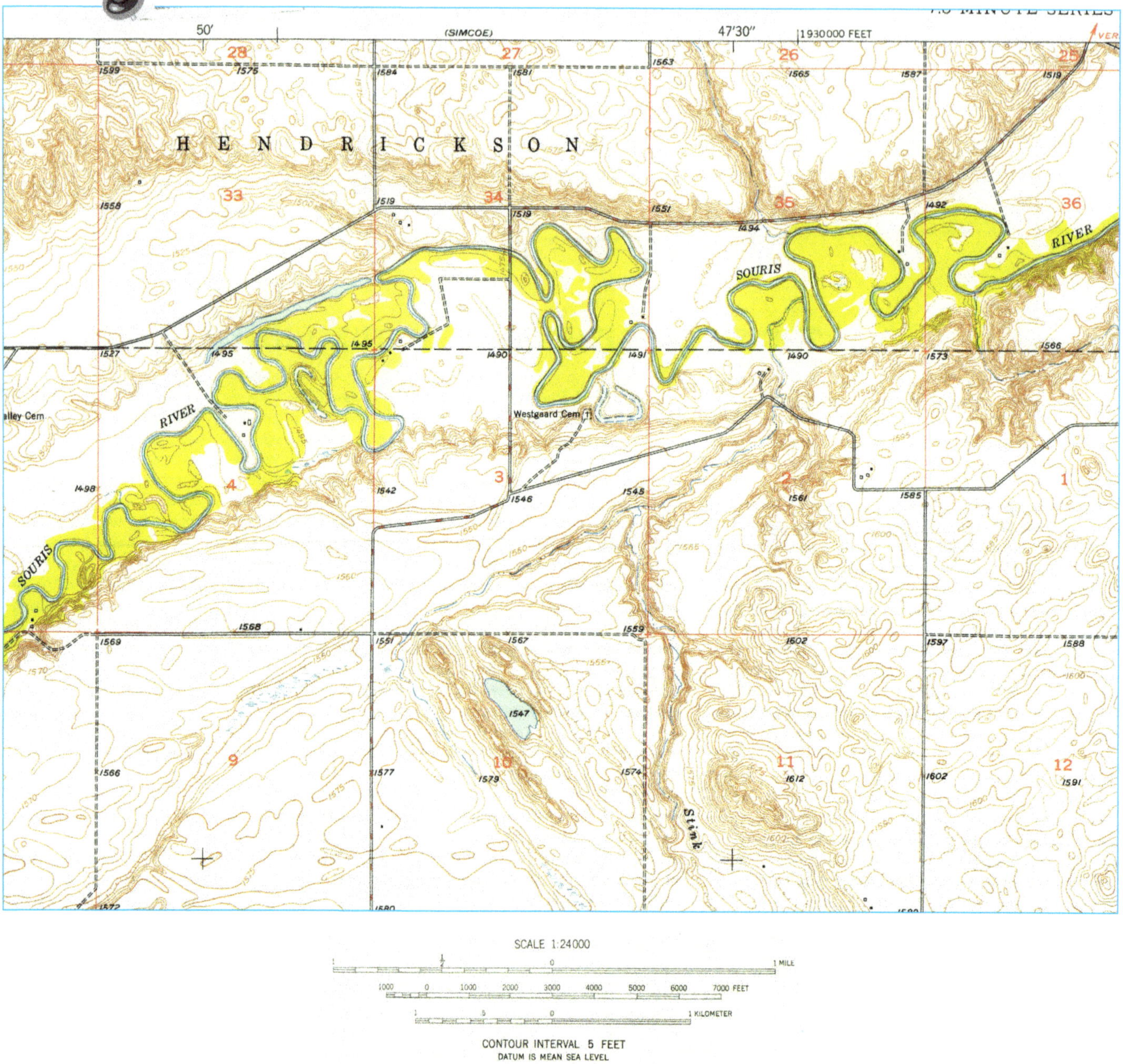

Map XVII – Dams Exercise

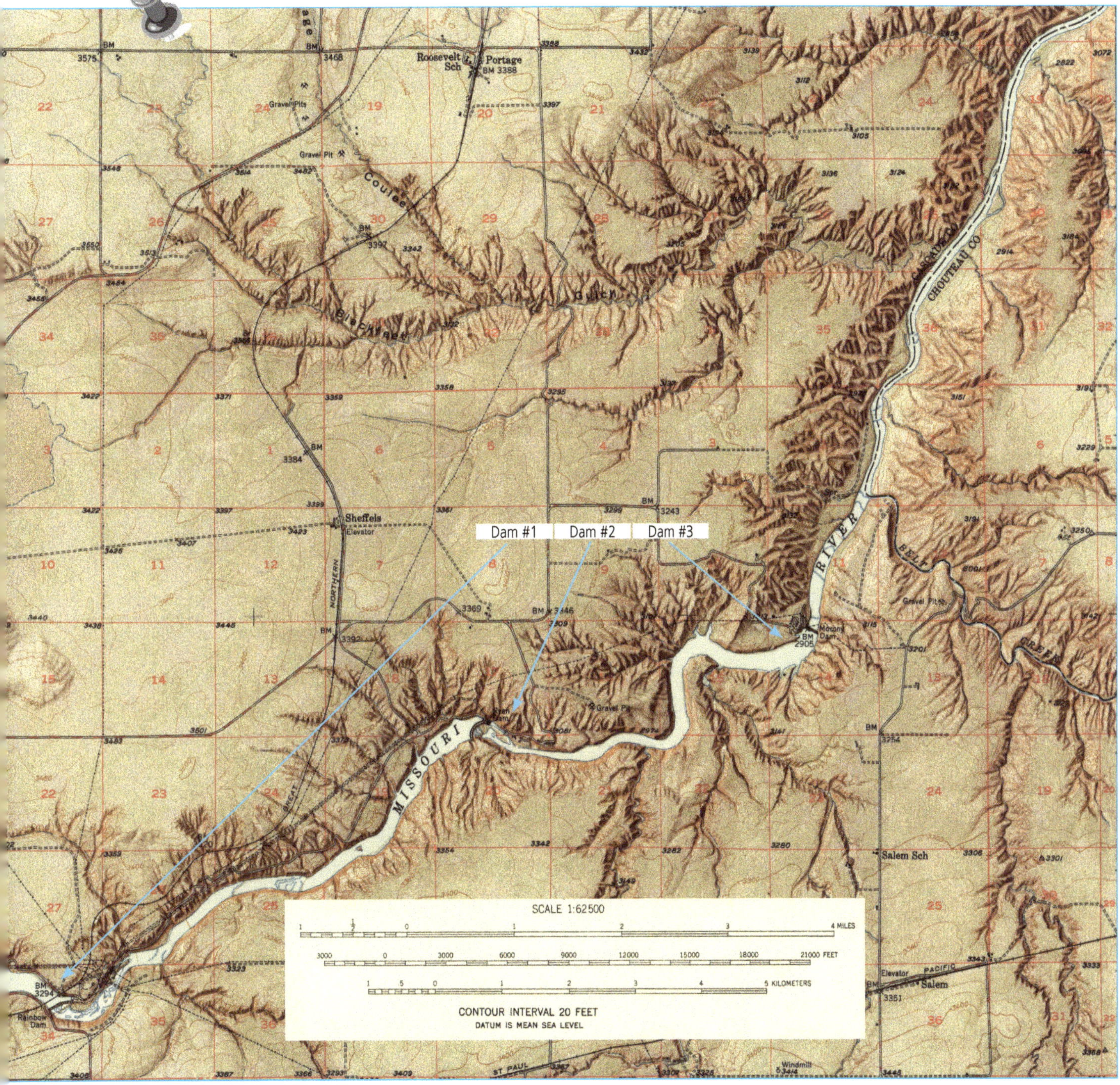

Laboratory Exercise XI

Groundwater

Introduction

The source of all fresh water on the planet is atmospheric precipitation. As part of the hydrologic cycle, the atmosphere evaporates salt water from the ocean and rains it back down onto the planet as fresh water, which then will: 1) become incorporated into the ice caps; 2) join the river systems; 3) be consumed by vegetation; or 4) filter into the ground and replenish the groundwater reservoir.

Groundwater is water contained below the surface within saturated sediment and rock. It arrives there by percolating into the ground and navigating downward through open spaces in the substrate. The area where spaces within the substrate are primarily occupied by air serves as a corridor through which the water must pass before reaching the groundwater reservoir. This relatively dry part of the subsurface is called the "zone of aeration" or the "unsaturated zone," as the openings are not fully filled with water. Due to the influence of gravity, the water will continue its downward descent until compaction of the sediment or the presence of an impermeable layer prevents additional movement. At this point, the water will pool and fill in open spaces between sediment grains and fractures within rock, generating an area known as the "zone of saturation." Water in this zone is called "groundwater" and the upper limit of this zone, which divides the zone of aeration from the zone of saturation, is called the "water table." See Figure 11.1.

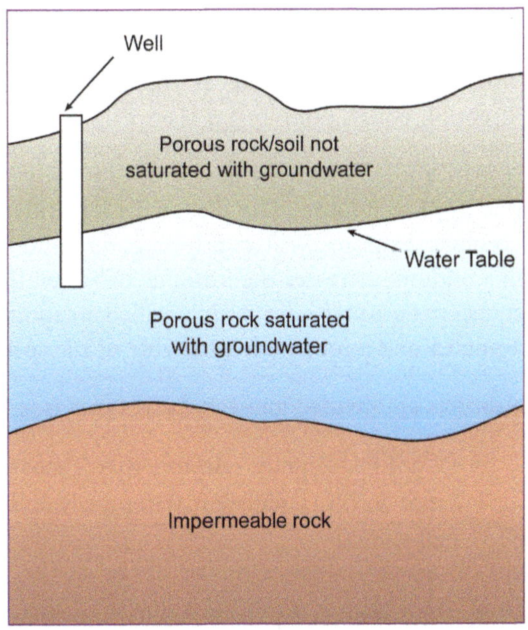

Figure 11.1 – Anatomy of the Groundwater System

Understanding where groundwater is housed, how and what rate it's transported, what techniques are available to bring it up to the surface, and how to responsibly use and preserve this precious resource is critically important—as groundwater is the primary supply of fresh water for human consumption.

OBJECTIVE

The objective of this lab is fourfold:

1) Investigate where groundwater is stored; 2) Develop an understanding of the factors that govern groundwater transport; 3) Examine the ways groundwater can be brought up to the surface and some ramifications of doing this irresponsibly, and 4) Explore the impact of groundwater on the landscape.

PART A: POROSITY, PERMEABILITY, AND AQUIFERS

Groundwater is housed and transported through "pores," which are fractures within rocks and between grains of unconsolidated material. "Porosity" is the volume of void spaces within and between the geologic material and is expressed as a percentage. While porosity defines the total amount of water that can be housed, not all the water in the pore is available for transport. When groundwater is subjected to the influence of gravity, most of the water is set in motion, but molecular forces and surface tension will retain some of the water in the pore. The volume of water that can be moved is called "specific yield" while the quantity of water retained in the pore is called "specific retention." Specific yield plus specific retention equals porosity; i.e. the total volume of water than can be housed in the voids.

"Permeability"- aka "hydraulic conductivity"- is a measure of the ease with which water moves from one pore to another. This, in turn is dependent on the quantity, size, and sinuosity of the conduits that connect the pores. Wider and straighter conduits promote higher permeability—while a higher quantity of conduits increases the amount of water that can be transported. A body of rock or sediment that transports groundwater easily is called an aquifer, one that grudgingly transports groundwater is called an aquitard, and a rock body incapable of transporting groundwater is called an aquiclude.

Aquifers are divided into two basic types (see Figure 11.2):

A. Confined Aquifer - Aquifer that is separated from the surface by an "aquitard" - sediments or rocks that do NOT transport water easily and therefore retard the motion of water (ex: mudstone or claystone)

B. Unconfined Aquifer – Aquifer in which there is no confining layer between the water table and the surface. – usually composed of "Alluvium" - the name given to sediments that have been weathered and eroded by some form of water then deposited in a non-marine setting. Typically made of sand and gravel.

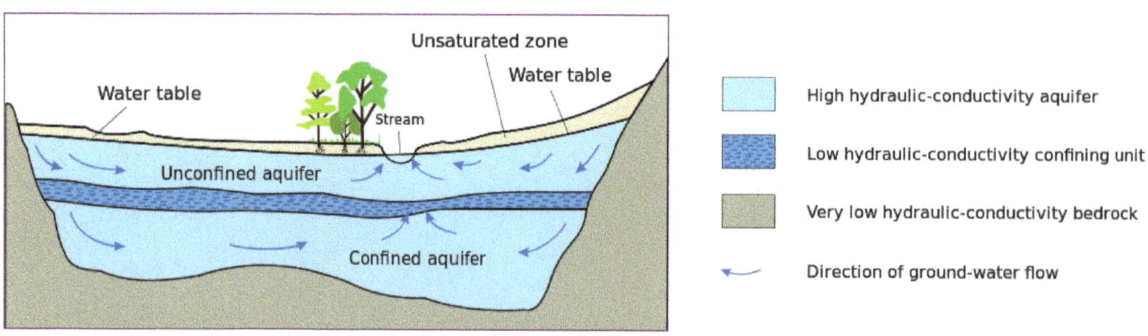

Figure 11.2 – Confined and Unconfined Aquifers

Bearing in mind that the ability of a geologic substrate to be a good aquifer is dependent primarily on its permeability, it's important to understand that there is a distinct difference between a material that is "porous" and a material that is "permeable." See Figures 11.3a and 11.3b..

- Porosity—The percentage of open pore spaces (empty space) in a given volume of rock or sediment that determines the total amount of water a material will hold. Porosity is largely influenced by particle size, shape, sorting, and compaction.

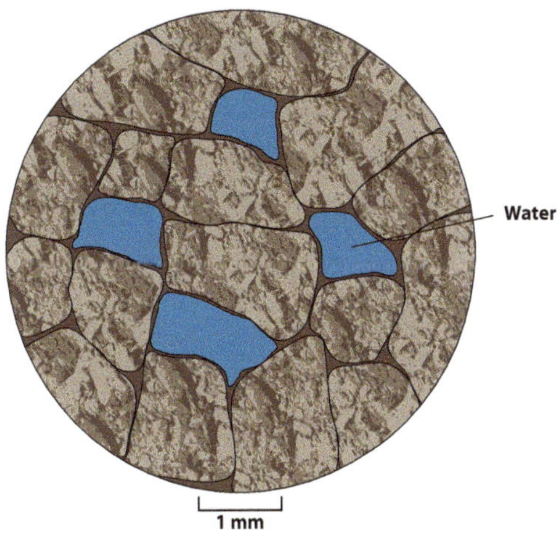

Figure 11.3a – Porosity

- Permeability—The ability of a material to allow fluids to pass through an **interconnected** network of pores.

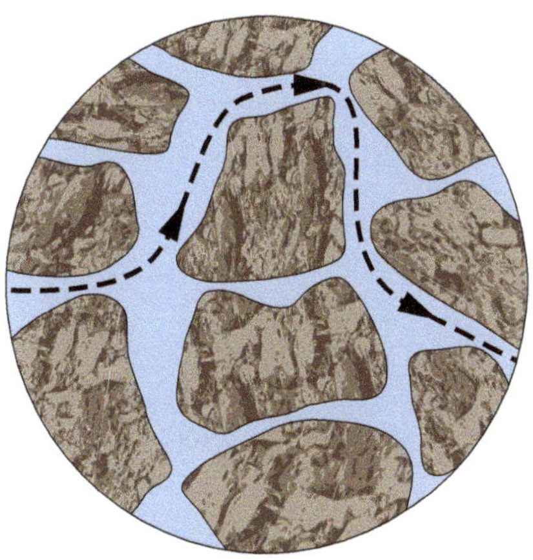

Figure 11.3b – Permeability

A material can be very porous, but if the pores are not interconnected, groundwater cannot flow through material, and the material will therefore be a poor aquifer. As such, in order for a material to make a good aquifer, it must be porous AND permeable. Porous materials that are not permeable are not aquifers.

Exercise: What substrate makes the best aquifer?

Table 11.A depicts a list of common geologic materials and their porosity and permeability. Refer to Table 11.A when answering the following questions:

	Geologic Material	Hydraulic Conductivity (lower limit) (m/yr)	Hydraulic Conductivity (upper limit) (m/yr)	Average Hydraulic Conductivity (m/yr)	Average % Porosity
Unconsolidated Sediment	Well-Sorted Gravel	10000	10000000		37.5
	Well-Sorted Sand	100	100000		37.5
	Silty Sand	10	10000		15
	Silt	0.01	100		42.5
	Glacial Till	0.00001	10		15
	Clay	0.00001	0.01		50
Consolidated Rock	Unfractured Igneous and Meta Rock	0.0000001	0.0001		1
	Shale	0.000001	0.01		5
	Sandstone	0.0001	10		29
	Fresh Limestone	0.01	10		5
	Karst Limestone	10	100000		n/a

Table 11.A

1. Calculate the average hydraulic conductivity for the geologic materials listed above and place your answers in the column provided.

2. Considering only the uncolidated samples that contain some quantity of medium and/or coarse grains (i.e. gravel, sand, silty sand, glacial till), based on the above data, which type of sediment makes a better aquifer—sorted or unsorted? ____________________. Explain why this is the case. ____

____

3. Using the average hydraulic conductivities, calculate how much more efficient Karst limestone is at transporting groundwater than fresh, unweathered limestone. ____________ x. Explain why this is the case ____

____

4. Graph 11.I depicts the porosity and permeability of some of the most common surface materials. Using that graph, answer the following questions.

   a. Is there any clearly quantifiable relationship between porosity and permeability of a given geologic material? Yes or No?

      Explain: ____

   b. Based on the graph and your answer to the question above, does a higher porosity necessarily mean higher permeability? Yes or No?

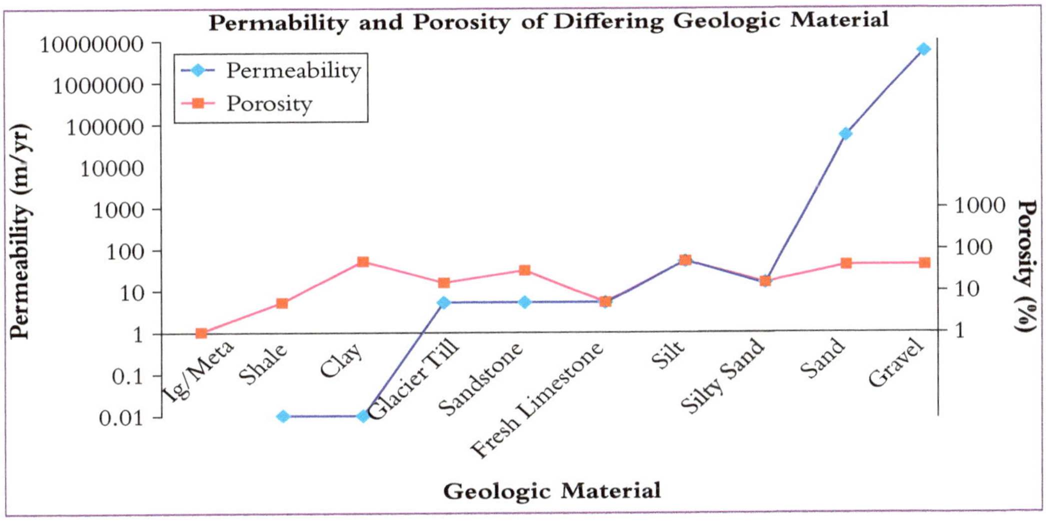

Graph 11.1

5. Given that there are infinite varieties of sediment and rock, each of which depicts different properties from the next, if you were a consulting geologist in search of a lucrative aquifer to supply a growing city, what is the primary property you would evaluate in the subsurface geology? ______________________

6. If an aquifer is 25 ft thick and has a specific yield of 15%, how much water is stored in a .5-mile square area? *Show your work* __________ ft^3

Now multiply by your answer by 7.48 to convert to gallons: __________ gallons

7. What additional information would you need in order to calculate if the aquifer in the question above is capable of supplying water to the surrounding community in a sustainable manner?
____
____

PART B: UNDERSTANDING GROUNDWATER TRANSPORT

The velocity of groundwater transport is calculated by multiplying the permeability of the material by the hydraulic gradient, then dividing by the specific yield. Oftentimes, the permeability is expressed in units such as ft/day or cm/sec, but the process of groundwater transport is easier to conceptualize when working with more familiar units. As such, the following equation permits you to calculate groundwater velocity using a permeability value (p) that is expressed in gallons per day per square foot (gpd/ft^2).

The hydraulic gradient (g) is the difference in water level per unit of distance in a given direction (e.g., the slope of the water table). Just like a slope on the surface, the hydraulic gradient can be measured directly from water-level maps in ft/ft or ft/mile. Using the fact that there are 7.48 gal per cubic foot, the following formula will allow you to calculate groundwater velocity in ft/day, where the specific yield (y) is given as a decimal (i.e., 10% = .10) and (g) is in ft/ft.

$$v = \frac{p \text{ x } g}{7.48y}$$

With respect to the direction of flow, as a general rule, the water table forms a gently sloping surface that mimics the land surface. As such, the water table can be mapped and contoured just like surface topography. In this case, however, control points are water elevations in wells, springs, lakes, or streams, as opposed to surface elevations. The direction of groundwater flow can be indicated by flow lines, **which are drawn perpendicular to water-level contour lines in the direction of flow**.

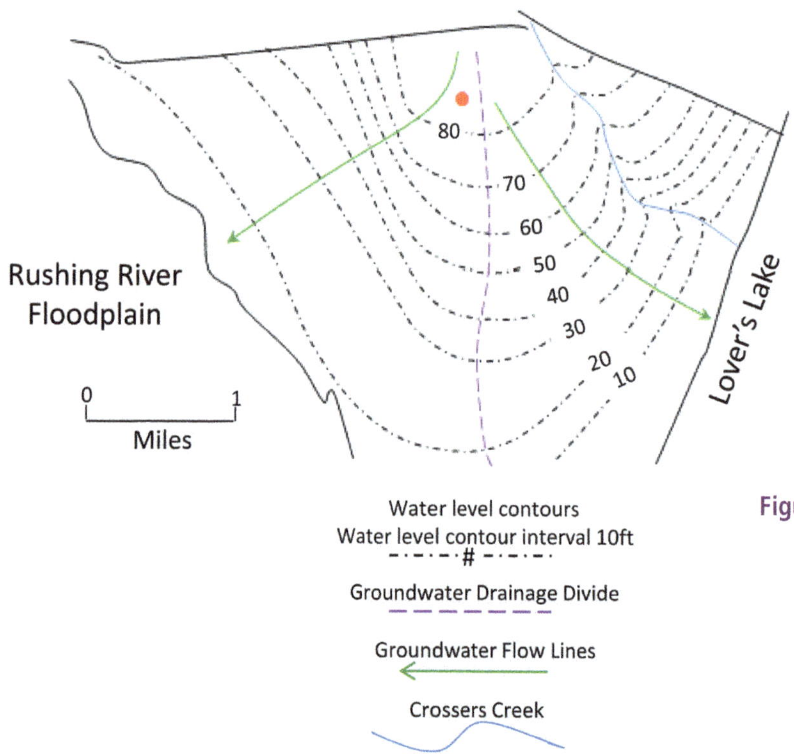

Figure 11.4 – Note: Elevations of water table are feet above sea level

Exercise: Groundwater flow—What Direction and at What Rate?

Use Figure 11.4 to answer the following questions. SHOW YOUR WORK.

1. Flow lines are used to depict the direction of groundwater flow and **are drawn at right angles to the contour lines**. Using a red pen or pencil, construct a flow line beginning at the red dot on Figure 11.4. In what direction is the water moving? ______________

2. If industrial contaminants spilled at the spot marked by the red dot, which body of water would it contaminate? ______________

3. What is the groundwater velocity of the flow line you just drew, given that the permeability of the aquifer is 750 gpd/ft^2 and the specific yield is 20%. *Show your work* _____________ ft/day

4. Based on the groundwater velocity calculated in the question above, approximately how long do scientists have to neutralize the pollution before it reaches the main body of water? *Show your work* _______________ days

PART C: GETTING WATER UP TO THE SURFACE

Groundwater moves slowly down gradient through pore spaces between the sediment until it reaches a point where the surface intersects the water table. There it can naturally discharge to the surface in the form of a stream, spring, lake, or wetland, while new atmospheric precipitation seeps into the ground at locations known as recharge sites and replenishes the groundwater lost to the surface. However, groundwater can also be prematurely withdrawn from the subsurface and brought up for human consumption through manmade devices called wells.

Wells are holes that people dig or drill such that water can be withdrawn as needed at a rate controlled by the demands of the population. The wells are designed to penetrate through the zone of aeration and into the zone of saturation—so water can be continuously withdrawn as long as the water table remains higher than the base of the well. Wells are classified into two basic categories: Ordinary and Artesian. See Figure 11.5.

Ordinary wells are constructed by digging a hole into the aquifer that infiltrates deeper than the water table. Water flowing through the aquifer then permeates the well and fills it, such that the water surface in the well is even with the water table. In order to get water from the well to the surface, it must be manually withdrawn, which can be done with anything from buckets to state-of-the-art pumps. The amount of water artificially discharged (withdrawn) from a well of this type must be carefully monitored to ensure there is adequate recharge replacing it, or the water table will drop and eventually cause the well to go dry, in addition to inducing a plethora of environmental problems.

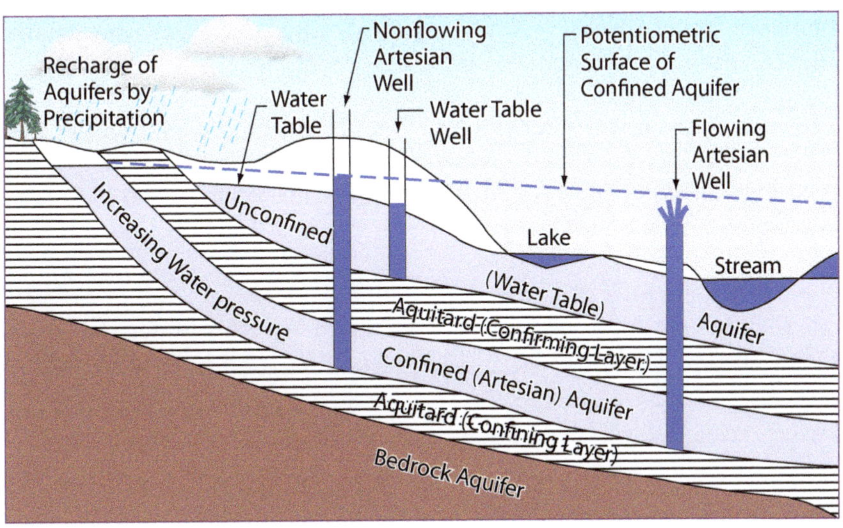

Figure 11.5 – Ordinary and Artesian Wells

Artesian wells differ from ordinary wells in that water is able to rise at least part of the way to the surface without any pumping or other forms of manual withdrawal. Artesian wells require very specific geologic conditions in which an aquifer lies beneath an aquitard and both are sloping away from the recharge area. A setup of this type puts the water in the aquifer under tremendous pressure and the pressure alone will push water up through any vertical opening, including a manmade well. In some cases, the pressure is sufficient to drive the water all the way to the surface, where it generates a "flowing artesian well." In other cases, the water is not pushed high enough to reach the surface and results in "non-flowing" artesian wells. Which one develops is dependent on the location of the "potentiometric surface"—an imaginary line where a given reservoir of fluid will equilibrate if allowed to flow. Once the ascending water reaches the potentiometric surface, the pressure equalizes and the water will go no higher.

Exercise 1: Ordinary Wells and the Water Table

Figure 11.6 depicts the topography of an area characterized by flowing surface water, which is underlain by a shallow, unconfined aquifer. Use Figure 11.6 to answer the following questions.

1. On the graph provided, construct a topographic profile of Fig 11.6 from the NE corner to the SW corner of the map.

2. Shallows holes were drilled throughout the landscape in order to determine the depth of the water table below the surface. The locations that were sampled are labeled 1–30 in Fig 11.6 and the depth to the water table for each location is listed in Table 11.B. For each of the sampled locations, calculate the water table elevation in feet above sea level by following the procedure below:

   a. Determine the surface elevation at samples sites 1–15 on Fig 11.6. (Note: record the surface elevation using 5-ft contour measurements, i.e., 455, 460, 465, etc.) Sample sites 16–30 have been filled in for you.

   b. Subtract the water depth from the surface elevation for all the sample sites to calculate the water table elevation above sea level.

   Record your answers for both the surface elevation and the elevation of the water above sea level in spaces provided in Table 11.B.

3. Next, transfer the water table elevations you just calculated onto Fig 11.6, such that each sample site depicts the elevation of the water table in that location.

4. Using a red pen, contour the water elevation lines in 20-ft intervals, beginning with 350.

5. On the same graph paper you used in #1 and using the same scale as before, construct a second topographic profile, from the NE corner to SW corner of Fig 11.6, this time using the water table contours.

6. What is the relationship between the elevation of the water table and the topography of the surface landscape? ________________________. Explain why this is the case.
   ____
   ____
   ____
   ____

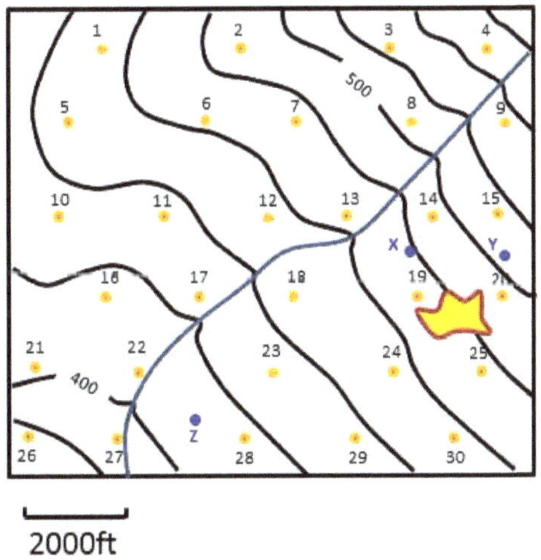

Figure 11.6 and corresponding graph

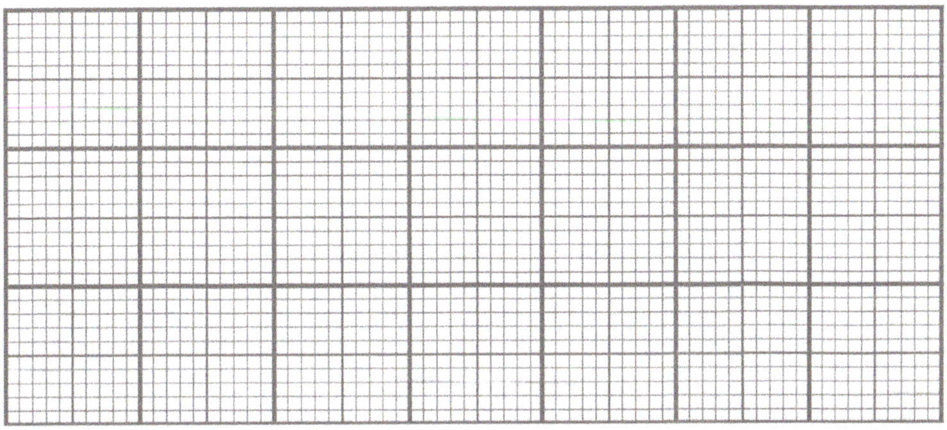

Site #	Surface Elevation	Depth to Water Table (Ft)	Water Table Elevation (ft above sea level)	Site #	Surface Elevation	Depth to Water Table (Ft)	Water Table Elevation (ft above sea level)
1	455	16	439	16	415	13	
2		14		17	425	14	
3		21		18	445	16	
4		24		19	470	18	
5		15		20	495	20	
6		16		21	405	12	
7		18		22	410	12	
8		20		23	430	14	
9		23		24	455	17	
10		15		25	475	19	
11		14		26	375	25	
12		17		27	390	21	
13		17		28	415	18	
14		20		29	435	15	
15		22		30	455	18	

Table 11.B Sample Site Water Level Data

7. Remember that groundwater flow occurs at right angles to the water elevation contour lines. Keeping this in mind, draw arrows on Fig 11.6 to indicate the direction of groundwater flow in the aquifer.

8. Calculate the average hydraulic gradient between the NE and SW corners of the map in ft/ft.
_______________ft/ft (*Show your work*)

9. Using the equation below and the values in Table 11.C, calculate the groundwater velocity in each of the four substrates represented in Figure 11.7. Record your answers in spaces provided in Table 11.C Note: you will need to calculate the hydraulic gradient for each individual substrate before you can calculate the velocity. Do NOT use the average rate you calculated in the question above.

$$V = \frac{\text{Hydraulic conductivity (ft/day) x hydraulic gradient (ft/ft)}}{\text{Specific Yield}}$$

Sediment Type	Hydraulic Conductivity (ft/yr)	Hydraulic Gradient (ft/ft)	Specific Yield (%)	Velocity (ft/day)
Gravel	219,000		22	
Sandy Gravel	365		25	
Coarse Sand	146,000		27	
Fine Sand	3,650		21	

Table 11.C Effect of Sediment Type on Groundwater Movement and Yield

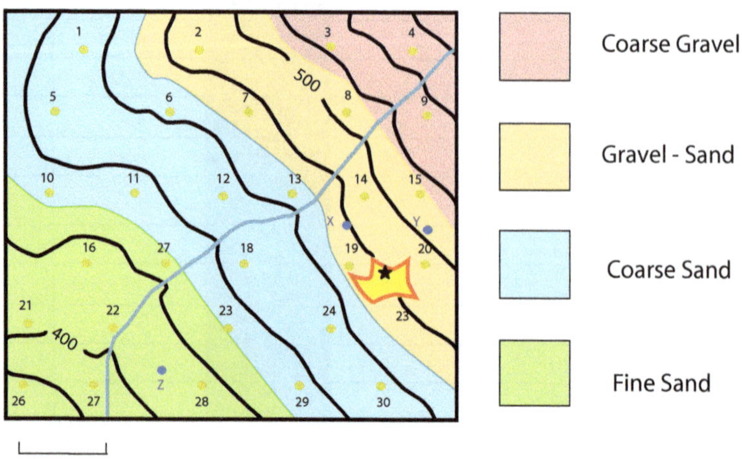

Figure 11.7 – Geological Map of the Aquifer

10. The yellow pond in Figure 11.7 marks the boundary of a landfill that has been leaking contaminants into the groundwater system from a spot marked by the black star. There are three wells throughout this area that are used by surrounding communities for their municipal water supply. Which if the three wells is at risk for being contaminated: X, Y, or Z? (circle one)

11. Using the values from question 9, calculate approximately how long will it take the contaminant to get from the contamination site to the well you chose in question 10? Show your work _________ days

Exercise 2: Artesian Wells and the Water Table

Recall that artesian wells are wells that are confined from above and are therefore under significant pressure. As such, when the confining layer is punctured by a well, the water naturally ascends until it reaches the potentiometric surface, where it will equilibrate, stabilize, and terminate any further ascent. Wells that have a potentiometric surface that is even with or higher than the land surface result in water that freely flows on the landscape and are called "flowing artesian wells." Wells that have a potentiometric surface lower than the land surface are called "non-flowing artesian wells," for while the water does rise above the water table, it does not reach the land surface and therefore does not freely flow.

Figure 11.8 depicts a landscape that overlays an artesian aquifer. Use Figure 11.8 to answer the following questions.

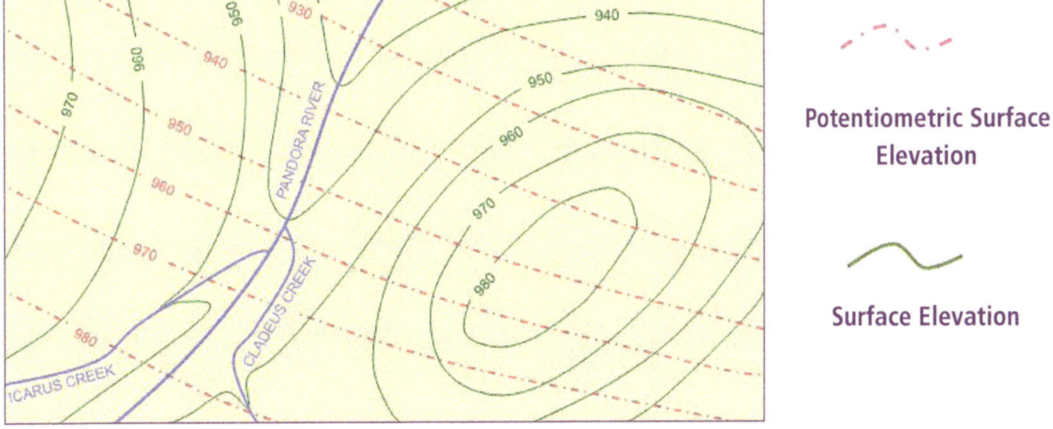

Figure 11.8 – Artesian Aquifer

1. Using a red pen or pencil, outline and shade in the area on Figure 11.8 that contains flowing artesian wells.

2. Considering the relationship between the potentiometric surface and the surface topography, put an "X" on Figure 11.8 in the most lucrative spot to drill a water well.

3. How many feet above the surface would water spurt at the site you chose for #2? ________ft

4. Place a "P" on the spot that would require the most pumping.

5. How many feet would the water need to be pumped at the site you chose for question #4? __________ ft

6. In what direction is the aquifer and confining layer sloping? ______________. How do you know?
____
____

7. Based on your answer to #6, in what direction is the recharge area located? ________________

PART D: OOPS—WE OVER-PUMPED

While wells are a necessary part of anthropogenic sustenance, care must be taken that the amount of water being withdrawn is replaced at a comparable rate or a plethora of negative changes will chronically plague the landscape. Examples of such problems include, but are not limited to, saltwater intrusion of coastal aquifers, drying up of the surface water, flow reversals, and ground subsidence. The following exercise is designed to demonstrate the drastic effects that over-pumping, even on a small scale, has on the landscape.

Exercise 1: Groundwater and Flow Reversals

Figure 11.9 is a schematic representation of an aquifer that is being tapped for groundwater to supply a nearby developing city. Figure 11.10 shows the same area after water was irresponsibly over-pumped from one of the wells. Use figures 11.9 and 11.10 to answer the following questions.

1. Using 5-ft contour intervals, contour the water table on Figure 11.9 *and* draw in two or three arrows to show the direction of flow.

2. What is the average hydraulic gradient of the water table? ________ft/ft

3. Approximately how many feet below the surface will they have to drill in order to reach the water table at the proposed well site? ___________ft

4. Should they stop digging upon reaching the water table? Yes or No? Explain.
____
____

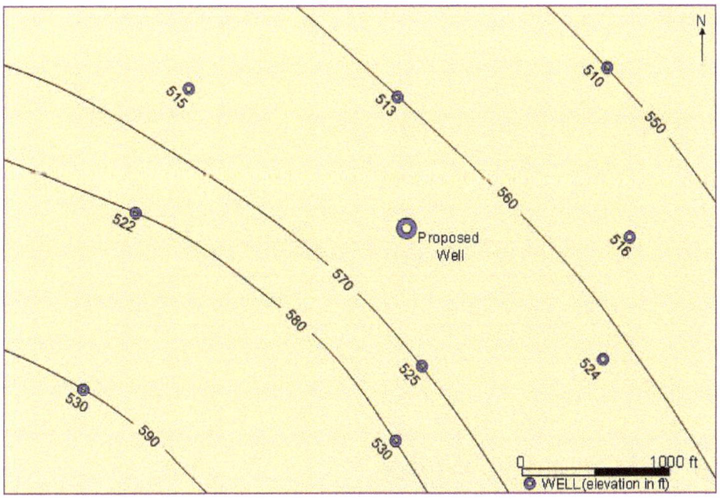

Figure 11.9 – Elevation of the water tables in the wells before over-pumping

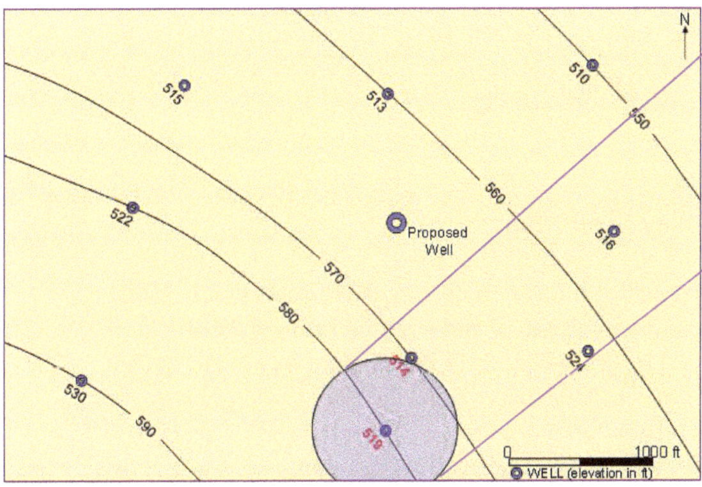

Figure 11.10 – Elevation of the water tables in the wells after over-pumping

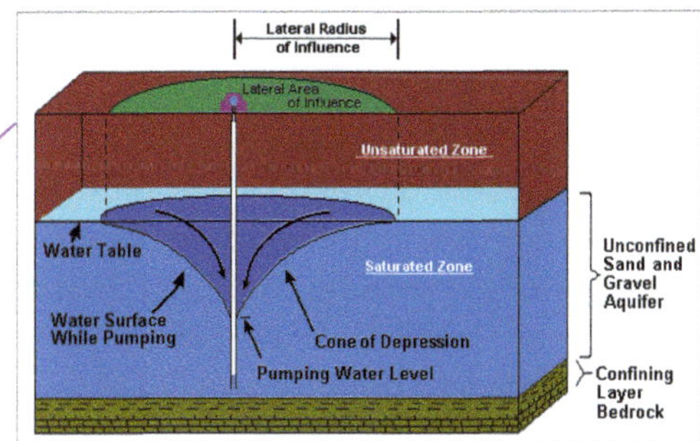

Note: The zone around a well in an unconfined aquifer is normally saturated, but becomes unsaturated during pumping. This occurs because as the water level in the well is lowered, a gradient is created between the water in the surrounding aquifer and the water in the well. Because water flows from high to low water levels, i.e., high to low pressure, this gradient results in water flowing from the surrounding aquifer into the well, resulting in an area where the water table dips down to form a cone shape known as the "cone of depression."

Figure 11.10 depicts water levels in the wells of the same aquifer after a well in the south central part of the aquifer was over-pumped. The water table in and around that well dropped 11 ft and affected the entire area within a radius of 500 feet from the well. Affected area is highlighted in purple.

5. Notice that a well to the NE of the over-pumped well also experienced a significant water table drop despite the fact it was not over-pumped. Explain why this occurred.

____

____

____

6. Using 5-ft contour intervals, contour the water table on Figure 11.10 and draw in two or three arrows to show the direction of flow.

7. Is the water still flowing in the same direction as before? Yes or No? If no, what's changed?
____

8. What is the average hydraulic gradient of the water table? _______ft/ft

9. How are the following changes going to affect the other wells in the aquifer?

   Change in Flow Direction: ____
   ____

   Change in Hydraulic Gradient: ____
   ____

Exercise 2: Groundwater and Ground Subsidence

Thus far we've discussed groundwater's use of the pore spaces within the sediment as a residence and transportation route, but it must also be recognized that the sediment relies on the groundwater as a source of support, which keeps the grains from compacting and prevents the surface from subsiding. The high fluid pressure that groundwater exerts adequately supports the materials above, such that the grains are essentially locked in place. If, however, fluid is removed from between the grains without replacement in sufficient time, support is reduced and subsidence will result. See Figure 11.11.

Santa Clara Valley, California

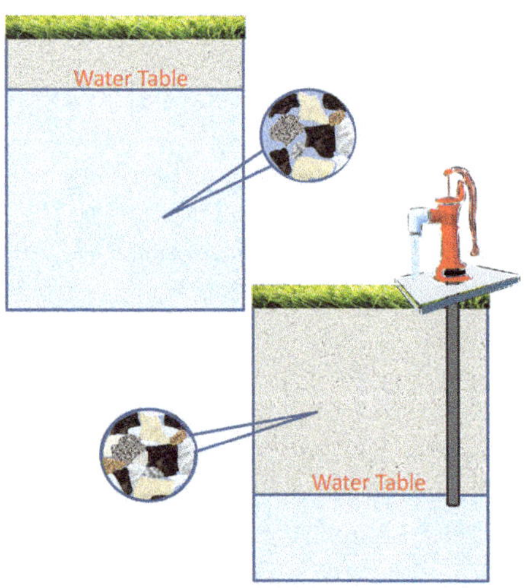

Grains compact when fluid is removed

Figure 11.11 – Ground subsidence

In the first half of the twentieth century, the Santa Clara Valley was intensively cultivated and was nicknamed "Garden of Eden" or "The Valley of Heart's Delight" for its myriad of fruit orchards and vegetable harvests. As the population grew, so did the demand for food; between 1920 and 1960, an average of about 100,000 acre-feet per year of groundwater was used to irrigate crops.

Note: Hydrologists frequently use the term "acre-foot" to describe a volume of water. One acre-foot is the volume of water that will cover an area of one acre to a depth of one foot. The term is especially useful where large volumes of water are being described. One acre-foot is equivalent to 43,560 cubic feet or about 325,829 gallons!

The period from the end of World War II to approximately 1970 was a time of rapid population growth that was associated with the transition from an agriculturally based economy to an industrial and urban one. Nonagricultural use of groundwater began to increase substantially during the 1940s, and by 1960 total groundwater withdrawals approached 200,000 acre-feet per year. In 1964 the water level in the USGS monitoring well in downtown San Jose had fallen to a historic low of 235 feet below the land surface, as groundwater was being used faster than it could be replenished.

Substantial land subsidence occurred in the northern Santa Clara Valley as a result of the massive groundwater overdrafts. In 1965 an integrated approach to groundwater recharge including surface water addition, storm flow water, and conservation of water during the rainy season led to substantial recovery of groundwater levels, and there has been little additional subsidence since about 1969.

The Santa Clara Valley was the first area in the U.S. where land subsidence due to groundwater overdraft was recognized. It was also the first area where organized remedial action was undertaken and subsidence was effectively halted. The groundwater resource is still heavily used, but importation of surface water has reduced groundwater pumping and allowed an effective program of groundwater recharge that prevents groundwater levels from approaching the historic lows of the 1960s.

The valley is a large structural trough filled with over 1500 ft of alluvium, including a minority of sand and gravel aquifers, and the majority composed of fine-grained alluvium. Below a depth of 200 ft, the groundwater is confined by layers of clay, except near the margins. Initially, wells as far south as Santa Clara flowed, but a decline in surface water pressure occurred due to pumping for agricultural purposes.

Refer to Figure 11.12 to answer the following questions.

Figure 11.12

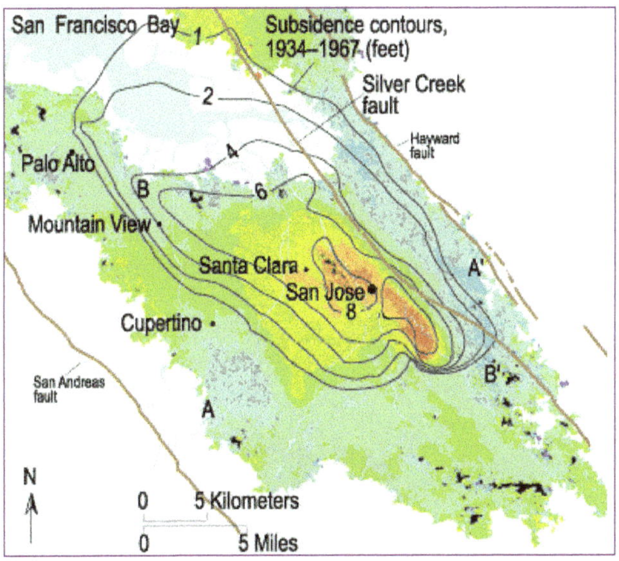

Colored area = alluvium
White area = consolidated rock

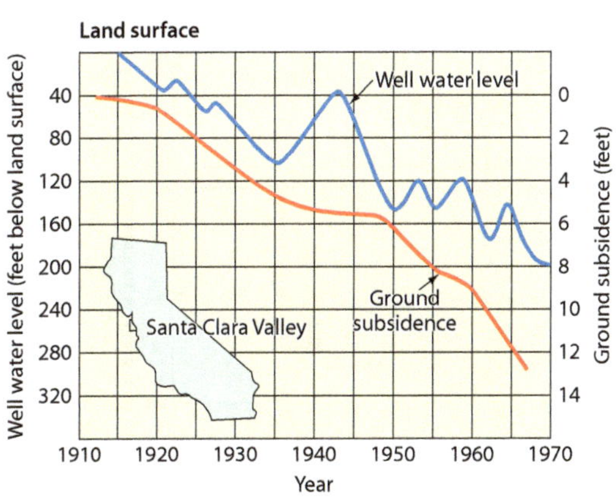

Graph 11.II – Ground subsidence in the Santa Clara Valley, 1934–1967

1. Based on Graph 11.II, what is the general relationship between water table elevation and ground subsidence? ____________________
____
____

2. Examine Figure 11.12. What geologic material was largely responsible for this subsidence? ____________________

   Explain why: ____________________
   ____
   ____

3. Assuming the ground subsidence problem was allowed to continue, would you expect the white area in Figure 11.12 to eventually become affected? Why or why not? ____________________
____
____

Use Figure 11.13 to answer the following questions.

4. In 1965 the importation of water to the valley began. Some imported water was used to recharge the groundwater through stream channels.

   a. Between what years was the pumpage rate highest in this region? ________ and ________.

   b. What was the pumpage rate during that time interval? ____________ thousands of acre-feet

   c. Substantial water importation began around 1965. What was the pumpage rate by 1980?______________ thousands of acre feet

   d. Did the water level (aka "artesian head") rise or fall in response to the importation of water?

      ____

   e. By what percentage did the pumpage rate decrease from 1965 to 1980?

      __________________%

   f. By what percentage did the import rate increase from 1965 to 1980?

      __________________%

   g. Was there a substantial change in the amount of rainfall the region received between 1965 and 1980? **Yes** or **No**?

   h. Which of the following three factors was most influential in causing the water levels to increase from their low in the 1960s: **decreased pumpage, increased precipitation**, or **water imports**?

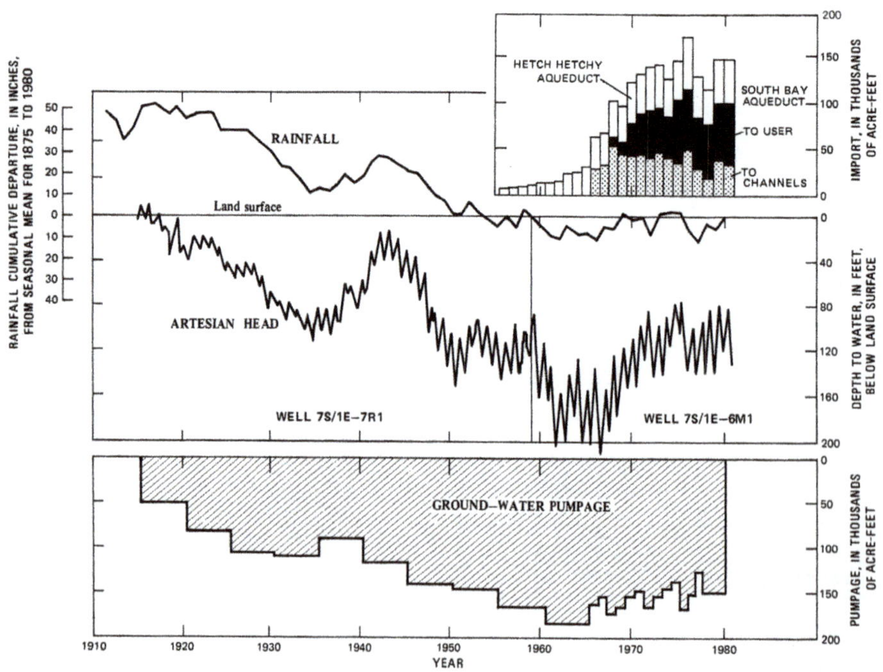

Figure 11.13 – Artesian head change in San Jose in response to rainfall, pumpage, and water imports

Ground Subsidence in San Joaquin Valley

A second severely subsidence-plagued area in California is the San Joaquin Valley; also the result of extensive ground water mining for agriculture. While the withdrawal of water has allowed the San Joaquin Valley to become one of the world's most productive agricultural regions, it has also contributed to one of the single largest land-surface alterations attributed to mankind (see Figure 11.14).

5. Figure 11.15 is a graph of compaction changes in response to fluctuations in the depth of the water table in the San Joaquin Valley. Considering the data presented in this graph, answer the questions that follow.

   a. What is the relationship between groundwater levels and ground subsidence? ____

   ____

   b. Would subsidence resume in the valleys if imports of water were reduced and pumping of groundwater was increased? Yes or No?_______________________

Figure 11.14 – Signs on pole show approximate altitude of land surface in 1925, 1955, and 1977. Maximum subsidence in the San Joaquin Valley occurred near Mendota, California, and was in excess of 28 ft.

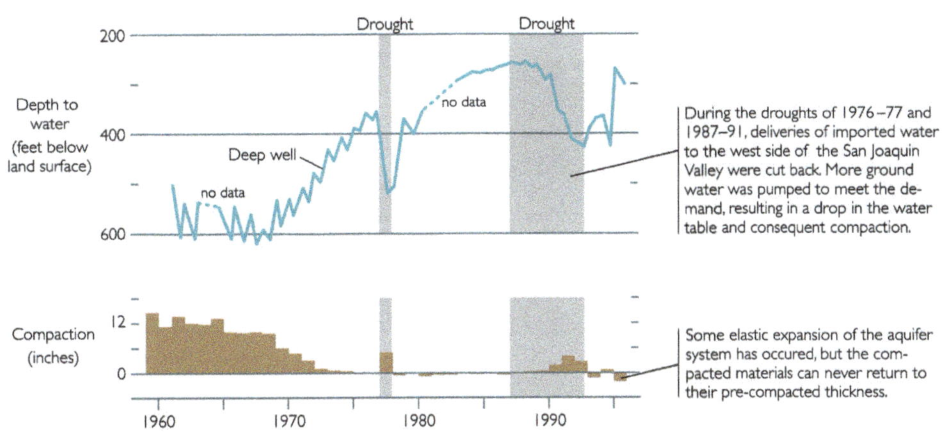

Figure 11.15 – Compaction changes in response to fluctuations in the depth of the water table in the San Joaquin Valley

PART E: TYING IT ALL TOGETHER

Exercise: Building a Complete Picture

Below is a hypothetical exercise demonstrating how geologists combine surface and subsurface data in order to gain a complete picture of the groundwater system and any human interaction with it.

1. Complete the cross-section at the bottom of Figure 11.16. The section is from points A to A' on the map above it. The topography map shows the surface topography along that line and the unconsolidated sediment that covers most of the area. Although not shown on the cross-section, the layers beneath the surface can be mapped using control points in water wells distributed throughout the area. To complete the cross-section: 1) Create a topographical profile of the A–A' line on the lower figure in 11.16; 2) fill in the data on the four well charts below, using the "example well" as a guide; and 3) use those data to construct a geologic cross-section of the underlying layers by drawing a point at the elevation of each contact, then lithostratigraphically correlating the layers.

Data Tables

Sample Well (Elev. = 880 ft)			
Sediment Type	Thickness	Depth	Elev.
Fine Sand	5	5	875
Shale	25	30	850
Gravel	325	355	525
Clay	25	380	500

Well A (Elev. = 900 ft)			
Sediment Type	Thickness	Depth	Elev.
Gravel	375		
Clay	25		

Well B (Elev. = 850 ft)			
Sediment Type	Thickness	Depth	Elev.
Fine Sand	50		
Shale	25		
Gravel	250		
Clay	25		

Well C (Elev. = 850 ft)			
Sediment Type	Thickness	Depth	Elev.
Fine Sand	175		
Shale	25		
Gravel	125		
Clay	25		

Well D (Elev. = 870 ft)			
Sediment Type	Thickness	Depth	Elev.
Fine Sand	200		

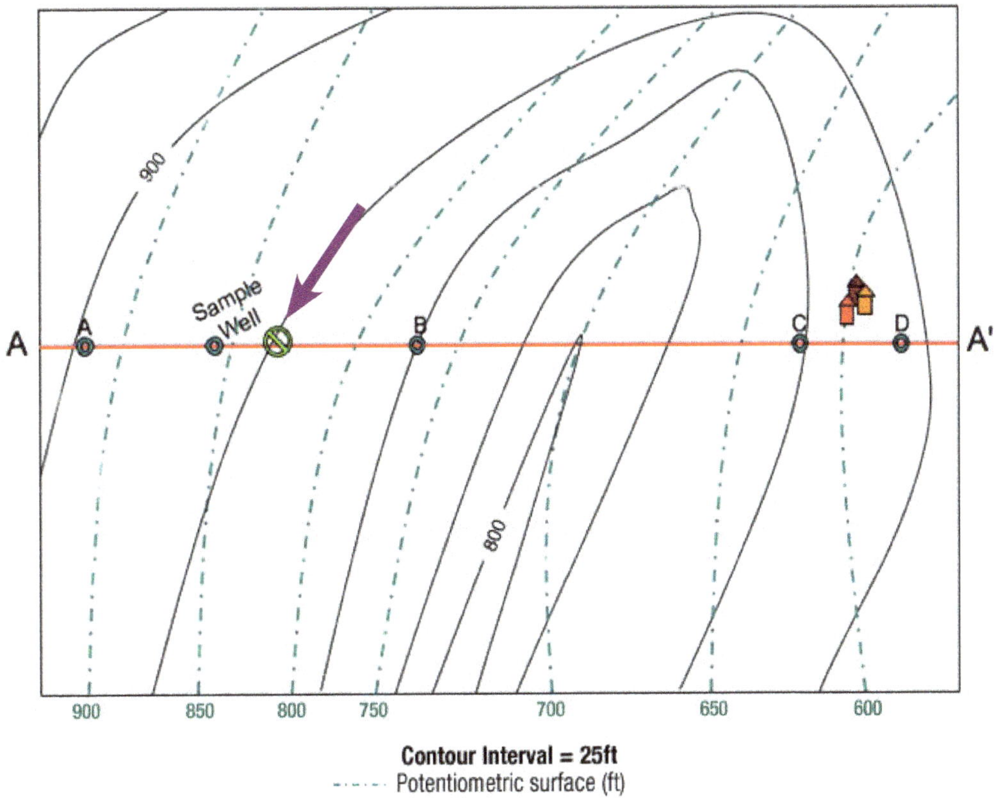
900
Sample Well
A
B
C
D
A'
800
900
850
800
750
700
650
600
Contour Interval = 25ft
Potentiometric surface (ft)

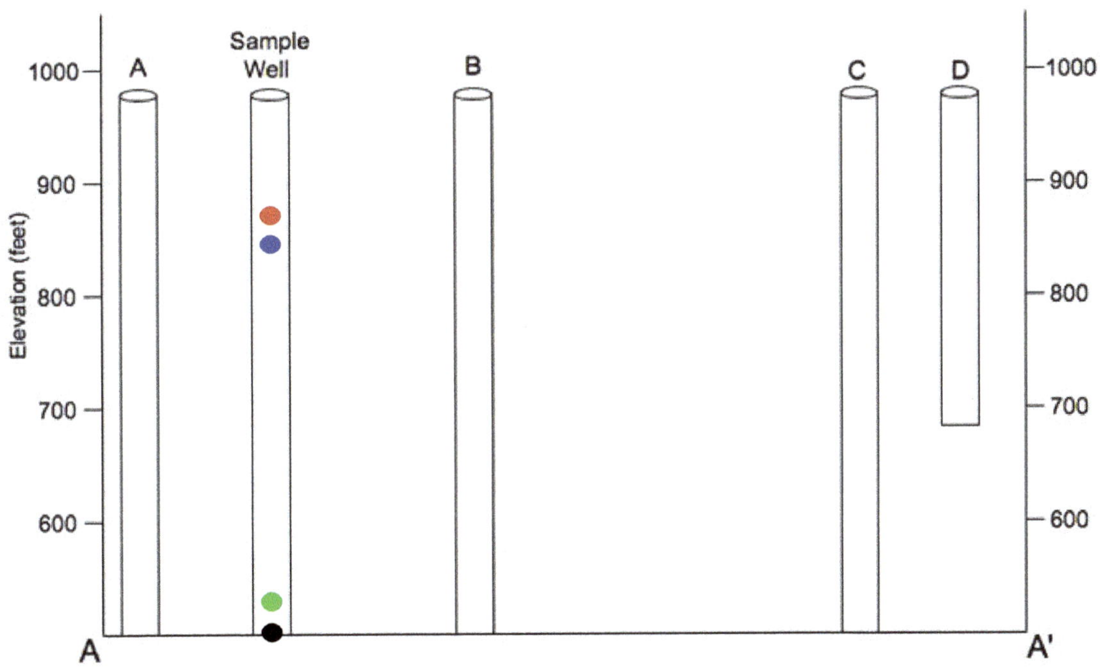
Sample Well
A
B
C
D
1000
900
800
700
600
Elevation (feet)
A
A'

Figure 11.16

Now use the cross-section to answer the questions that follow:

2. The shallow well in the unconsolidated sediment at points B and C fill with water to within 30 feet below of the surface. Using this information, sketch the probable position of the water table in the unconsolidated sediment.

   What is the main source of recharge water for the river? ______________________________

3. Given your answer to #2, is the water quality in the river representative of the ground water quality? Yes or No? _______

4. Knowing that the confining shale bed is impermeable and that because the wells are lined, no water can enter the holes where they pass through the unconsolidated sediment, to what elevation would water from the *gravel* aquifer naturally rise in each of the wells?

   *Why was well D not included in the above question? ______________________________

5. Are these deeper wells flowing artesian, non-flowing artesian, or non-artesian? Explain.____

   ____

   Well A _______________; Well B_______________; Well C_______________

6. In the late 1970s, a small metal-processing plant was located at the position marked by a purple arrow. The company illegally disposed of their chemical waste by drilling a shallow well into the unconsolidated sediment and dumping the chemical waste into the ground, but they were soon discovered because the contamination reappeared. Where do you think the contamination discharged: the surface in well B, the river, or both? Explain.

   ____

   ____

7. A small residential community was built near the site of well C. In order to supply the community with water for domestic needs, a shallow well (~150 ft deep) was drilled about 2500 ft to the east of well C. Compare this well to well A.

   a. Which well is likely to supply more water? ______________________________

   b. Which well is likely to have to a lower contamination issue? ______________________________

Figure 11.17 – Dissolution of carbonate rocks

PART F: GROUNDWATER AND THE LANDSCAPE

The term 'karst' describes a landscape shaped by the dissolution of a layer or layers of soluble bedrock, usually carbonate rock such as limestone. The limestone dissolves because groundwater, after reacting with the rocks and minerals on the surface, becomes mildly acidic, permitting it to dissolve the calcite mineral crystals that make up the limestone. See Figure 11.17.

Much of the drainage in karst areas occurs underground rather than as surface streams. Rainwater seeps into the ground among fractures in the bedrock, whereupon the acidic water dissolved the limestone around it. The cracks widen into narrow caves, which may eventually widen into huge cave galleries. See Figure 11.18.

The systems of fractures and caves that typically develop in limestone are what make limestone good aquifers.

Figure 11.18 – Underground karst cave network

Karst regions are generally recognizable from the surface as they depict the following features.

- **Sinkholes**: Develop where the ceilings of cave galleries collapse, and lakes or ponds form wherever water fills the sinkholes. See Figure 11.19.

Figure 11.19 – Sinkholes

Figure 11.20 – Springs

Figure 11.21 – Disappearing Streams

- **Springs**: Places where water flows naturally discharge to the surface and flow from the ground. See Figure 11.20.

- **Disappearing Streams**: Terminate abruptly by seeping into the ground. See Figure 11.21.

Exercise: Mammoth Cave, Kentucky

1. Examine the Park City Quadrangle (Map 11.I). This map depicts a karst region within Mammoth Cave, Kentucky that is composed of limestone. Sandstone overlies the limestone only in the area north of Park City.

   a. List two ways you can identify which part of the map has limestone at the surface and which part of the map has sandstone at the surface.
      - ______________________________
      - ______________________________

   b. Using a red pen, draw the boundary between the karst topography region and the area dominated by sandstone.

2. You've decided to purchase a plot of land, where primarily sandstone is at the surface, for the purpose of establishing a farm. You select an area on one of the highest hills (~850 ft) located north of Park City, and make plans to drill a well into the limestone aquifer on your property for water to irrigate your crops. How deep would you have to drill your well (through the sandstone) to obtain water from the limestone aquifer? ____________ ft

   Explain your reasoning: ______________________________

   ____

3. Why are there so few sinkholes developed in the SE part of this map area (even though limestone crops out there)? *Hint:* Drawing a quick sketch could be very useful here.

   ____

   ____

   ____

4. Compare this map to the quadrangle of Interlachen, Fl (Map 11.II). How do the two locales differ with respect to:

   a. The size of the sinkholes? ______________________________

   b. The height of the water table? (hint: are the sinkholes primarily filled or empty?)

      ____

      ____

3. Which of these two landscapes is more evolved? ______________________________

4. Would you recommend purchasing property in either one? Yes or No? Explain

   ____

   ____

   ____

Map 11. I

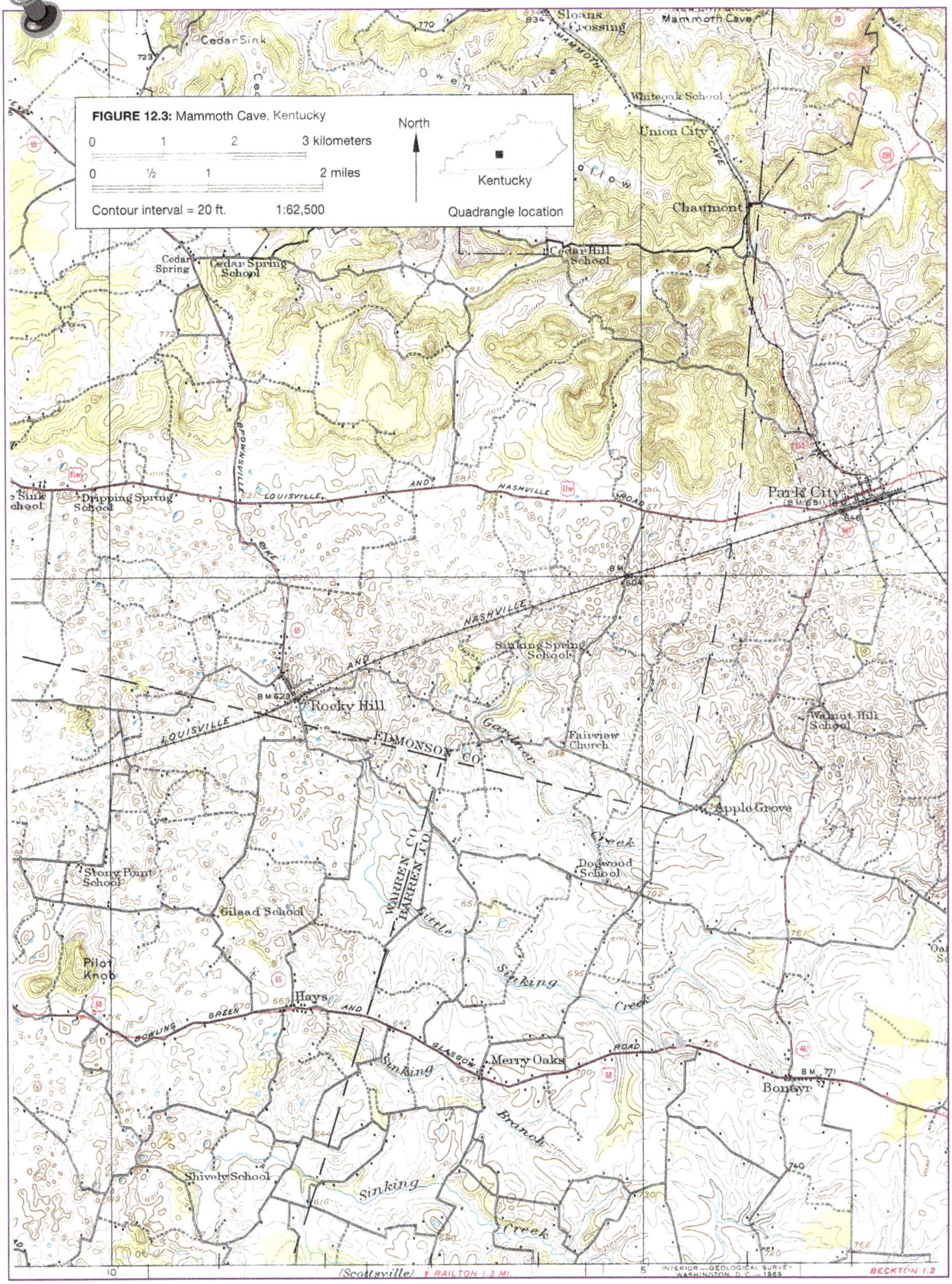
FIGURE 12.3: Mammoth Cave, Kentucky
0 1 2 3 kilometers
0 ½ 1 2 miles
Contour interval = 20 ft.
1:62,500
North
Kentucky
Quadrangle location
Cedar Sink
Sloans Crossing
Mammoth Cave
Whiteoak School
Union City
Chaumont
Cedar Hill School
Cedar Spring
Cedar Spring School
Dripping Spring School
Louisville and Nashville Road
Park City
Sinking Spring School
Rocky Hill
Louisville and Nashville
Edmonson Co
Fairview Church
Walnut Hill School
Apple Grove
Gardner Creek
Dogwood School
Stony Point School
Gilead School
Warren Co
Barren Co
Little Sinking Creek
Pilot Knob
Hays
Bowling Green and Glasgow Road
Merry Oaks
Bonayr
Sinking Branch
Shively School
Sinking Creek
(Scottsville)
Railton 1.3 mi
Beckton 1.2

Map 11. II

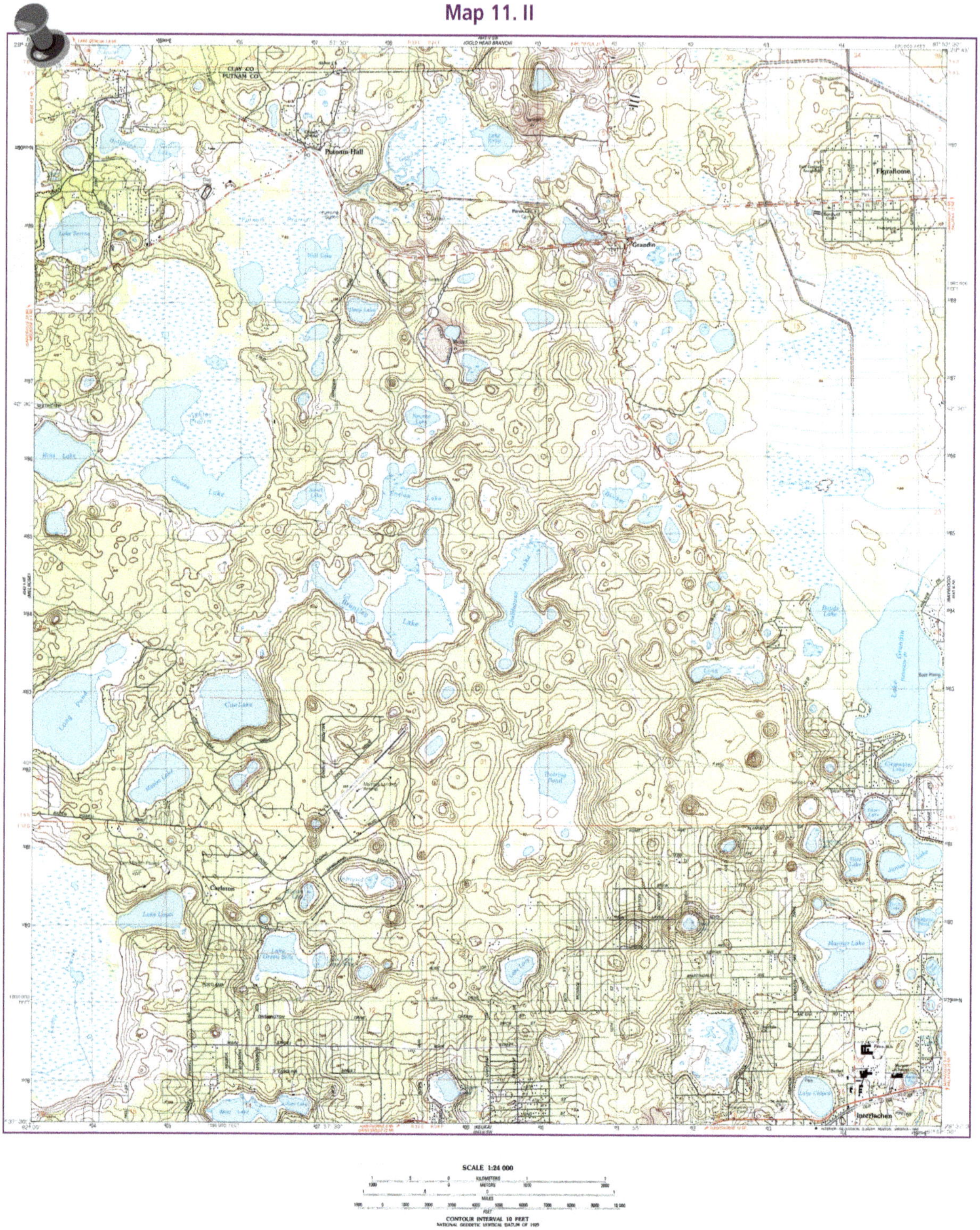
CLAY CO
PUTNAM CO
Putnam Hall
Florahome
Grandin
Deep Lake
Cue Lake
Carleton
Interlachen
SCALE 1:24 000
CONTOUR INTERVAL 10 FEET

Laboratory Exercise XII
Coastal Processes

Introduction

The coastal environment is a unique landscape, as this is the realm where the land processes meet the sea processes, generating areas of rapid change. While most geologic modifications, and particularly those involving landscape evolution, require slow, steady, long-term, incremental processes, the coastal environments are exceptional in that visible changes can be seen on a daily and sometimes hourly basis, making the coasts one of the fastest-evolving terrains on the surface.

The rapidity with which the coasts alter has significant implications that extend far beyond geologic interests. From a societal perspective, the coasts have always been a desirable place to be—whether for living, vacationing, or recreation. Currently, the United States' largest cities (Los Angeles, Miami, New York, etc.), along with three quarters of the US population, are located in coastal states, while the associated restaurants, strip malls, hotels, and beaches generate billions of dollars in revenue per year. Because putting permanent structures in these dynamic, unstable environments is essentially inviting disaster, accelerated colonization of the coastal regions has resulted in an escalating attempt to stabilize the coasts through interference of natural sediment transport processes. This interruption of normal coastal activities not only amplifies the instability by creating more problems than it solves, but it also generates an erroneous perception that the coasts are now stable and safe. Belief in these incorrect ideologies then entices more coastal development—ultimately creating a parallel trend of accelerating damage.

While the coasts have a wealth of benefits to offer, it must be understood that they cannot be used indiscriminately and the best uses of coastal areas are those that are compatible with change. The more we understand about the myriad processes that take place on the coast, the better equipped we are to forecast changes, and thereby make informed decisions that will help us safely adapt to the changes that will inevitably be produced.

Objective

The objective of this lab is fourfold:

1) Acquaint you with types of coasts and the variables that influence coastal landscapes;

2) Examine the role of sediment transport and the features it produces;

3) Explore the relationship between coastline dynamics and plate tectonics; and

4) Investigate the consequences of inappropriate anthropogenic use of the coasts.

Part A: Sediment Supply: Where does the sediment along a coastline come from?

Coastlines are built, maintained, destroyed, and modified through the addition and removal of sediment. While the erosional, transportation, and depositional processes along a coastline will be discussed in depth shortly, it's important to first address the source(s) of sediment on a coast—as the composition, size, sorting, and quantity of sediment are major determiners in the character of a coastline.

Sediment can be added to the coast by a variety of methods, including but not limited to:

1. Coastal Erosion: Eroding cliffs along the coast drop boulders onto the coastline that are further weathered into transportable-sized particles. See Figure 12.1.

Figure 12.1 – Coastal errosion

2. Seafloor Erosion: As waves approach the coast and come in contact with the ocean bottom, they scour the seafloor, uplifting sand, silt, and local biological debris. This material is then delivered to the coast as the wave swashes onto the land. When the wave backwashes, the finer silt and mud are returned to the ocean, while the coarser sand and biological debris remain on the coast. See Figure 12.2.

Figure 12.2 – Seafloor errosion

3. Aeolian Erosion: Wind selectively picks up sand-sized particles and carries them to the coast. Wind-driven additions are a major source of coastal sediment in arid environments, such as the shorelines in the Middle East and Western South America. See Figure 12.3.

The Lange Wand, or Long Wall. Namib Desert, Namibia

Puerto Madryn, Patagonia

Satellite View, Pakistan Coast

Figure 12.3 – Aeoloian errosion

4. River Erosion: While the methods just discussed supplement the sediment supply along a coast, the primary supplier of sediment to any coastline is the river systems. As the rivers travel from the highlands to the plains and through the continental interior on their way to the coasts, tons of continental crust is weathered and transported. Ultimately, when the river empties into the ocean, a colossal amount of sediment is deposited in the delta, where it can be picked up by ocean transport processes and taken down the coastline. See Figure 12.4.

Mahanadi River Delta, India

Skokomish River Delta; Skokomish Indian Reservation, Washington State, USA

Krishna River, India

Figure 12.4 – River errosion

Exercise: Sediment Source and the Character of a Coastline

1. Compare figures 12.5a and 12.5b to answer the following questions.

   a. Examine the two photographs. Which is an area of erosion (where sediment enters the ocean transport system) and which is an area of deposition (where sediment taken from elsewhere exits the ocean transport system)?

   Erosion: ___________; Deposition: ___________

   b. Examine Figure 12.5a. Based on what you know about the sources of sediment to a coastline, would you expect the sediment in this part of the coast to have the same composition as the cliffs behind it? Yes or No? Explain

   ____

   c. Examine Figure 12.5b. Based on what you know about the sources of sediment to a coastline, would you expect the sediment in this part of the coast to have the same composition as the cliffs behind it? Yes or No? Explain

   ____

Figure 12.5a

Figure 12.5b

2. Compare figures 12.6a and 12.6b to answer the following questions.

   a. Examine the two photographs. What is the most obvious difference between the sand on these two coastlines? ____

   b. Given the locality of the coast in Figure 12.6a, what is the general mineralogical composition of the sand? ____

c. Given the locality of the coast in Figure12.6b, what is the general mineralogical composition of the sand? ______

d. Why do these two coastlines have such drastically different sediment types? That is, what is the main sediment supply in each case and how does that explain why they differ? ______

Figure 12.6a – Punaluu Beach, Big Island of Hawaii

Figure 12.6b – Third Beach of Stanley Park, Vancouver, BC, Canada

3. While weathered rock makes up a significant percentage of coastal sediment, indigenous, loose debris is also a large contributing component and causes local variations in sediment composition throughout a contiguous coast.

   a. Examine figures 12.7a and 12.7b. If you were to pick up a handful of sediment from either of these coastal areas, what additional "ingredient" (besides weathered rock) would you likely find in abundance? ______

Figure 12.7a

Figure 12.7b

b. Examine figures 12.8 and 12.9. If you were to pick up a handful of sediment from either of these coastal areas, what additional "ingredient" (besides weathered rock) would you likely find in abundance? ______________________

Figure 12.8

Figure 12.9

4. Considering that the primary controls on the rate that sediment is added to a coast are the quantity of sediment available and the ease of transportation, which environment(s) would you expect to have the fastest influx of sediment? Tropical/Temperate, Glacial, or Aeolian (Circle all that apply)

5. What factors did you consider when drawing your conclusion to question 4 above? i.e., justify your answer. ____

____

6. Examine figures 12.10–12.12. These are satellite images of the three coastal environments discussed in #4. Do these images support or contradict the answer(s) you chose for #4? ______________________

If they contradict, revisit numbers 4 and 5 and reassess the questions.

Tropical/Temperate Environments

Figure 12.10b – Connecticut River, CT, USA

Figure 12.10a – Northeast coast of Brazil, midway between the mouth of the Amazon River and the coastal city of São Luis

Figure 12.10c – Ganges River, India

Glacier Environments

River sediment

Figure 12.11a – Gulf of Alaska

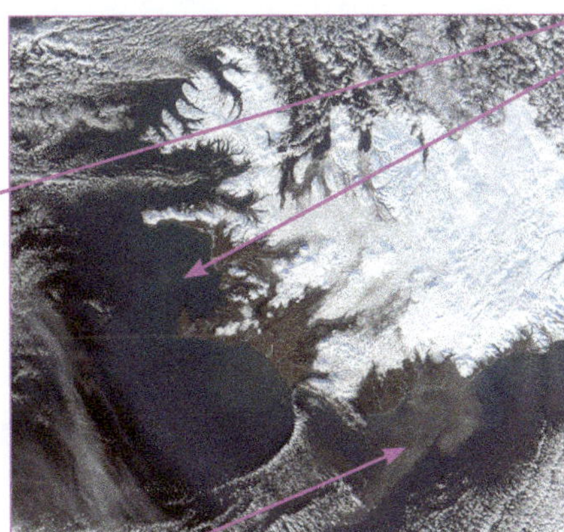

Figure 12.11b – Iceland

Figure 12.11c – Yukon Delta, Alaska

Windblown glacial dust

Aeolian Environments

Figure 12.12a – Desert coast of Western Sahara and Morocco

Figure 12.12c – Coast of Namibia

Figure 12.12b – United Arab Emirates

7. Milliman and Mead, 1983, did an extensive study on the annual quantities of suspended load that exit the mouth of major river systems worldwide. The results of that study are presented in Table 12.A.

Table 12.A– Calculated Total Suspended Sediment Discharge From the Continents

River System	Sediment Discharge (10^6t yr^{-1})
North America	
St. Lawrence	4
U.S. Atlantic Coast	13
Gulf Coast	256
Colorado	.1
Columbia	8
Rest of W. U.S.	62
Canada West Coast	61
S. Alaska (glacial)	340
S. Alaska (non-glacial)	104
N. Alaska	42
MacKenzie	100
N. NE Canada	30
SUBTOTALS	1020
Central America	
Mexico	210
Remainder	323
SUBTOTALS	422
South America	
Northwest	150
Magdalena	220
Northern	1218
Eastern	28
Southern	154
Western & South	18
SUBTOTALS	1788

Europe	
Western	31
Alpine	66
Black Sea	133
SUBTOTALS	230
Eurasian Arctic	
West of 140°	59
East of 140°	25
SUBTOTALS	84
Asia	
Northeast	100
NE China/Korea	658
Yellow (Huangho)	1080
Rest of China	930
SE Asia and Himalayas exclusive of Indus)	3128
India	286
Indus	100
Asia Minor	67
SUBTOTALS	6349
Africa	
Northwest	110
West	113
Southwest	17
East	240
Zambesi	20
Tana	30
Nile	0
SUBTOTALS	530
Australia	
East, North	62
Oceanic Islands	3000
TOTALS	**13,505**

a. According to Table 12.A, what is the total estimated quantity of suspended sediment load that enters the oceans from the major worldwide river systems annually?
_______________________ t/yr

b. Given that this study deals exclusively with suspended sediment load, does it account for all the river sediment that actually enters the ocean? Yes or No? Explain
____
____

8. As demonstrated above, when allowed to work unhindered, rivers transport enormous quantities of sediment to the coasts. However, over the past few decades, anthropogenic interference with the river systems, in the form of various engineering structures, has drastically decreased the quantity of sediment actually reaching the coastline. Consider the following scenario:

a. Originally, the Colorado River transported approximately 135 x 10^6 tons of sediment per year to the Gulf of California. Following the installation of multiple dams for reservoirs, soil conservation, bank stabilization, and flood-control projects, the Colorado River now delivers an estimated 0.1 x 10^6 tons of sediment per year to the Gulf of California. By what percentage did the quantity of sediment delivered by the Colorado River decrease after the engineering structures were emplaced? (Show your work) _______________%

b. What affect does damming up the rivers have on the coastal environments?
____
____
____

Part B: Additions, Deletions, and Transport of Sediment Down the Coast

While Part A examined the various ways sediment can be added to the shoreline sediment-transport system, it's important to understand that once on the coast, the dominant force that controls both sediment transport and coastal landscape evolution is the **wave**. See Figure 12.13.

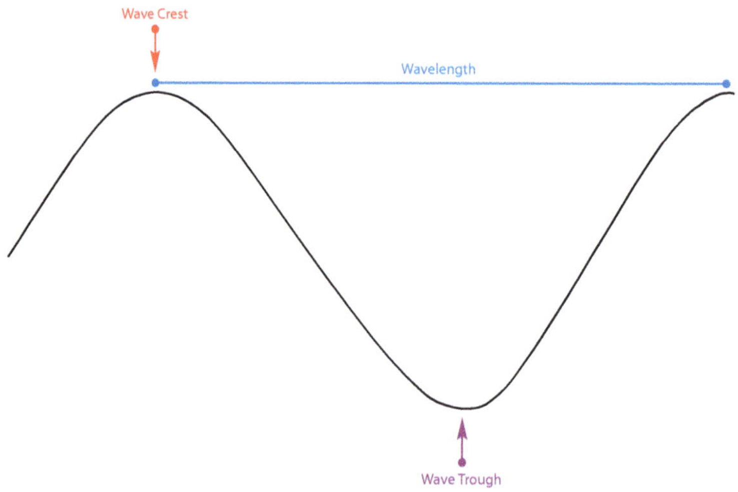

Wavelength – distance between successive crests

Figure 12.13 – Wave Features

Waves are generated by offshore storms. Because the wind is blowing faster than the water is moving, a frictional layer develops between the two fluids and energy is transferred from the wind to the water in an attempt to equilibrate the imbalance. This energy manifests itself in the form of a wave.

Wave energy exhibits a series of circular energy belts that extend vertically below the ocean's surface and propagate forward in circular motions. The top energy belt is largest, with the diameter equaling the amplitude of the wave. With increasing depth, the diameter of the energy belt decreases until the "wave base" is reached, after which there is no wave movement at all. See Figure 12.14.

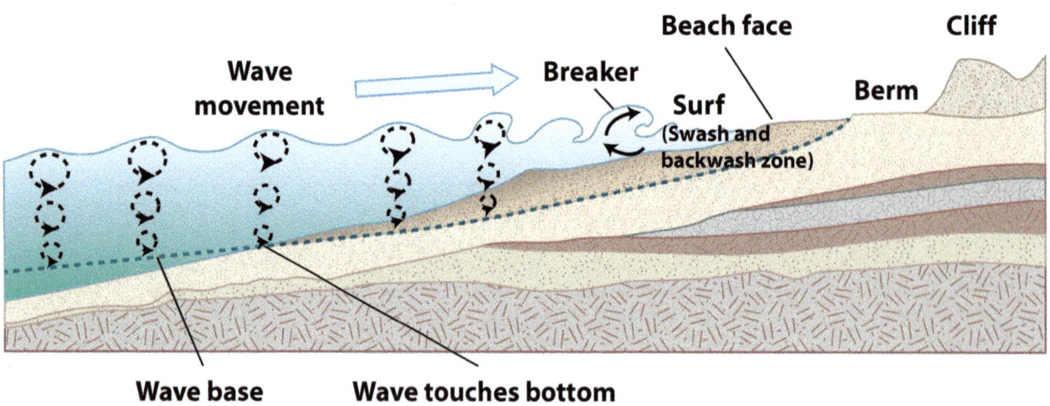

Figure 12.14 – As a wave approaches the coast

As long as the wave base remains above the sea floor, the wave energy propagates unhindered through the ocean—the equivalent of a bowling ball rolling beneath a blanket. However, once the wave energy enters shallow water, it encounters the seafloor, a phenomenon described as "feeling bottom." In addition to stirring up and suspending seafloor sediments and debris that will be brought to the shore with the wave, the friction between the wave and the sea floor also decreases the wave velocity, resulting in a shortening of the wavelength and subsequent increase in wave height. Eventually, the wave crest destabilizes, and falls over or "breaks" onto the coast, expending all the energy along the shoreline.

The amount of energy transported in the form of waves is astronomical. It has been estimated that were we able to efficiently harness wave energy for power generation, there exists the potential to produce up to 80,000 x 10^{12} watts of electricity per year. All coasts are regularly subjected to various intensities of wave energy, which serves the purpose of reworking the sediment, as well as reshaping the coastline. Depending on a number of factors, including, but not limited to wave frequency, intensity, and sediment availability—stretches of coastline will be predisposed to erosion, deposition, or a steady-state sediment budget. As such, all coasts are constantly in flux—endlessly evolving due to the incessant interaction with wave energy.

Section I: Erosion Along a Coastline: A Natural Process—Not a Hazard

Erosive shoreline environments form where wave processes are strong enough to wash out all available sediment to deeper off-shore waters, leaving behind a bare and rocky coast (see Figure 12.15). These erosive shorelines are typically associated with strong waves, but are also found in shoreline locales that have reduced sediment input, such as desert coasts and shorelines downstream from the mouth of rivers with low sediment loads.

Figure 12.15 – Rocky coastline

Figure 12.16 depicts some common coastal landforms that characterize erosive shorelines.

- Wave-Cut Notch—An indentation cut into a sea cliff at water level by wave action.
- Wave-Cut Cliff—Eventually, the overhang above a wave-cut notch becomes unstable due to lack of support at the base, then shears off, leaving behind a nearly vertical cliff face.
- Wave-Cut Bench/Platform—As the cliff gradually retreats inland due to the repeated creation of wave-cut notches followed by collapse, a flat or gently sloping bench/platform is left in its path, which becomes visible at low tide.
- Sea Caves, Sea Arches, Sea Stacks, and Tombolos

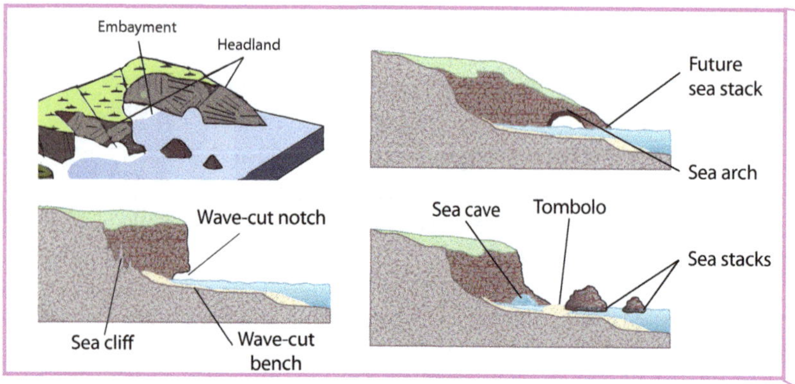

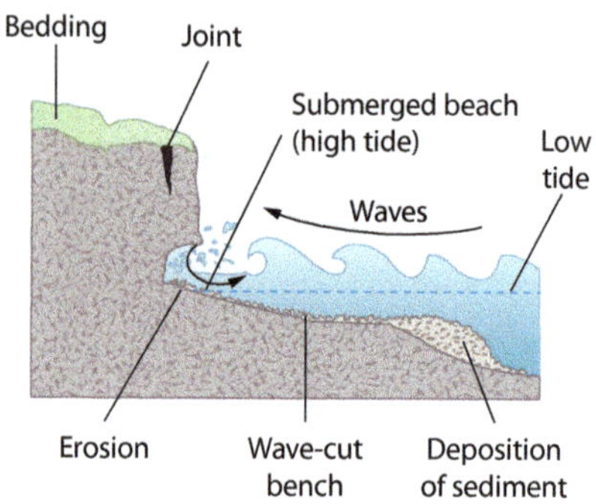

Figure 12.16 – Features of Erosive Shorelines

As a wave approaches an irregular coast, the wave refracts and changes shape such that it mimics the shape of the coastline. The reason the wave refracts at irregular coastlines is that the part of the wave closest to the headlands touches bottom first and slows down due to friction, while the parts farther offshore continue to move at their original velocity toward the embayment. Due to this refraction, there is a convergence of the waves at the headlands, which promotes intense erosion and a divergence of waves at the embayment, which facilitates deposition (see Figure 12.17). With the exception of coarse-grained sand and gravel beaches formed in the embayment, extensive sediment deposits are uncommon on erosive coasts, as any deposited sediment is quickly swept away by waves, tides, and the oceanic sediment-transport system. The net result is a straightening of the coastline. See Figure 12.18.

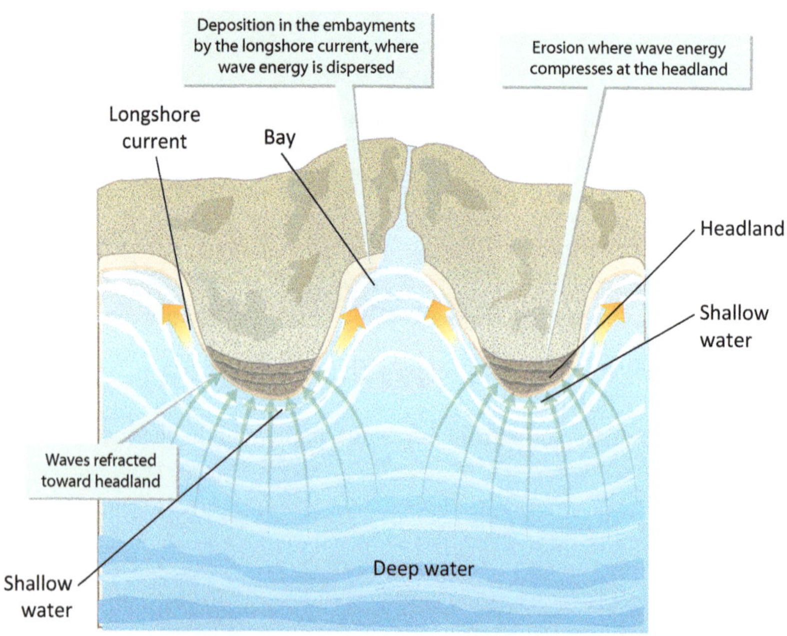

Figure 12.17 – Wave Refraction

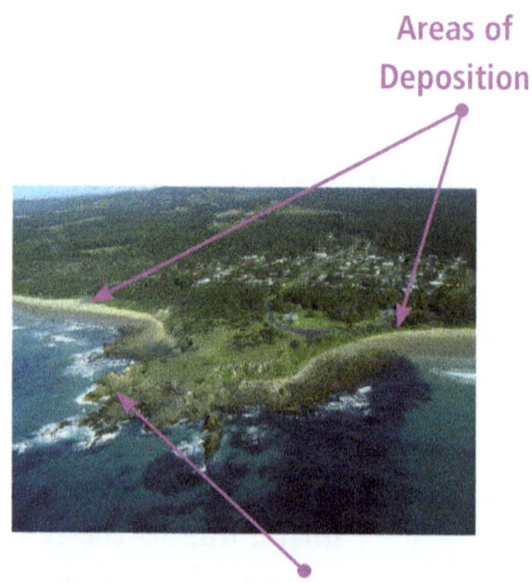

Figure 12.18 – Erosion + Deposition straighten the coastline

Because of the refraction, the waves curve and attack the sides of a headland. This process generates the following features (see Figure 12.19):

- Sea Cave—Large tunnel through the interior of a headland
- Sea Arch—Over time, erosion wears down the headland until all that remains is a slender arch, which is connected to the mainland by a narrow bridge
- Sea Stack—Eventually the arch collapses, leaving an isolated stack of rocks just offshore

Once formed, a sea stack protects the adjacent shore from waves, generating a quiet area where sediment deposition can take place. Therefore, sand collects in the strip between the sea stack and shore, building a "tombolo"—a narrow ridge of sand that links the sea stack to the mainland. See Figures 12.20a and 12.20b.

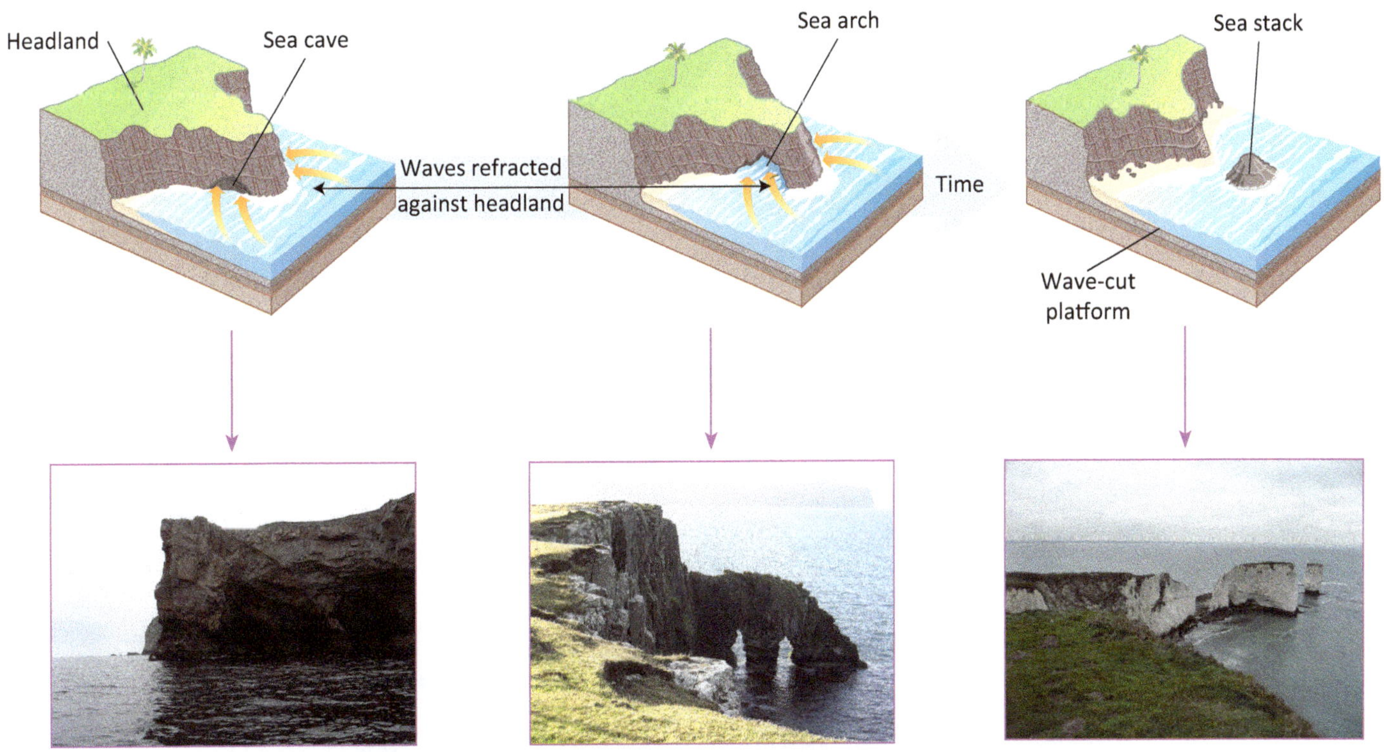

Figure 12.19 – Sea Caves, Sea Arches, and Sea Stacks

Figure 12.20a – Tombolo at St Ninian's Isle, United Kingdom

Figure 12.20b – Tombolo at Cape Paximadhi, Southern Greece

Exercise: Features and Processes of Erosive Coastlines

1. Examine figures 12.21–12.27. On the lines provided, identify the coastal feature depicted in each figure as one of the following: Wave-cut notch, Wave-cut cliff, Wave-cut platform, Sea Caves, Sea Arch, Sea Stack, or Tombolo.

____

Figure 12.21a – 12.21d

____

Figure 12.22a – 12.22b

Figure 12.23a – 12.23c

Figure 12.24a – 12.24b

Figure 12.25a – 12.25c

Figure 12.26a – 12.26b

Figure 12.27a – 12.27d

2. Examine Map 12.I located at the end of this chapter. La Push, Washington is an example of an erosive coastline. On the map are several arrows, each labeled with a letter. In the space provided below, identify the landscape feature associated with each arrow.

   A. ______________________________

   B. ______________________________

   C. ______________________________

   D. ______________________________

   E. ______________________________ (note the steep slope)

3. Examine Map 12.II, located at the end of this chapter. Is Cape Sebastian, Oregon an erosive coastline? Yes or No? Justify your answer.

   ____

   ____

   ____

   ____

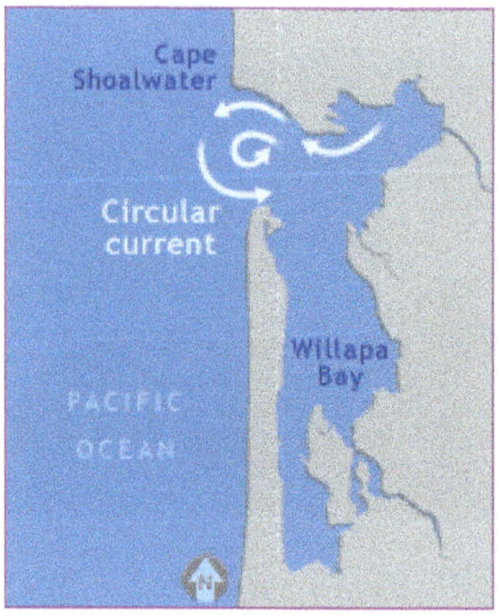

Figure 12.28

4. It's extremely important to note that erosion of sea cliffs is a normal, natural, and foreseeable shoreline process that becomes a hazard only if we get in its way. Unfortunately, failure to acknowledge the dynamic nature of the coasts and the rapidity with which those changes take place has resulted in significant development on eroding coastlines and an astronomical amount of property damage. Nowhere is this more apparent than in Cape Shoalwater, Washington.

Currently nicknamed "Washaway Beach," Cape Shoalwater boasts the most rapid erosion rate on the U.S. Pacific Coast. Cape Shoalwater is a massive spit, through which a tidal channel is deepening and migrating northward. As the channel migrates, it cuts into the shore, eroding the coast with gargantuan speed. See Figure 12.28.

Figure 12.29 shows the progressive changes along this coastline from 1871 to 1999.

Using the first frame, answer the following questions.

a. Approximately how many feet did the coast line erode from 1871 to 1999? ________ft

b. What is the average rate of erosion per year from 1871 to 1999 at the spot marked by the purple line? ______ ft/yr (Show your work)

c. Would you recommend permanent development on coast of this type? Yes or No? Explain

____

____

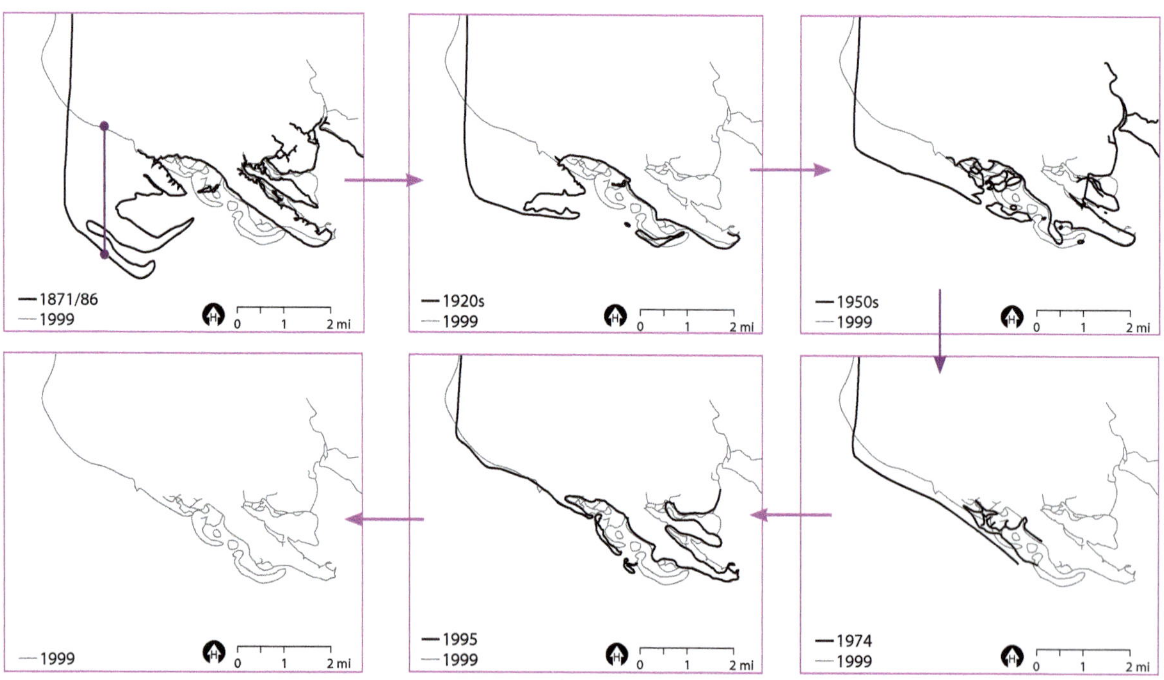

Figure 12.29 – Changes along the Shoalwater coastline from 1871 to 1999

Side Note

During the 1920s, over thirty homes were relocated or claimed by erosion . In the years that followed, erosion destroyed a lighthouse, a life-saving station, a clam cannery, a school, and a Grange Hall. Erosion also forced the relocation of a cemetery and State Highway 105. In recent decades, erosion has destroyed twenty homes, private property, and part of the Willapa National Wildlife Refuge.

In 1995, erosion threatened to undermine State Highway 105. There were also concerns that sea water could invade and damage cranberry bogs worth millions of dollars. To protect State Highway 105, the Washington State Department of Transportation constructed a $27 million submerged groin and beach fill area—but still...the struggle continues.

Rapid erosion created cliffs along the beach in the Cape Shoalwater area, April, 1992. Dune ridges on Cape Shoalwater were once 49 feet high.

Section II: Transportation and Deposition of Sediment

Regardless of whether the sediment originates from the rivers, the wind, the ocean, or erosive coastlines, once it reaches the shore, it joins what is known as the "littoral transport system"—nature's way of moving sediment down the coast. The backbone of the littoral transport system is called the "longshore current," which develops as a result of wave interaction with the coastline.

The longshore current is capable of redistributing sediment down the coast as a result of the wave refraction process discussed above. Because the waves refract upon their approach to the shoreline, the water swashes onto the coast at an angle. However, when the water spilled on the coast backwashes to return to the ocean, it does so under the influence of gravity and therefore travels perpendicular to the coast. Thus, when the wave hits the coast, sediment is propelled forward at an angle, and when the water returns to sea, the sediment is pulled straight back. As such, the sediment travels down the coast in a saw-tooth pattern, due to continuous wave action. See Figure 12.30.

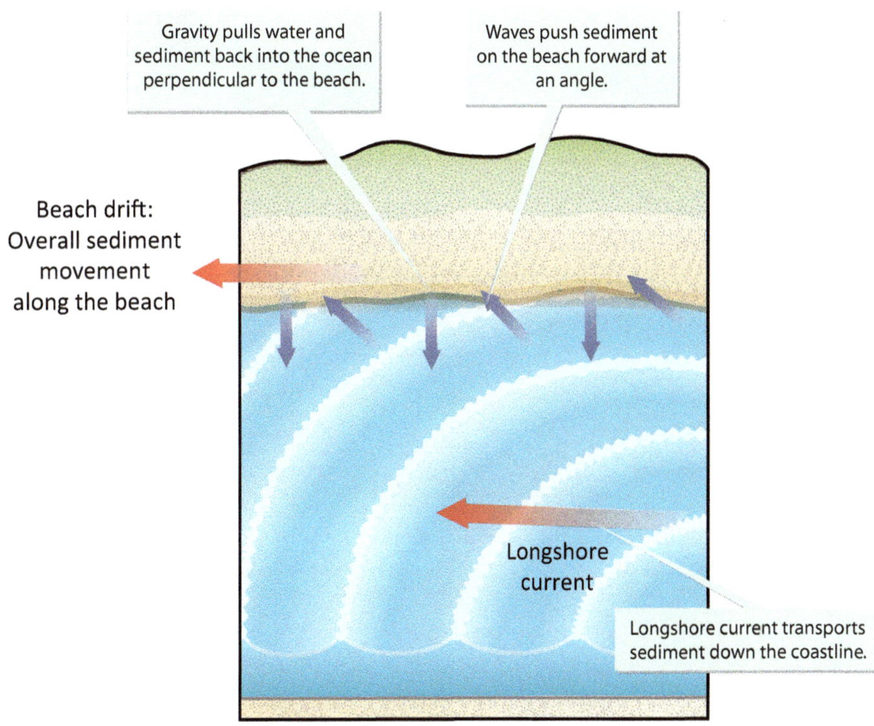

Figure 12.30 – The Longshore Current

Side Note

Think about what happens when you're at the beach and you walk a few feet into the ocean to bob in the waves and cool off. After you've bobbed around for a few minutes and are ready to soak up some more sun on the beach, do you exit the ocean at the same spot you entered it? Yes or No? If no, why not?

____

____

____

Figure 12.31 – Plunging Breakers

Figure 12.32 – Spilling Breakers

Just as the longshore current picks up and transports sediment away from a given location, it also deposits sediment at areas down-current from the sediment source. As such, what is known as the "sediment budget" along a defined piece of coastline is a direct result of how much sediment is added to the area from rivers, wind, the ocean, and rocky coasts; how much is transported away by the longshore current; and how much is deposited by the longshore current.

As discussed earlier, erosive shoreline environments form where wave processes are strong enough to wash all available sediment off the coast. These areas are generally characterized by steep offshore topography, which generate aggressive, plunging breakers that sweep out sediment native to the locale, in addition to removing any sediment that was deposited when the wave hit the coastline. See Figure 12.31.

Depositional shorelines form where wave processes are gentle and moderated. These areas are generally characterized by mild offshore topography, which generate tranquil, spilling breakers that don't have enough power to remove sediment upon backwashing. As such, any deposited sediment remains on the coast in these areas. See Figure 12.32.

Therefore, coastlines with intense wave activity tend to lose more sediment to the littoral transport system then they gain, thereby resulting in erosion; while calmer shorelines gain more sediment from the littoral transport system than they lose, thereby resulting in deposition.

Figures 12.33 and 12.34 depict some common coastal landforms that characterize depositional shorelines.

- Wide Beaches—A "beach" is a landform consisting of loose material that has accumulated at the shoreline due to wave action.

- Spits and Baymouth Bars—Where the coastline indents landward, sediment-laden longshore currents will continue depositing into open water, stretching out the beach to create a "sand spit." Some sand spits grow across the opening of a bay or estuary to form a "baymouth bar."

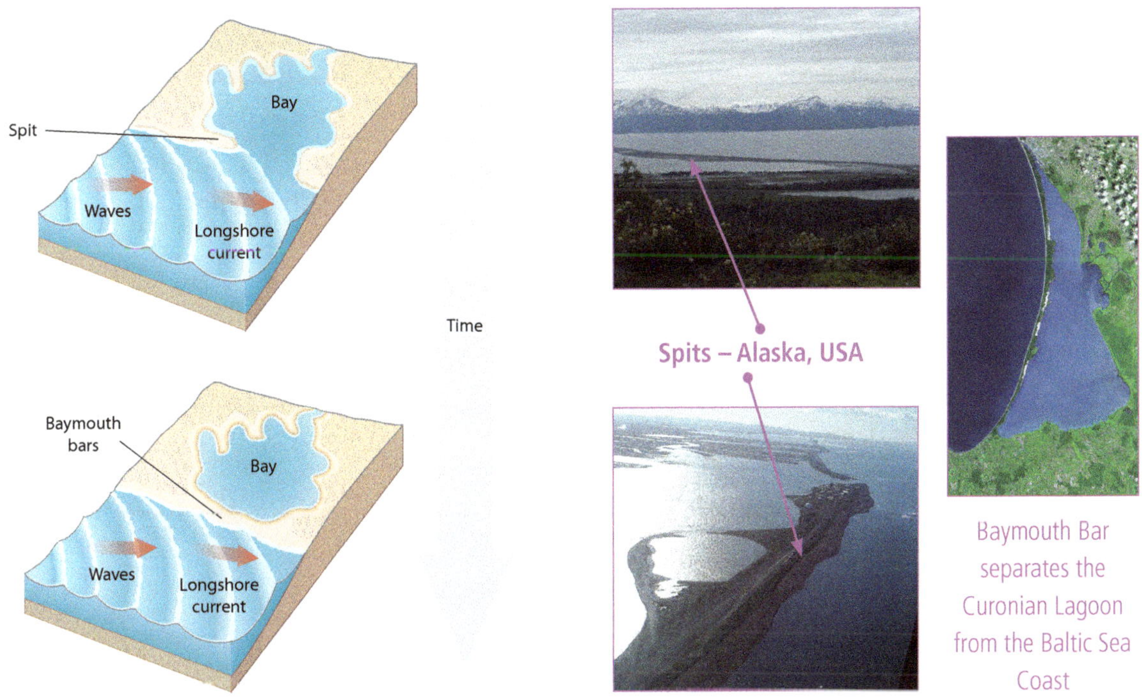

Figure 12.33 – Spits and Baymouth Bars

- Barrier Islands, Tidal Inlets, and Lagoons—"Barrier Islands" are relatively narrow strips of sand that are parallel to the mainland coast. They usually occur in chains and, excepting the tidal inlets that separate the islands, may extend uninterrupted for over a hundred kilometers. A "lagoon" is a shallow body of water separated from a larger body of water by barrier islands or reefs. The lagoon side of the barrier island is protected from wave action, whereas the ocean side is exposed to perpetual wave processes. The net result is deposition on the landward side of the island and erosion on the ocean side of the island, generating a slow migration of the barrier island toward the coast.

- Tidal Flats—Tidal flats are level muddy surfaces, alternately submerged and exposed to the air by changing tidal levels. They form when mud is deposited by tides or rivers bordering an estuary.

- See Figure 12.34, 12.35 and 12.36.

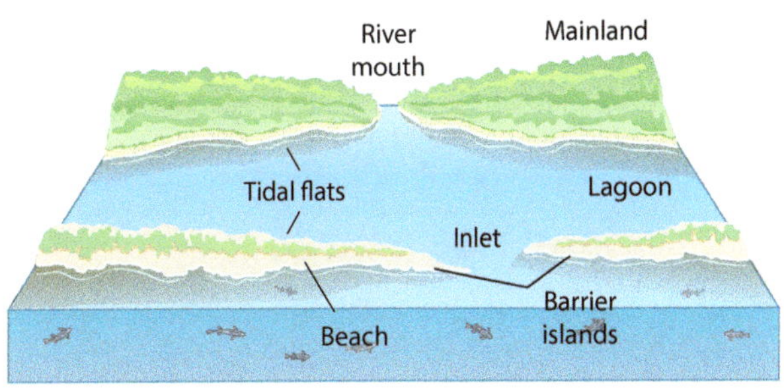

Figure 12.34 – Barrier Islands, Lagoons, and Tidal Flats

Figure 12.35 – Tidal flat

Figure 12.36 – Barrier island and lagoon

Exercise: Features and Processes of Depositional Shorelines

1. Use Map 12.III of Chatham, Massachusetts, located at the end of this chapter, to answer the following questions.

   a. List at least three landscape features present on this map that indicates it's an area of deposition.

   - ____________________
   - ____________________
   - ____________________
   - ____________________

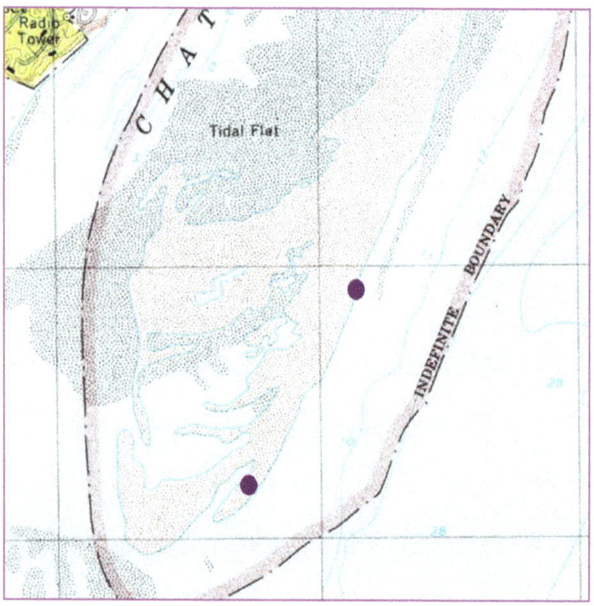

Figure 12.37

   b. Which direction is the longshore current moving?

   ____

   How do you know?

   ____
   ____
   ____

   c. If the longshore current were carrying a grain of sand between the two purple dots on Figure 12.37, draw in the EXACT path and pattern the grain of sand would travel.

2. Use Map 12.IV of Morehead City, North Carolina , located at the end of this chapter, to answer the following questions.

   a. What direction is the longshore current traveling along the eastern coast? ______________

   b. What are the segmented strips of sand surrounding the mainland? ______________

   c. What is the body of water between the feature you specified in question b on the previous page and the mainland called? ______________

   d. Would you expect the coast of the mainland to be characterized by intense or gentle wave activity? ______________. Justify your answer ______________

   ____

   e. Would you expect the *landward* coast of the surrounding island chain to be characterized by intense wave activity? Yes or No?

   f. Would you expect the *seaward* coast of the surrounding island chain to be characterized by intense wave activity? Yes or No?

   g. Given how the segmented islands surrounding the mainland are affected by the waves, if you were to envision a map of how this area might look in 100 years, where would that island chain be with respect to the mainland? (i.e., closer to, farther away from, in the same spot, or completely eroded away?) *Justify your answer.*

   ____

   ____

   ____

SECTION III—TYING IT ALL TOGETHER

Use Map 12.V of Drakes Bay, California, located at the end of this chapter, to answer the following questions.

1. Examine the shoreline to the northwest of Drakes Bay.

   a. Is this a "rocky" coast or a "sandy" coast? ______________

   b. Would you expect plunging breakers or spilling breakers along this shoreline?

   ____

   c. Is this part of the coast erosional or depositional? ______________

   d. What evidence can you cite to justify your answer to question 1c above? ______________

   ____

   e. Is this area classified as a headland or embayment? ______________

   f. What direction is the longshore currently moving along this coast? ______________

   g. What evidence can you cite to justify your answer to question 1f above? ______________

   ____

2. Examine Point Reyes in the southwest corner of the map.
   a. Is this a "rocky" coast or a "sandy" coast? ________________________
   b. Would you expect plunging breakers or spilling breakers along this shoreline?
   ____
   c. Is this part of the coast erosional or depositional? __________________
   d. What evidence can you cite to justify your answer to question 1c above? ______________
   ____
   e. Is this area classified as a headland or embayment? _________________

PART C: SUBMERGENT AND EMERGENT COASTLINES

While sediment abundance, wave intensity, and tidal behavior control coastal morphology in the geologic short-term, it must also be recognized that there are underlying long-term controls on shoreline evolution, the crux of which is sea-level change.

Sea-level changes can be referred to as either "Eustatic" or "Relative."

- Eustatic sea-level fluctuations refer to a global rise or fall in sea level. If one considers Earth's ocean basin as one large container, global sea level changes can be accomplished either by adding or removing water (i.e., major glacier advances and retreats) or by changing the size of the container (i.e., increasing or decreasing rifting rates). In both cases, the water level of the ocean will physically increase or decrease with respect to the world's landmasses.
- Relative sea-level change involves the appearance of the sea level fluctuating, as opposed to it actually doing so. Just like sea level is capable of rising and falling, the continents, too, can be tectonically uplifted or forced to subside. Uplift will generate a landscape where the shoreline is raised relative to an unchanged sea level, giving the appearance of a sea-level fall; while land subsidence lowers the shoreline relative to an unchanged sea level, giving it the appearance of a sea-level rise. Unlike eustatic changes, relative sea-level changes occur only in the vicinity of the tectonic phenomena, and therefore the affects are regional as opposed to global.

Regardless of whether the sea-level change is eustatic or relative, the affected area(s) will experience a landward or seaward shift in the position of the shoreline.

- Sea-level fall and tectonic uplift produce an "emergent" coast, where the shoreline progressively extends further out to sea.
- Sea-level rise and land subsidence produce a "submergent" coast, where the shoreline progressively retreats toward the continent.

Features of Emergent Coastlines

Marine Terraces: Series of steps, each consisting of a wave-cut platform, sea cliff, and a mixture of marine and terrestrial deposits, created from the emergence of older shorelines as relative sea level drops. See Figure 12.38.

1. Low tide cliff/ramp
2. Modern wave-cut platform
3. Modern wave-cut notch
4. Modern sea cliff
5. Paleo wave-cut platform I
6. Paleo wave-cut notch I
7. Paleo sea cliff I
8. Terrace cover—marine deposits and terrestrial mass wasting deposits
9. Alluvial fan
10. Decayed and covered Paleo sea cliff and wave-cut platform II
11. Paleo–sea level I
12. Paleo–sea level II

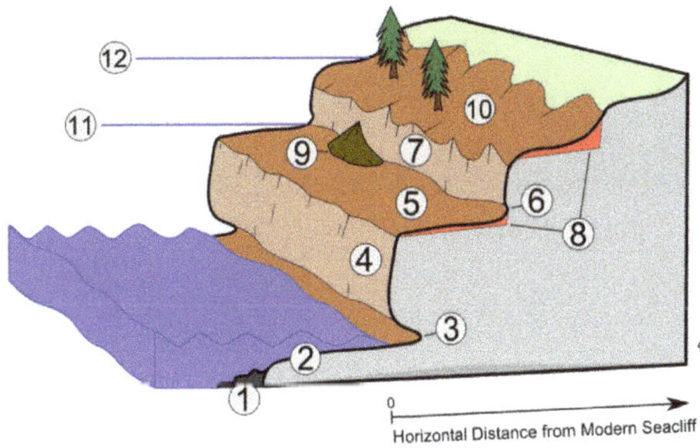

Figure 12.38 – Marine Terraces

Features of Submergent Coastlines

Estuary—A river valley that has been flooded by sea water. See Figures 12.39a and 12.39b.

An estuary is a partly enclosed coastal body of water with one or more rivers or streams flowing into it, and with a free connection to the open sea. Estuaries form a transition zone between river environments and ocean environments and are subject to both marine influences, such as tides, waves, and the influx of saline water; and riverine influences, such as flows of fresh water and sediment.

Figure 12.39a – Columbia Estuary, USA

Figure 12.39b – Exe Estuary, UK

Fjord—A glacial-cut U-shaped valley that is flooded by seawater. See Figures 12.40a and 12.40b.

Figure 12.40a – Fjord in Sognefjord, Norway

Figure 12.40b – Fjord in Greenland

Wetlands– A vegetated, flat-lying stretch of coast that floods with shallow water but does not feel the impact of strong waves.

Wetlands are characterized as having a water table that stands at or near the land surface for a long enough period each year to support aquatic plants. The water found in wetlands can be saltwater, freshwater, or brackish.

- Temperate wetlands include:
    - Bogs (wetlands dominated by mosses and shrubs) - see Figure 12.41a
    - Marshes (wetlands dominated by grasses) - see Figure 12.41b
    - Swamps (wetlands dominated by trees) - see Figure 12.41c
- Tropical and semi-tropical wetlands include:
    - Mangrove swamp–dominated wetlands

Figure 12.41a – Bogs in Cape May, NJ

Figure 12.41b – Marshes, Morden Hall Park, UK

Figure 12.41c – Swamp Island Ponds, UK

Exercise 1: Emergent Coastlines

As was discussed above, emergent coastlines develop due to global sea-level fall or localized tectonic uplift. Given that the west coast of the U.S. is on the Ring of Fire subduction zone and riddled with transform faults, it should be no surprise that the coastline between Santa Cruz and Point Año Nuevo, California is home to some extremely prominent emergent coastline features. See Figure 12.42.

1. Examine Map 12.VI located at the end of this chapter.

   a. Notice that the topography along the coastline in this region is composed of staircase-like benches. What are these features? ____________________

   b. How and why did they form? (Be specific with respect to what's happening in this particular region) ____________________

   ____

Investigation into the elevations and ages of these paleo-shorelines divulges important information about the tectonic history of this region. Since the initial study in 1930, the exceptionally well-preserved Ben Lomond Mountain marine terrace sequence has been the subject of numerous studies and reinterpretation. To summarize, the terrace sequence consists of six marine terraces cut into the slowly rising coastline during the Pleistocene. Terrace names, elevations, and estimated ages from one of the studies are shown in Table 12.B, while distribution of the terraces throughout the coastline is visible in Figure 12.43.

Figure 12.42 – Aerial photograph of the coastline north of Santa Cruz, California

   c. Fill in the remaining two columns in Table 12.B (on the following page) by calculating the amount of uplift between terraces and the rate at which that uplift took place. (Note: Sea level was not always where it is now ... keep this in mind when determining the amount of uplift between terraces.)

   d. Which terrace shows the highest rate of tectonic uplift? ____________ .

   e. Which shows the slowest? ____________

   f. As is clear from the data, not only was the continent uplifting, but sea level was also significantly fluctuating. Knowing that these processes have been dated to the middle Pleistocene, what do you think was responsible for the sea-level fluctuations? ____________________

Table 12.B – Data from Ben Lomond Mountain Marine Terrace Sequence

Marine Terrace	Elevation (m)	Estimated Age (yrs)	Paleo-sea level (m)	Tectonic Uplift (m)	Uplift Rate (cm/1000 yrs)
Santa Cruz★					
DAV	17	80,000	-19		
Hy 1	32	105,000	-9		
Cement	58	125,000	+6		
Western	92	213,000	-5		
Wilder (p)★★	140	320,000	0		
Blackrock	190	430,000	0		
Quarry	240	545,000	0		

★ The lowest emergent terrace (Santa Cruz terrace) contains two separate wave-cut platforms and shoreline angles named, from youngest to oldest, the Davenport and Highway 1 (note: only one terrace, but two abrasional surfaces).

★★ p, shoreline angle elevation projected

2. Examine Figure 12.43. Notice that of the six terraces identified on this coastline, only five are visible at any one given area.

   a. What terrace is missing on the west side of Sand Hill Bluff? ______________________

   b. What 2 terraces are missing on the east side of Sand Hill Bluff?
   ______________________ and ______________________

   c. Why are these older terraces not contiguous throughout the coast? ______________________

   d. Excluding absolute dating, what techniques did geologists likely use to construct a complete terrace sequence along this coastline? ______________________ and ______________________

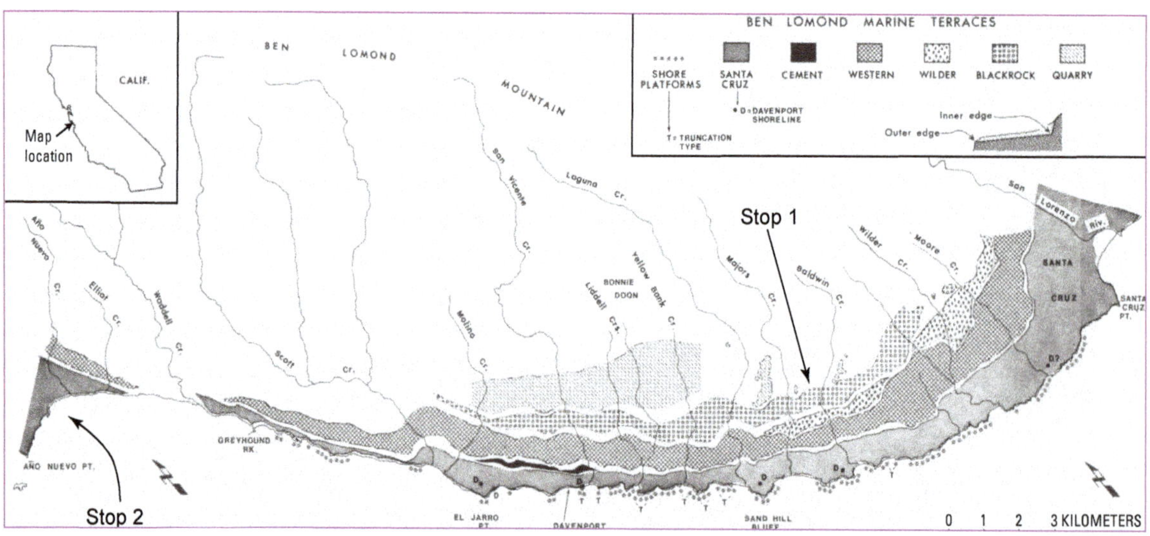

Figure 12.43 – Distribution of Marine Terraces in the Ben Lomond Mountain Marine Terrace Sequence

Exercise 2: Submergent Coastlines

1. Re-examine Map 12.III of Chatham, Massachusetts. List four landscape features that indicate this is a submergent coastline.

   - ______________________________
   - ______________________________
   - ______________________________
   - ______________________________

2. Figures 12.44a and 12.44b are images of a coastal segment bordering Virginia and Maryland. What feature here indicates this is a submergent coastline? ______________________________

Figure 12.44a – Satellite View

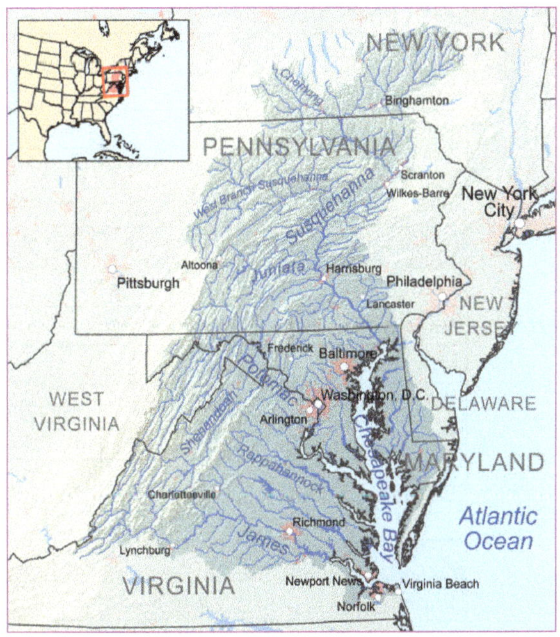

Figure 12.44b – Map View

3. Revisit Map 12.IV of Morehead City, North Carolina. Is this an emergent or submergent coastline?______________________________

4. Justify your answer to number 3 above. ______________________________
____

5. Now that you've explored a good amount of the U.S. Eastern Seaboard, would you say it's an emergent or submergent coastline? ______________________________

Exercise 3: Tying It All Together

Examine Map 12.VII. This map depicts the relief of the ocean floor in the vicinity surrounding the 48 contiguous states. The boundary between the continental shelf and the deep ocean (aka shelf-slope break) has been outlined for you in red and lies at approximately 600 feet below sea level. Use this map to answer the following questions.

1. Fill in the chart below by measuring the width of the continental shelf at each of the locations specified, then calculating the average slope of the continental shelf for the western, eastern, and southern coasts of the United States.

Eastern Shoreline		
Location	Width of shelf (mi)	Slope (ft/mi)
Long Beach, NJ		
Norfolk, VA		
Savannah, GA		
Portland, ME		
Average		

a. Is this margin passive or active? (circle one)
What tectonic activity, if any, is taking place along this coastline?
____

Southern Shoreline		
Location	Width of shelf (mi)	Slope (ft/mi)
Brownsville, TX		
Galveston, TX		
Biloxi, MS		
Perry, FL		
Average		

b. Is this margin passive or active? (circle one)
What tectonic activity, if any, is taking place along this coastline?
____

Western Shoreline		
Location	Width of shelf (mi)	Slope (ft/mi)
Newport, OR		
San Francisco, CA		
Morrow Bay, CA		
Average		

c. Is this margin passive or active? (circle one)
What tectonic activity, if any, is taking place along this coastline?
____

2.

a. What is the relationship between the tectonic processes occurring along a margin and the width of the continental shelf? ____

b. What is the relationship between the tectonic processes occurring along a margin and the slope of the continental shelf? ____

3. Refer to Figure 12.45 to help answer the following questions:

   a. Why do active margins produce narrow, steep sloped continental shelves, while passive margins produce wide, flatter continental shelves?

   ____

   ____

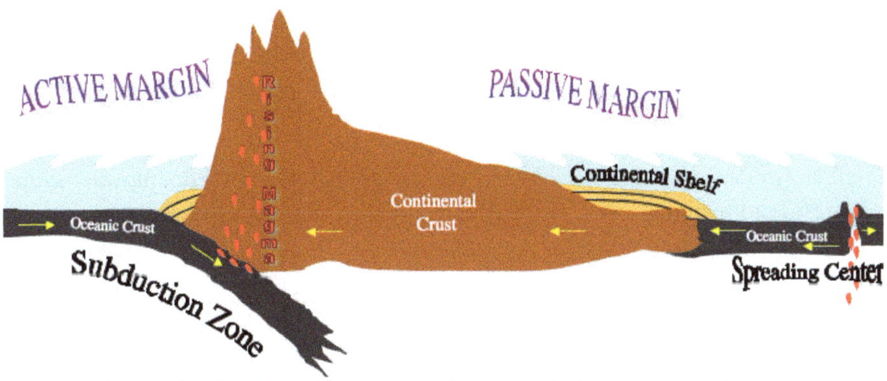

Figure 12.45 – Tectonic Process and emerging/submerging coastlines

   b. Fill in the following chart with respect to the types of shoreline processes that characterize each type of coast.

	Continental Slope	Wave Type
Active Margin	Steep or mild?	Plunging or spilling?
Passive Margin	Steep or mild?	Plunging or spilling?

4. Based on the data above, is the western coast of the United States:

   a. Rocky or sandy?

   b. Erosive or depositional?

   c. Emergent or submergent?

5. Based on the data above, is the eastern and southern coasts of the United States:

   a. Rocky or sandy?

   b. Erosive or depositional?

   c. Emergent or submergent?

Part D: Human Interaction with the Coasts

Section I: Engineering Structures

Due to the lucrative nature of the coasts, federal, state, and municipal governments are generally unwilling to accept the inherent dynamic nature of the coastline and insist on developing areas that are not designed for permanency. Most commonly, developers attempt to "stabilize" the coastline through the use of engineering structures that are meant to interfere with the littoral transport system and ultimately prevent erosion in places nature intended for destruction and force unnatural amounts of deposition in prime coastal real estate localities.

Because coastal erosion is a fairly predictable process, society has invested enormous amounts of time and money in an attempt to terminate or control affected areas. Examples of engineering structures used for this purpose include but are not limited to:

- Seawall: Vertical or near-vertical shore-parallel structures designed to prevent sea-cliff erosion. See Figure 12.46.

Figure 12.46 – Seawalls

- Groin: Linear structures placed perpendicular to the shore. See Figure 12.47.

  Each groin is designed to trap sand being carried down the coast by the longshore current, thus building an irregular but wider beach.

Figure 12.47 – Groins

- Jetty: Linear structures, constructed in pairs at the mouth of an inlet to lagoons, estuaries, or bays. See Figure 12.48.

  Designed to stabilize the channel, reduce deposition in the channel, and protect it from large waves

Figure 12.48 – Jetties

- Breakwater: Linear Structures, constructed parallel to the shore. See Figure 12.49.

  Designed to intercept incoming waves and provide a protected area for anchored ships

Figure 12.49 – Breakwater

Engineering structures may bring artificial stability in the short term (a few years at most), all the while enticing further development; but over time, unmanageable side effects inevitably develop, severely compromising the safety of the area they are trying to protect.

Exercise: Side effects of engineering structures

1. Answer the following questions concerning the side effects of using seawalls.

   a. Examine Figure 12.50. Observe the amount of energy the seawall is exposed to regularly from incoming ocean waves. What is going to progressively happen to the integrity of the seawall over time?

   ____

   ____

   ____

Figure 12.50

b. Notice the quantity of paving and guttered buildings the seawall is protecting. How might all the paving and guttering affect the amount of water that pools on the surface of the seacliff?

____

____

c. How does all that land-based water reach the ocean?

____

____

d. In what two ways might the land-based water contribute to accelerating the degradation of the seawall?

- ______________________________________
- ______________________________________

e. The combination of prolonged pounding waves, land-based waterfalls, and ocean storms will jointly degrade the seawalls. As such, the seawalls require regular and expensive maintenance and repair. Whether from lack of funds or insufficient attention, many seawalls are not maintained and end in catastrophic collapse (see figures 12.51a and 12.51b). How does collapse of a seawall generate a much larger and immediate erosion threat than what was present before the seawall was built?

____

f. How does putting up a seawall affect landscapes down current?

____

____

____

Figure 12.51a – Norfolk, UK

Figure 12.51b

Maryland's Atlantic coast is a popular year-round recreational destination. Maryland's barrier island beaches, found on Fenwick and Assateague Islands (see Figure 12.52), are readily accessible to more than 30 million people and Ocean City, the state's only coastal resort, hosts over eight million visitors annually. Because the state beaches are primary economic and recreational resources in Maryland, state and local governments have attempted to protect and maintain the state beaches—until recently, using short-sighted solutions that have caused more problems than they solved.

Ocean City, Maryland and the surrounding landscape exhibits one of the most dramatic case studies with respect to demonstrating the negative effects associated with inappropriate coastal development and use of engineering structures. The following questions are designed to guide you through the series of mistakes that were made in this location, help you assess the unfortunate results, and investigate more sustainable solutions to taking advantage of the coastal resources.

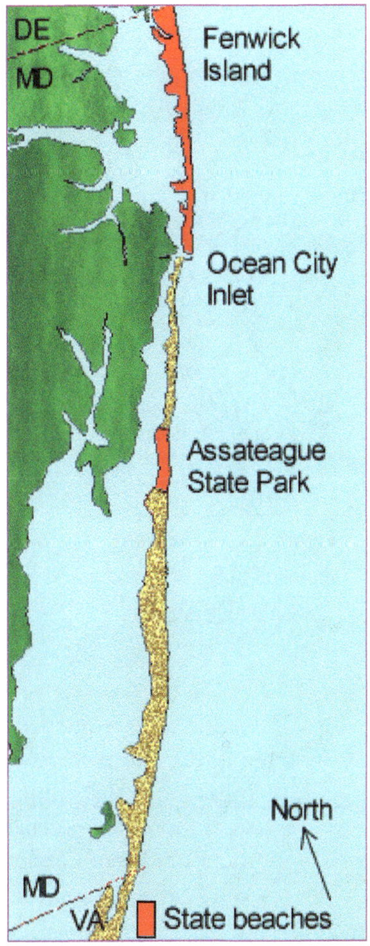

Figure 12.52 – Ocean City public beaches occupy the entire 8-mile length of Fenwick Island within Maryland. The remaining 2 miles of state-managed beaches are confined to Assateague State Park. The rest of Assateague Island falls under the jurisdiction of the National Park Service.

Background:

Prior to the 1930s Fenwick and Assateague Islands were joined, thereby existing as one large barrier island. Given that the inlet was not yet open, the eastern side of the island was not easily accessible by water, and therefore no serious development was attempted. In 1933, a large hurricane split the island in two, creating what is now Assateague Island in the south and Fenwick Island in the north. Although natural sediment transport processes would have closed the gap between the two islands, the new inlet was kept open through the installation of Jetties in 1934–1935 in order to provide better navigation to the Ocean City area of Maryland.

7.

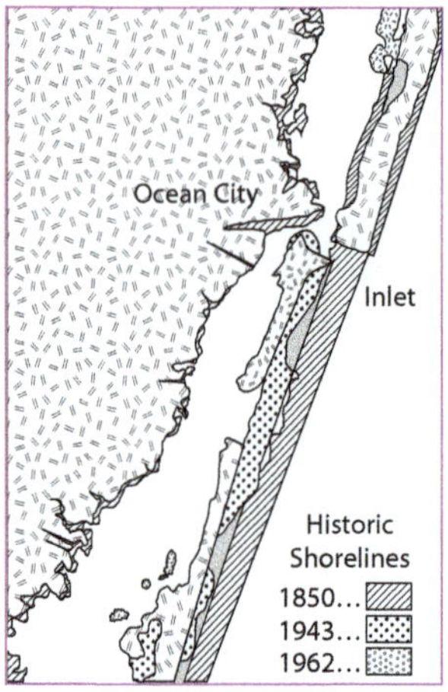

Figure 12.53 – Historic Shoreline of Fenwick and Assateague Islands

a. Examine Map 12.VIII. Find the Ocean City Inlet and the surrounding jetties. Which way is the longshore current traveling in this area? ______________________________

b. On what did you base your conclusion to letter a above?

____

c. In less than a decade after the jetties were installed, Assateague Island began a landward migration at an unprecedented rate (See Fig. 12.53). While barrier islands are by nature mobile structures that migrate landward over time, explain why Assateague Island began migrating so much faster than Fenwick Island after the installation of the jetties?

____
____
____
____
____

d. While undeveloped, the barrier islands underwent a seasonal dune cycle. During the winter months, storm waves caused sand to migrate off the beach and into sand bars beyond the breakers. In the summer, gentle waves moved the sand back onto the beach, where it was stored in dunes—generating a large, well-nourished beach (see Figure 12.54). When development of Ocean City began, building was permitted up to and beyond the primary dune line. What affect would this development have on the seasonal dune cycle?

____

e. How would the loss of the seasonal dune cycle affect the beach at Ocean City?

____
____
____
____

f. After the loss of the dune cycle, groins were installed on Fenwick Island in order to trap sand and enlarge the beach. What affect did these likely have on the already sand-starved Assateague Island?

____
____
____
____

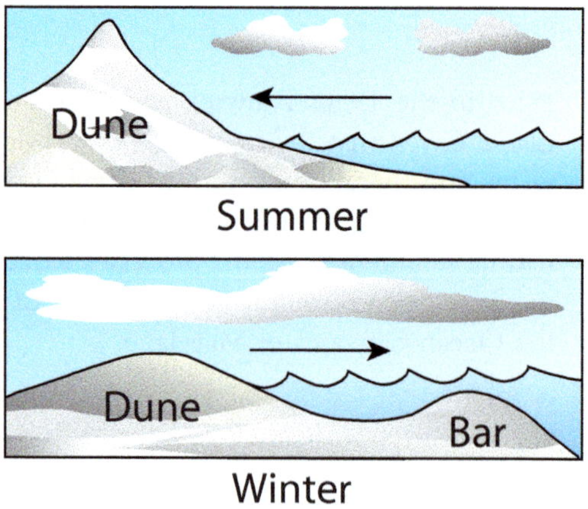

Figure 12.54 – Seasonal Dune Cycle

g. Hurricanes that threaten the eastern coast of the United States generally approach from the southeast (see Fig. 12.55). How does the accelerated migration of Assateague Island put Ocean City at exceptionally greater risk for hurricane damage? ______________________________
____
____

Figure 12.55 – Common Hurricane Paths on the East Coast

Although the state's coastal beaches are immensely valuable resources, they are also potentially an expensive liability. In Maryland, rapid shoreward erosion of these islands jeopardizes both property and economy. Realizing that the engineering structures are augmenting the problem (see Figure 12.56), like many Atlantic coast communities, Ocean City has turned to natural erosion-prevention methods by opting for beach nourishment (see Figures 12.57a and 12.57b) and dune management as the primary maintenance strategy.

A 1980 study by the Army Corps of Engineers concluded that the most practical and efficient method to prevent further erosion to the beaches and costly damage to Ocean City homes and businesses, as well as retard the migration of Assateague Island, would be to widen the beach and create an artificial dune, with sand dredged from shoals located within state waters. This study led to the Ocean City beach replenishment project, which called for periodic replenishment of beach sand with offshore sands every four years to maintain the new beach geometry.

After reading the "side note" discussing some details of Ocean City's Beach Nourishment Program, answer the questions that follow.

Side Note

Beach nourishment projects demand that sand resources meet certain physical, economic, and environmental criteria. Sand used for replenishment must be of an optimum grain size, which are specific for each region. The volume of sand required for restoration is also dependent on local geologic processes. The proximity of sand sources to the target beach is an important economic factor. Considering these factors, native offshore sands are the most desirable materials for beach nourishment projects in Maryland.

Currently used resources are located north of Ocean City Inlet, within the three-mile limit of state jurisdiction. These sands are committed to the reconstruction and periodic nourishment of Ocean City beaches. Demand for offshore sands is increasing as more shore communities opt for shoreline replenishment. In addition, an increase in the frequency of strong storms has accelerated erosion of the restored beaches. These factors place increasing demands on the sand resources within state waters. From the period 2014 to 2044, at least 7.6 million cubic meters of sand will be required to maintain Ocean City beaches, costing close to 100 million dollars. New sand sources are needed outside state waters to meet increased demand, making access to sand resources in federal waters necessary to continue shoreline restoration projects.

Figure 12.56 – Erosion continues on Fenwick and Assateague Islands

Right: Figure 12.57a and 12.57b – Beach nourishment: Massive development on Fenwick Island has both increased the urgency for and complicated the process of beach restoration.

3. Looking into the future, how might this long-term plan run into serious problems regarding:
   - Practicality? ____________________
   - Politics? ____________________
   - Sustainability? (Hint: consider the increasing storm activity) ____________________
   - Finances? ____________________

4. While certainly safer and more eco-friendly than the use of engineering structures, is beach nourishment a practical long-term solution to coastal development? Yes or No? Explain ____________________

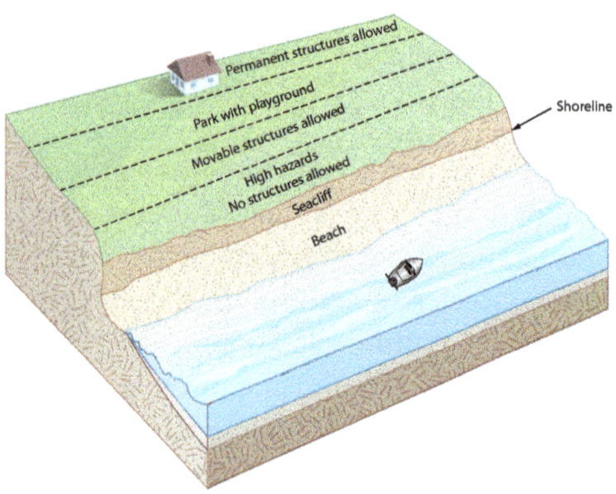

Figure 12.58

5. Coastal features are temporary land forms, but they are often developed as though they were permanent features. Urbanization of these fragile locales enhances their inherent instability by interrupting the natural storage and flow of sand in and around the shoreline. The dynamic nature of this environment, exaggerated by development and accelerated by sea-level rise, is nothing short of a recipe for disaster and ultimately poses a serious threat to regional economic and cultural stability. Given what you've learned thus far, suggest a way to take advantage of shoreline resources without compromising the safety of our livelihoods, property, or the environment. (Hint: See Figure 12.58). ____________________

SECTION II: SEA-LEVEL RISE

It has been well documented that the accelerated rise in average global temperature since the onset of the industrial revolution falls outside the range of natural variability and is believed to be caused in part by excessive anthropogenic burning of fossil fuels. The progressively escalating temperatures that have been documented over the last century and a half have numerous problematic impacts, among which is sea-level rise.

Rising sea levels can pose difficulties for society on a number of levels, all of which manifest themselves at the coasts. Just a small rise in sea level, on the order of 1.5 ft, would create complications for river drainage, which in turn would result in severe continental flooding. In addition, with less sediment being brought to the ocean and sea water encroaching further onto the coastline, the existing coastal erosion problems can be expected to increase.

Scientists are very concerned that the accelerated rise in global temperature will soon result in complete melting of the Antarctic and Greenland ice sheets, generating insurmountable problems for coastal cities.

Exercise: Continental Coasts: There They Are, There They Go!

1. The Antarctic Ice Sheet extends approximately 5.4 million mi^2 and on average, is about 1.3 miles thick. What is the volume of water that would be released if this entire ice sheet melted? ____________ mi^3 (show your work).

2. The Greenland Ice Sheet extends about 656,000 mi^2, and on average, is about 1.4 miles thick. What is the volume of water that would be released if this entire ice sheet melted? ____________ mi^3 (show your work).

3. What is the total volume of water that would be released into the ocean if both these ice sheets melted? ____________ mi^3 (show your work).

4. Given that the world's oceans cover a surface are of approximately 135 million mi^2, how many feet would sea level rise due to the addition of all the water from both the melted ice sheets? ________________ ft (show your work).

5. Use Figure 12.59 below to answer the following questions:
   a. Using a dashed blue line, draw in on figure 12.59 below, the sea level elevation if only half of both ice sheets melted
   b. Using a dashed red line, draw in on figure 12.59 below, the sea level elevation if all of both ice sheets melted
   c. List some inland cities that would become coastal

6. Earth currently consists of 148 million Km^2 of landmass. Were both ice sheets to melt, approximately 29,600,000 km^2 of the existing landmass would be submerged. Given this data, were both ice sheets to melt, what percentage of Earth's landmass would be lost? (show your work) _________ %

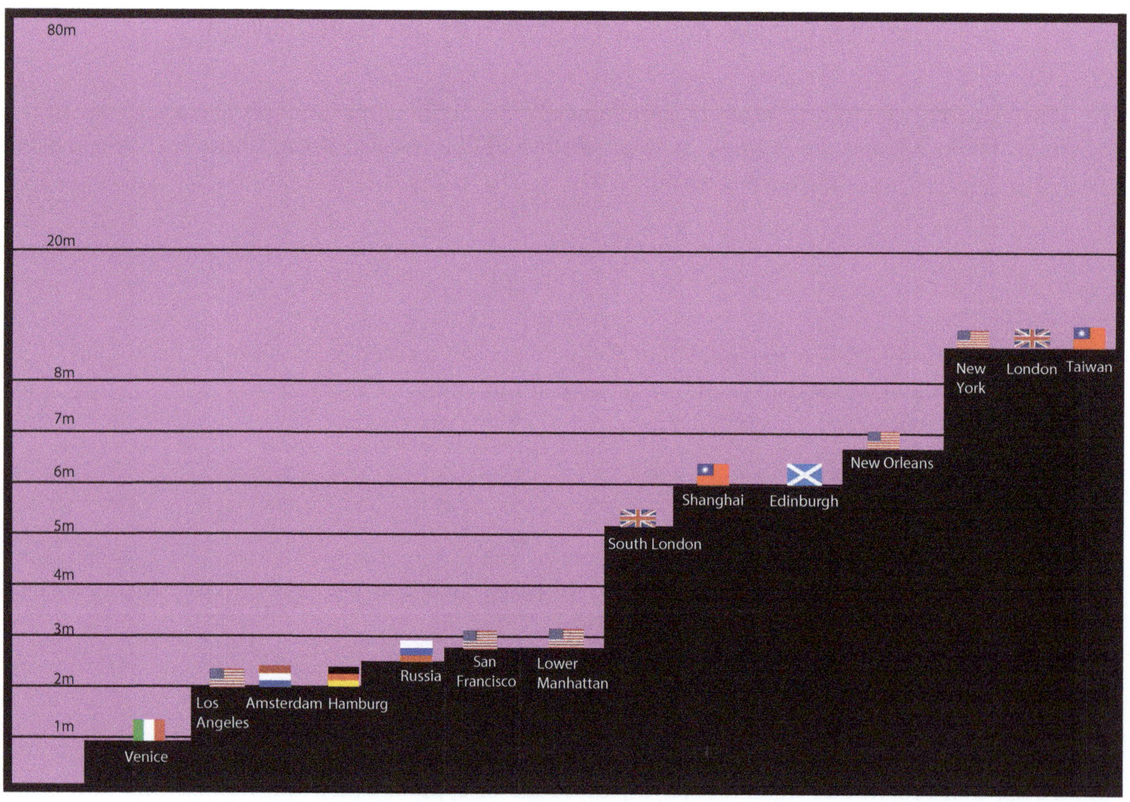

Figure 12.59

Map 12.I – La Push, WA

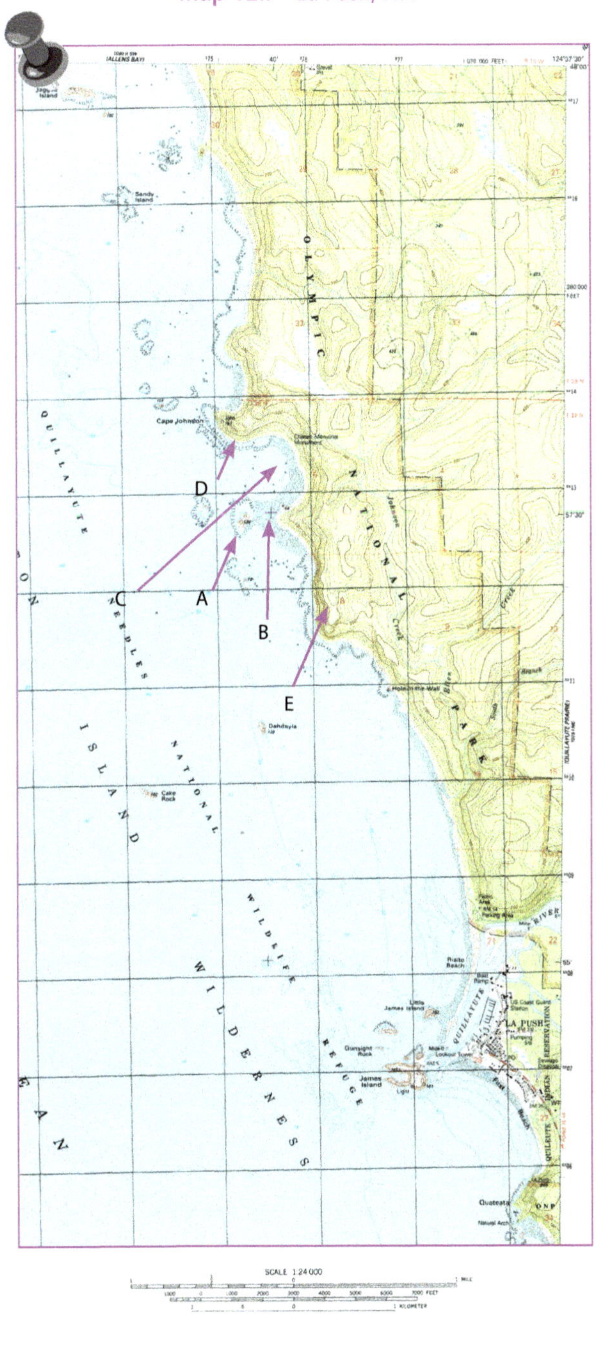

Map 12.II – Cape Sebastian, OR

Map 12.III – Chatham, MA

Map 12.IV – Morehead City, NC

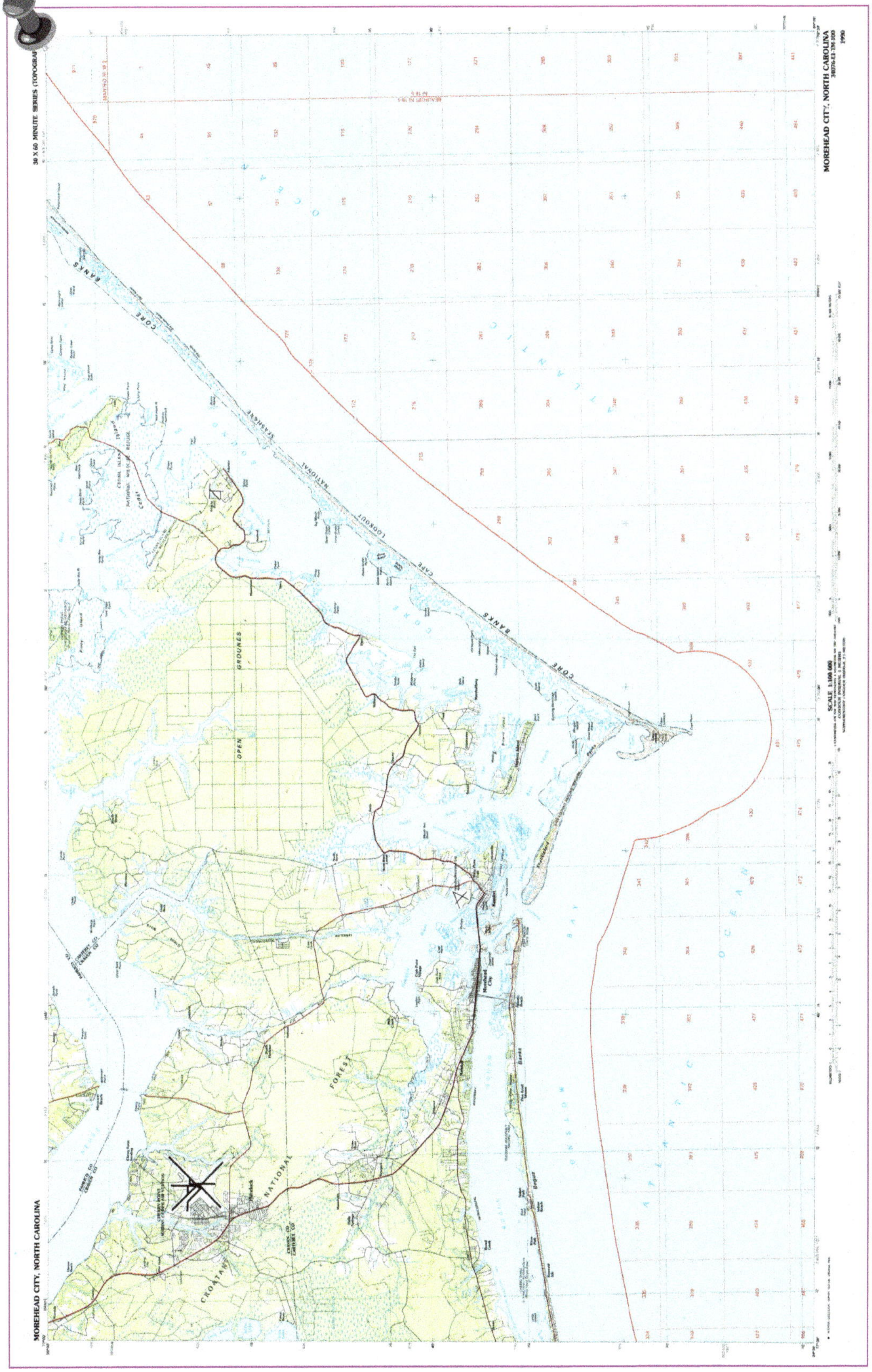

Map 12.V – Drake's Bay

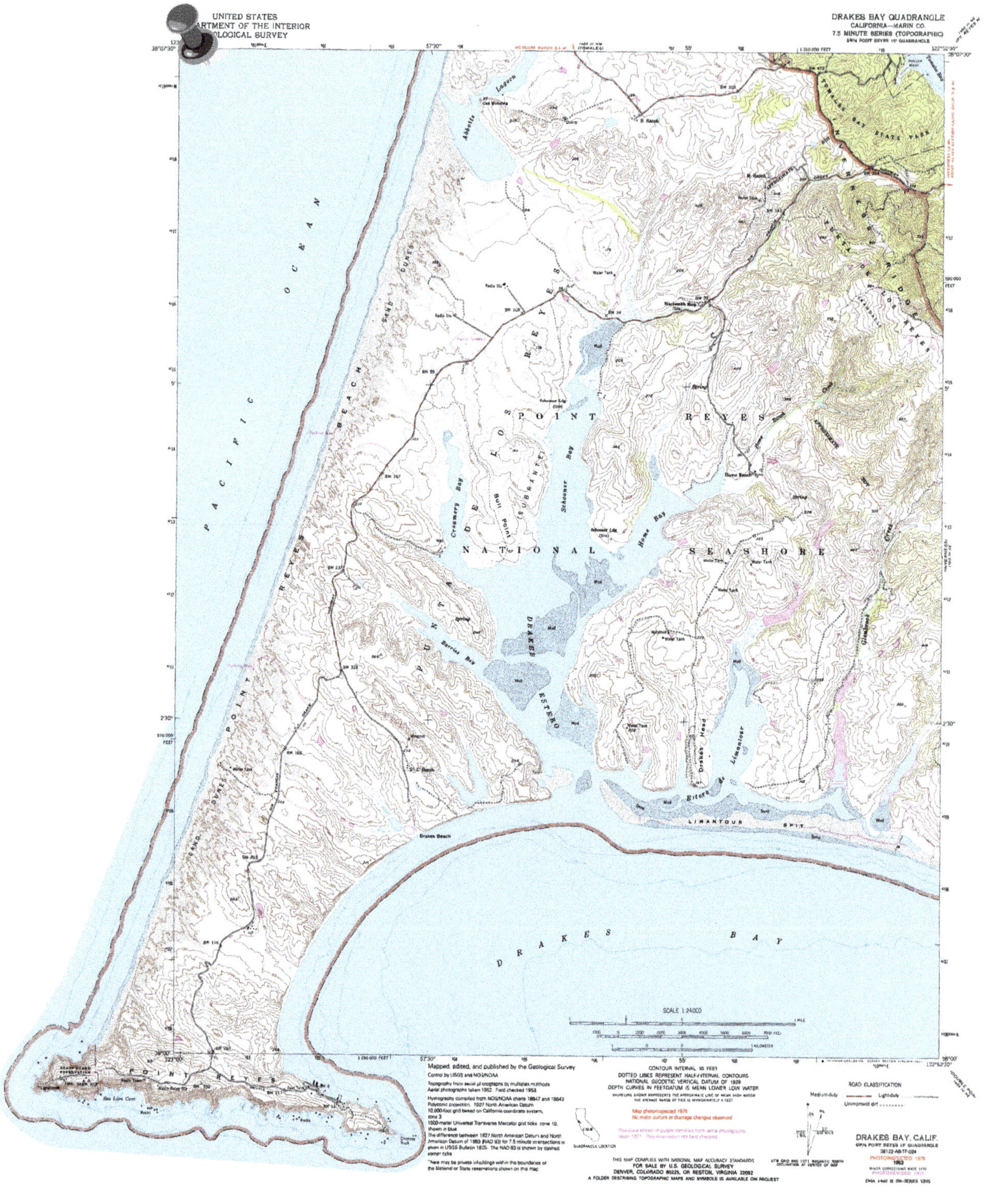

Map 12.VI – Davenport, CA

Map 12.VII

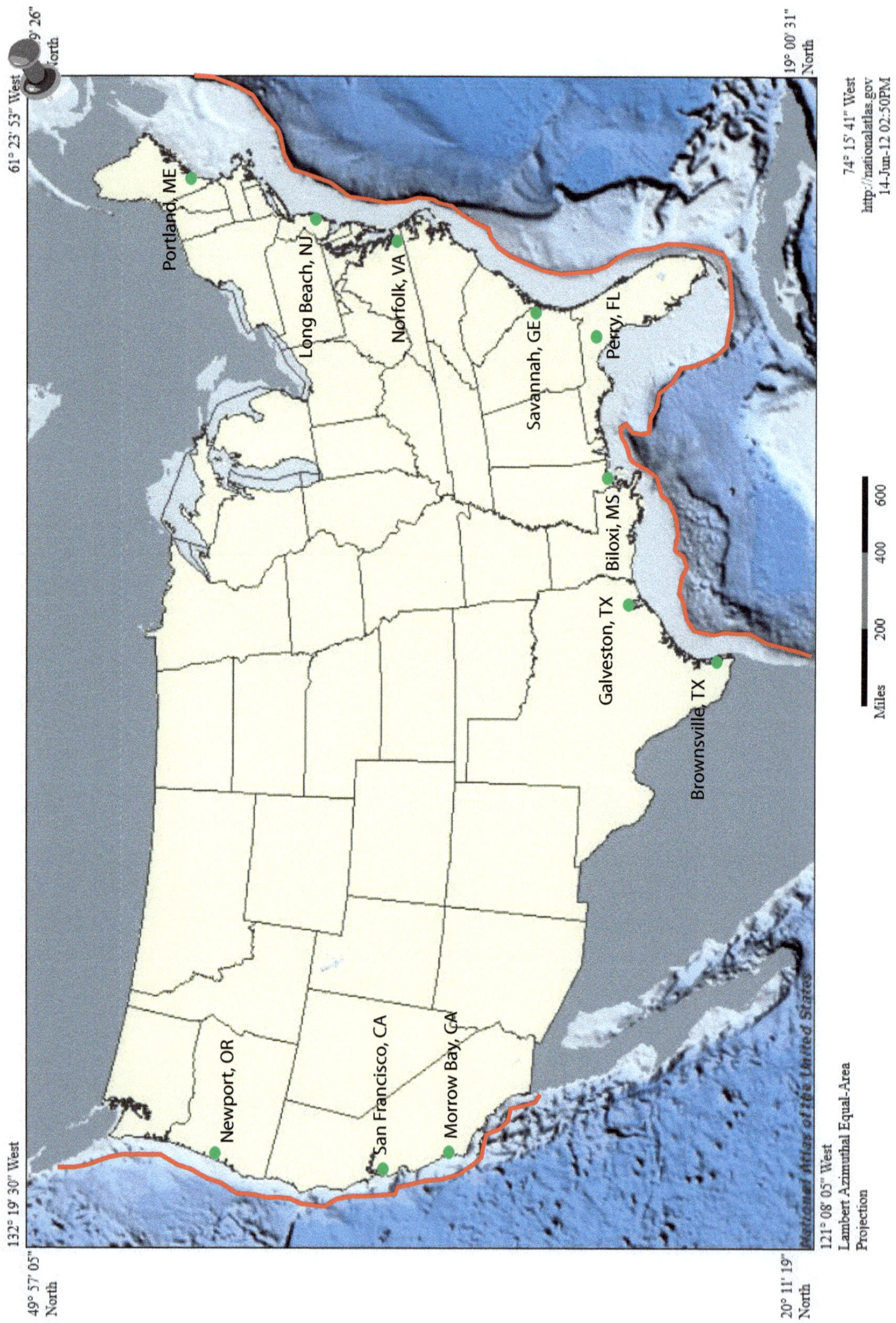
Portland, ME
Long Beach, NJ
Norfolk, VA
Savannah, GE
Perry, FL
Biloxi, MS
Galveston, TX
Brownsville, TX
Newport, OR
San Francisco, CA
Morrow Bay, CA
61° 23' 53" West
19° 00' 31" North
74° 15' 41" West
http://nationalatlas.gov
14-Jun-12 02:50PM
Miles
200
400
600
132° 19' 30" West
49° 57' 05" North
20° 11' 19" North
121° 08' 05" West
Lambert Azimuthal Equal-Area
Projection
National Atlas of the United States

Map 12.VIII – Ocean City, MD

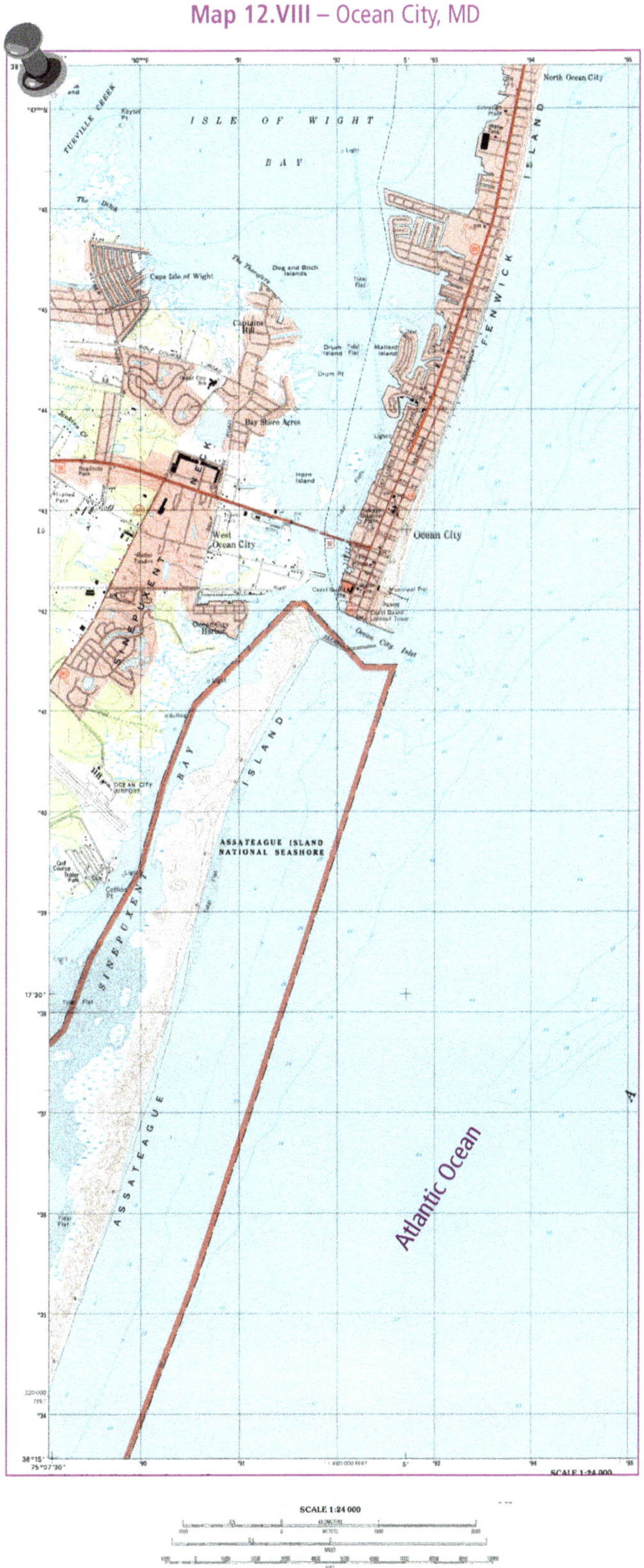

Credits

Cover Images

1. Copyright © 2009 by User:Troxel / Wikimedia Commons / CC BY 3.0.
2. Lucas Löffler / Public Domain.
3. Copyright © 2005 by Dirk Beyer / CC BY-SA 3.0.
4. Copyright © 2009 by Brocken Inaglory / CC BY-SA 3.0.
5. D. Harlow / USGS / Public Domain.
6. Copyright © 1992 by User:Rolfcosar / Wikimedia Commons / CC BY-SA 2.5.
7. Copyright © 2012 by Chuanlun Zhang. Reprinted with permission.
8. U.S. Fish and Wildlife Service / Public Domain.
9. Copyright © 2011 by Depositphotos / muha04.
10. Copyright © 2012 by Depositphotos / filmstroemstock.
11. Copyright © 2011 by Depositphotos / Hackman.
12. Copyright © 2010 by Depositphotos / Igor_Gubarev.
13. Copyright © 2007 by Vic Brincat / Wikimedia Commons / CC BY 2.0.
14. Copyright © 2010 by Depositphotos / kamchatka.

Interior Images

1. Copyright © 2010 by Rob Lavinsky / iRocks.com / CC BY-SA 3.0.
2. Copyright © 1980 by Rock Currier / CC BY 3.0.
3. Copyright © 2010 by Rob Lavinsky / iRocks.com / CC BY-SA 3.0.
4. Copyright © 2010 by Rob Lavinsky / iRocks.com / CC BY-SA 3.0.
5. Copyright © 2010 by Rob Lavinsky / iRocks.com / CC BY-SA 3.0.
6. Copyright © 2010 by Rob Lavinsky / iRocks.com / CC BY-SA 3.0.
7. Copyright © 2010 by Rob Lavinsky / iRocks.com / CC BY-SA 3.0.
8. Copyright © 2010 by Chong Fat / CC BY-SA 3.0.
9. Copyright © 2010 by Rob Lavinsky / iRocks.com / CC BY-SA 3.0.
10. Copyright © 2010 by Rob Lavinsky / iRocks.com / CC BY-SA 3.0.
11. Copyright © 2004 by Eurico Zimbres / CC BY-SA 2.0 BR.
12. Copyright © 2010 by Rob Lavinsky / iRocks.com / CC BY-SA 3.0.
13. Copyright © 2010 by Rob Lavinsky / iRocks.com / CC BY-SA 3.0.
14. Copyright © 2010 by Rob Lavinsky / iRocks.com / CC BY-SA 3.0.
15. Copyright © 2010 by Rob Lavinsky / iRocks.com / CC BY-SA 3.0.
16. Copyright © 2010 by Rob Lavinsky / iRocks.com / CC BY-SA 3.0.
17. Ingo Wölbern / Public Domain.
18. Copyright © 2010 by Rob Lavinsky / iRocks.com / CC BY-SA 3.0.
19. Copyright © 2005 by Eurico Zimbres / CC BY-SA 2.0 BR.
20. Copyright © 2010 by Rob Lavinsky / iRocks.com / CC BY-SA 3.0.
21. Copyright © 2010 by Rob Lavinsky / iRocks.com / CC BY-SA 3.0.
22. USGS / Public Domain.
23. User:Alkivar / Wikimedia Commons / Public Domain.
24. Copyright © 2008 by Hannes Grobe / CC-BY-SA-2.5.

25. Copyright © 2007 by User:Casse-cailloux / Wikimedia Commons / CC BY-SA 3.0.
26. Copyright © 2010 by Rob Lavinsky / iRocks.com / CC BY-SA 3.0.
27. Copyright © 2010 by Rob Lavinsky / iRocks.com / CC BY-SA 3.0.
28. Copyright © 2007 by User:Itub / Wikimedia Commons / iRocks.com / CC BY-SA 3.0.
29. Copyright © 2004 by User:Anton / Wikimedia Commons / CC BY-SA 3.0.
30. Copyright © 2007 by iStockphoto LP / Orchidpoet.
31. Copyright © 2010 by Rob Lavinsky / iRocks.com / CC BY-SA 3.0.
32. Copyright © 2010 by Rob Lavinsky / iRocks.com / CC BY-SA 3.0.
33. Copyright © 2006 by Daniel Schwen / CC BY-SA 2.5.
34. Copyright © 2011 by User:Ra'ike / Wikimedia Commons / CC BY-SA 3.0.
35. Fred Anzley Annet / Public Domain.
36. Copyright © 2010 by Rob Lavinsky / iRocks.com / CC BY-SA 3.0.
37. Copyright © 2010 by Rob Lavinsky / iRocks.com / CC BY-SA 3.0.
38. Copyright © 2010 by Rob Lavinsky / iRocks.com / CC BY-SA 3.0.
39. Copyright © 2010 by Rob Lavinsky / iRocks.com / CC BY-SA 3.0.
40. Copyright © 2010 by Rob Lavinsky / iRocks.com / CC BY-SA 3.0.
41. USGS / Public Domain.
42. Copyright © 2007 by User:Ra'ike / Wikimedia Commons / CC BY-SA 3.0.
43. Copyright © 2005 by Luis Miguel Bugallo Sánchez / CC BY-SA 3.0.
44. Copyright © 2008 by Hannes Grobe / CC BY-SA 2.5.
45. Copyright © 2010 by Didier Descouens / CC BY 3.0.
46. Copyright © 2006 by User:Ra'ike / Wikimedia Commons / CC BY-SA 3.0.
47. Copyright © 2005 by Immanuel Giel / CC BY-SA 3.0.
48. Copyright © 2010 by Rob Lavinsky / iRocks.com / CC BY-SA 3.0.
49. Copyright © 2010 by Rob Lavinsky / iRocks.com / CC BY-SA 3.0.
50. Copyright © 2010 by Rob Lavinsky / iRocks.com / CC BY-SA 3.0.
51. Myrddin / Wikimedia Commons / Public Domain.
52. USGS / Public Domain.
53. Copyright © 2008 by Piotr Sosnowski / CC BY-SA 3.0.
54. Copyright © 2007 by User:Friman / Wikimedia Commons / CC BY-SA 3.0.
55. Copyright © 2011 by High Contrast / CC BY 3.0 DE.
56. Copyright © 2011 by User:Roll-Stone / Wikimedia Commons / CC BY-SA 3.0.
57. USGS / Public Domain.
58. USGS / Public Domain.
59. User:Locutus Borg / Wikimedia Commons / Public Domain.
60. Copyright © 2007 by User:Natursicilia / Wikimedia Commons / CC BY-SA 3.0.
61. Copyright © 2007 by Roy Parkhouse / CC BY-SA 2.0.
62. Copyright © 2009 by User:Kevmin / Wikimedia Commons / CC BY-SA 3.0.
63. USGS / Public Domain.
64. Copyright © 2005 by Patrick-Emil Zörner / CC BY-SA 2.0 DE.
65. Pearson Scott Foresman / Public Domain.
66. Copyright © 2006 by Halvard Hatlen / CC BY-SA 3.0.
67. Copyright © 2011 by Parent Géry / CC BY-SA 3.0.
68. Copyright © 2011 by User:Geologian / Wikimedia Commons / CC BY-SA 3.0.
69. USGS / Public Domain.
70. Copyright © 2006 by User:Lanzi / Wikimedia Commons / CC BY-SA 3.0.
71. Copyright © 2007 by User:Geologian / Wikimedia Commons / CC BY-SA 3.0.
72. Copyright © 2005 by User:Geologian / Wikimedia Commons / CC BY-SA 3.0.

73. Copyright © 2010 by Rob Lavinsky / iRocks.com / CC BY-SA 3.0.
74. Mark A. Wilson / Public Domain.
75. Copyright © 2010 by Michael C. Rygel / CC BY-SA 3.0.
76. Copyright © 2006 by Eurico Zimbres / CC BY-SA 2.5.
77. Copyright © 2011 by User:Geologian / Wikimedia Commons / CC BY-SA 3.0.
78. Copyright © 2010 by User:Der Messer / Wikimedia Commons / CC BY 3.0.
79. Copyright © 2005 by Pablo Alberto Salguero Quiles / CC BY-SA 3.0.
80. Copyright © 2008 by User:Nillerdk / Wikimedia Commons / CC BY-SA 3.0.
81. User:Scheinwerfermann / Wikimedia Commons / Public Domain.
82. Copyright © 2011 by Depositphotos / svl861.
83. Thomas Högner / Public Domain.
84. Copyright © 2009 by Ashley Dace / CC BY-SA 2.0.
85. Copyright © 2006 by Nigel Chadwick / CC BY-SA 2.0.
86. Copyright © 2007 by Mike Norton / CC BY-SA 3.0.
87. Copyright © 2008 by Lisa Jarvis / CC BY-SA 3.0.
88. Copyright © 2011 by Karl and Ali / CC BY-SA 3.0.
89. James L. Stuby / Public Domain
90. Copyright © 2008 by Hugh Meechan / CC BY-SA 3.0.
91. Copyright © 2009 by Piotr Sosnowski / CC BY-SA 3.0.
92. User:Roquai / Wikimedia Commons / Public Domain.
93. Copyright © 2005 by Kevin Walsh / CC BY 2.0.
94. Copyright © 2011 by Depositphotos / myper.
95. Copyright © 2011 by Roll-Stone / Wikimedia Commons / CC BY-SA 3.0.
96. Reiner Flassig / Public Domain.
97. User:Myrddin / Wikimedia Commons / Public Domain.
98. Copyright © 2011 by Anders Damberg / Geological Survey of Sweden / CC BY 2.0.
99. User:Woudloper / Wikimedia Commons / Public Domain.
100. User:Omphacite / Wikimedia Commons / Public Domain.
101. Copyright © 2005 by Kevin Walsh / CC BY 2.0.
102. Copyright © 2008 by Piotr Sosnowski / CC BY-SA 3.0.
103. Reiner Flassig / Public Domain.
104. User:Omphacite / Wikimedia Commons / Public Domain.
105. Copyright © 2007 by Quinn Dombrowski / CC BY-SA 2.0.
106. Copyright © 2008 by User:名古屋太郎 / Wikimedia Commons / CC BY-SA 3.0.
107. User:西城越 / Wikimedia Commons / Public Domain.
108. Copyright © 2008 by Jennifer Tu. Reprinted with permission.
109. Willie Scott / USGS / Public Domain.
110. USGS Cascade Volcano Observatory / Public Domain.
111. M. Hollunder / Public Domain.
112. Copyright © 2010 by Don Nelson Page / CC BY-SA 3.0.
113. User:G310Luke / Wikimedia Commons / Public Domain.
114. USGS / Public Domain.
115. Jack Lockwood / USGS / Public Domain.
116. USGS / Public Domain.
117. USGS / Public Domain.
118. Copyright © 2006 by Siim Sepp / CC BY-SA 3.0.
119. Copyright © 2013 by Depositphotos / Siim Sepp.
120. Copyright © 2006 by User:Chmee2 / Wikimedia Commons / CC BY-SA 3.0.
121. Copyright © 2006 by Siim Sepp / CC BY-SA 3.0.
122. Copyright © 2009 by Beatrice Murch / CC BY 2.0.
123. Copyright © 2010 by User:Avenue / Wikimedia Commons / CC BY-SA 3.0.
124. Copyright © 2008 by User:Ra'ike / Wikimedia Commons / CC BY-SA 3.0.
125. User:Woudloper / Wikimedia Commons / Public Domain.
126. User:Wilson44691 / Wikimedia Commons / Public Domain.
127. Copyright © 2006 by Daniel Mayer / CC BY-SA 3.0.

128. Copyright © 2011 by User:Ji-Elle / Wikimedia Commons / CC BY-SA 3.0.
129. Copyright © 2010 by Rob Lavinsky / iRocks.com / CC BY-SA 3.0.
130. Copyright © 2005 by Siim Sepp / CC BY-SA 3.0.
131. Copyright © 2006 by Gescanntes Handstück von Grabenstedt / CC BY-SA 3.0.
132. USGS / Public Domain.
133. Roll-Stone / Wikimedia Commons / Public Domain.
134. M. Murphy / Public Domain.
135. Copyright © 2005 by Shirley Grant / CC BY-SA 2.0.
136. Copyright © 2009 by Alexander Klink / CC BY 3.0.
137. Suzanne M. MacLeod / Public Domain.
138. Copyright © 2011 by User:Seteun / Wikimedia Commons / CC BY-SA 3.0.
139. Copyright © 2007 by Toby / CC BY-SA 2.0.
140. Copyright © 2007 by Paul Glazzard / CC BY-SA 2.0.
141. Copyright © 2010 by Michael Graham / CC BY-SA 2.0.
142. Copyright © 2007 by iStockphoto LP / Razvan.
143. Copyright © 2009 by Richard Dorrell / CC BY-SA 2.0.
144. Copyright © 2005 by User:יקד / Wikimedia Commons / CC BY-SA 3.0.
145. Copyright © 2006 by User:Nationalparks / Wikimedia Commons / CC BY-SA 2.5.
146. Copyright © 2005 by User:Zoharby / Wikimedia Commons / CC BY-SA 3.0.
147. User:Duk / Wikimedia Commons / Public Domain.
148. Copyright © 2010 by User:Jiekeren / Wikimedia Commons / CC BY-SA 3.0.
149. Copyright © 2008 by SMJ / CC BY-SA 2.0.
150. Copyright © 2006 by Jim Champion / CC BY-SA 2.0.
151. Copyright © 2010 by Philip Halling / CC BY-SA 2.0.
152. Copyright © 2007 by Yoo Chung / CC BY-SA 3.0.
153. Copyright © 2010 by Eric Jones / CC BY-SA 2.0.
154. Copyright © 2008 by Anne Burgess / CC BY-SA 2.0.
155. Copyright © 2008 by Anne Burgess / CC BY-SA 2.0.
156. Copyright © 2008 by Ulrich Lange / CC BY-SA 3.0.
157. Copyright © 2005 by User:Ducky / Wikimedia Commons / CC BY-SA 3.0.
158. Copyright © 2007 by Jim Champion / CC BY-SA 2.0.
159. Copyright © 2007 by Jonathan Billinger / CC BY-SA 2.0.
160. Copyright © 2007 by Mick Garratt / CC BY-SA 2.0.
161. Copyright © 2006 by Jim Bain / CC BY-SA 2.0.
162. Mark A. Wilson / Public Domain.
163. Copyright © 2010 by User:Ji-Elle / Wikimedia Commons / CC BY-SA 3.0.
164. Mark A. Wilson / Public Domain.
165. Copyright © 2007 by Hannes Grobe / CC BY-SA 2.5.
166. Copyright © 2003 by Daniel Mayer / CC BY-SA 3.0.
167. Copyright © 2010 by User:Wiki-Shacke / Wikimedia Commons / CC BY-SA 3.0.
168. Copyright © 2006 by Ulrich Koberg / CC BY-SA 2.5.
169. Copyright © 2010 by User:Qfl247 / Wikimedia Commons / CC BY-SA 3.0.
170. Copyright © 2008 by User:Lvklock / Wikimedia Commons / CC BY-SA 3.0.
171. Copyright © 2005 by Manfred Mader / CC BY-SA 3.0 DE.
172. Copyright © 2007 by User:Mizu_Basyo / Wikimedia Commons / CC BY-SA 3.0.
173. USGS / Public Domain.
174. Copyright © 2011 by Matthew Gordon / CC BY-SA 3.0.
175. Mark A. Wilson / Public Domain.
176. Mark A. Wilson / Public Domain.
177. Copyright © 2005 by Dave Dunford / CC BY-SA 2.0.
178. Copyright © 2004 by Val Vannet / CC BY-SA 2.0.
179. Copyright © 2011 by iStockphoto LP / malerapaso.
180. Copyright © 2009 by Hannes Grobe / AWI / CC BY 3.0.
181. USGS / Public Domain.

182. Copyright © 2007 by Kai Point Coal Mine / CC BY-SA 3.0.
183. Copyright © 2011 by User:Cliff.rigg / Wikimedia Commons / CC BY 3.0.
184. Copyright © 2005 by Eurico Zimbres / CC BY-SA 2.5.
185. Copyright © 2010 by Nic Bullivant / CC BY-SA 2.0.
186. Copyright © 2007 by Hannes Grobe / AWI / CC BY-SA 2.5.
187. Copyright © 2003 by Daniel Mayer / CC BY-SA 3.0.
188. Copyright © 2008 by Brian Stansberry / CC BY 3.0.
189. User:Woudloper / Wikimedia Commons / Public Domain.
190. Copyright © 2006 by Eirian Evans / CC BY-SA 2.0.
191. User:Kanohara / Wikimedia Commons / Public Domain.
192. Copyright © 2011 by Anne Burgess / CC BY-SA 2.0.
193. Copyright © 2009 by Anne Burgess / CC BY-SA 2.0.
194. Copyright © 2010 by User:Finetooth / Wikimedia Commons / CC BY-SA 3.0.
195. Copyright © 2008 by Alison Rawson / CC BY-SA 2.0.
196. Copyright © 2008 by Evelyn Simak / CC BY-SA 2.0.
197. Copyright © 2004 by Michael C. Rygel / CC BY-SA 3.0.
198. Copyright © 2006 by Michael C. Rygel / CC BY-SA 3.0.
199. Copyright © 2006 by Michael C. Rygel / CC BY-SA 3.0.
200. Copyright © 2011 by User:Nwhit / Wikimedia Commons / CC BY-SA 3.0.
201. Copyright © 2011 by User:Noonat / Wikimedia Commons / CC BY-SA 3.0.
202. NASA / Public Domain.
203. Copyright © 2008 by Richard Webb / CC BY-SA 2.0.
204. User:G. Thomas / Wikimedia Commons / Public Domain.
205. Copyright © 2011 by User:PePeEfe / Wikimedia Commons / CC BY-SA 3.0.
206. Copyright © 2009 by Andrew Curtis / CC BY-SA 2.0.
207. Copyright © 2008 by User:Kalima / Wikimedia Commons / CC BY-SA 3.0.
208. Copyright © 2010 by User:Qfl247 / Wikimedia Commons / CC BY-SA 3.0.
209. Copyright © 2009 by Daniel Mayer / CC BY-SA 3.0.
210. Copyright © 2007 by User:Anky-man / Wikimedia Commons / CC BY-SA 3.0.
211. NASA / Public Domain.
212. Copyright © 2009 by User:Qfl247 / Wikimedia Commons / CC BY-SA 3.0.
213. Copyright © 2011 by User:Kol35 / Wikimedia Commons / CC BY-SA 3.0.
214. Copyright © 2001 by Michael C. Rygel / CC BY-SA 3.0.
215. Copyright © 2011 by User:Kol35 / Wikimedia Commons / CC BY-SA 3.0.
216. User:Vargklo / Wikimedia Commons / Public Domain.
217. User:Rippleyripples / Wikimedia Commons / Public Domain.
218. Copyright © 2006 by Michael C. Rygel / CC BY-SA 3.0.
219. Copyright © 2005 by Bob Shires / CC BY-SA 2.0.
220. Copyright © 2006 by Michael C. Rygel / CC BY-SA 3.0.
221. James L. Stuby / Public Domain.
222. Angela K. Kepler / Public Domain.
223. Copyright © 2011 by iStockphoto LP / lore.
224. Copyright © 1989 by Janine Forbes / CC BY-SA 2.0.
225. Copyright © 2008 by Daniel Ventura / CC BY-SA 3.0.
226. Copyright © 2008 by Daniel Ventura / CC BY-SA 3.0.
227. Copyright © 2011 by Jonathan Wilkins / CC BY-SA 2.0.
228. Jose F. Vigil / USGS / Public Domain.
229. Copyright © 2007 by David Monniaux / CC BY-SA 3.0.
230. USGS / Public Domain.
231. USGS / Public Domain.
232. Copyright © 2013 by Remi Mathis / CC BY-SA 3.0.

233. USGS / Public Domain.
234. Copyright © 2008 by Piotr Sosnowski / CC BY-SA 3.0.
235. Copyright © 2011 by Depositphotos / Kavring.
236. USGS / Public Domain.
237. Copyright © 2005 by Siim Sepp / CC BY-SA 2.5.
238. USGS / Public Domain.
239. Copyright © 2008 by Piotr Sosnowski / CC BY-SA 3.0.
240. User:Fed / Russian Wikipedia / Public Domain.
241. Copyright © 2006 by Eurico Zimbres / CC BY-SA 2.5.
242. NOAA / Public Domain
243. Lyn Topinka / USGS Cascade Volcano Observatory / Public Domain
244. USGS / Public Domain.
245. Lyn Topinka / USGS Cascade Volcano Observatory / Public Domain
246. Copyright © 2006 by Eurico Zimbres / CC BY-SA 2.5.
247. USGS / Public Domain.
248. USGS / Public Domain.
249. USGS / Public Domain.
250. USGS / Public Domain.
251. USGS / Public Domain.
252. Will R. Stettner / USGS / Public Domain.
253. User:Woudloper / Wikimedia Commons / Public Domain.
254. NASA / Public Domain.
255. Copyright © 2008 by Alan Bowring / CC BY-SA 2.0.
256. Copyright © 2006 by Sascha Brück / CC BY-SA 3.0.
257. Copyright © 2004 by User:Maveric149 / Wikimedia Commons / CC BY-SA 3.0.
258. Copyright © 2006 by User:Ratte / Wikimedia Commons / CC BY-SA 2.5.
259. Copyright © 2008 by Alan Bowring / CC BY-SA 2.0.
260. Copyright © 2008 by ceridwen / CC BY-SA 2.0.
261. Copyright © 2009 by Ruth Sharville / CC BY-SA 2.0.
262. Black Tusk / Public Domain.
263. Copyright © 2011 by Nigel Chadwick / CC BY-SA 2.0.
264. Copyright © 2008 by Jonathan Wilkins / CC BY-SA 2.0.
265. Copyright © 2008 by Bruno Navez / CC BY-SA 3.0.
266. Copyright © 2009 by Ashley Dace / CC BY-SA 2.0.
267. Copyright © 2009 by Charles Gordon Clark / CC BY-SA 2.0.
268. Copyright © 2011 by User:יקד / Wikimedia Commons / CC BY-SA 3.0.
269. Copyright © 2006 by Rob Farrow / CC BY-SA 2.0.
270. Copyright © 2007 by Rose and Trey Clough / CC BY-SA 2.0.
271. Copyright © 2010 by User:J.smith / Wikimedia Commons / CC BY-SA 3.0.
272. Copyright © 2007 by User:Petr Brož / Wikimedia Commons / CC BY-SA 3.0.
273. Copyright © 2008 by Anne Burgess / CC BY-SA 2.0.
274. User:Lamprus / Wikimedia Commons / Public Domain.
275. Copyright © 2008 by Anne Burgess / CC BY-SA 2.0.
276. Copyright © 2004 by Patrick Mackie / CC BY-SA 2.0.
277. Copyright © 2008 by Peter Knudssen / CC BY-SA 2.0.
278. Copyright © 2008 by Anne Burgess / CC BY-SA 2.0.
279. USGS / Public Domain.
280. Copyright © 2004 by Stefan Kraft / CC BY-SA 3.0.
281. Dénes Emőke / Public Domain.
282. Copyright © 2011 by David Roseborough / CC BY 2.0.
283. Copyright © 2006 by User:Froggydarb / Wikimedia Commons / CC BY-SA 3.0.
284. Copyright © 1980 by Ryan Somma / CC BY-SA 2.0.
285. Copyright © 2004 by Patrick Coin / CC BY-SA 2.5.
286. Copyright © 2008 by Franco Andreone / CC BY-SA 2.5.
287. Copyright © 2008 by Brocken Inaglory / CC BY-SA 3.0.
288. Tim Vickers / Public Domain.

289. Copyright © 2010 by Ester Inbar / http://commons.wikimedia.org/wiki/User:ST.
290. Copyright © 2011 by Donald Macauley / CC BY-SA 2.0.
291. Copyright © 2011 by edenpictures / Flickr / CC BY 2.0.
292. Copyright © 2012 by William Starkey / CC BY-SA 2.0.
293. Copyright © 2012 by William Starkey / CC BY-SA 2.0.
294. Copyright © 2006 by Aris Katsaris / CC BY-SA 3.0.
295. USGS / Public Domain.
296. Copyright © 2007 by User:debivort / Wikimedia Commons / CC BY-SA 3.0.
297. Copyright © 2006 by Dmitry Bogdanov / CC BY-SA 3.0.
298. Copyright © 2007 by User:debivort / Wikimedia Commons / CC BY-SA 3.0.
299. Copyright © 2007 by Nobu Tamura / CC BY 3.0.
300. Copyright © 2007 by User:Apokryltaros / Wikimedia Commons / CC BY 3.0.
301. User:Daderot / Wikimedia Commons / Public Domain.
302. Copyright © 2008 by Nobu Tamura / CC BY 3.0.
303. Copyright © 2008 by Nobu Tamura / CC BY 3.0.
304. Copyright © 2008 by Nobu Tamura / CC BY 3.0.
305. Copyright © 2005 by User:Luna04 / Wikimedia Commons / CC BY-SA 3.0.
306. Copyright © 2008 by Nobu Tamura / CC BY 3.0.
307. Copyright © 2008 by Nobu Tamura / CC BY 3.0.
308. Copyright © 2008 by Nobu Tamura / CC BY 3.0.
309. Source: The IRIS Consortium.
310. Copyright © 1999 by Geology Labs On-Line. Reprinted with permission.
311. Copyright © 1999 by Geology Labs On-Line. Reprinted with permission.
312. Copyright © 1999 by Geology Labs On-Line. Reprinted with permission.
313. Copyright © 1999 by Geology Labs On-Line. Reprinted with permission.
314. Copyright © 2008 by User:Bidgee / Wikimedia Commons / CC BY 3.0.
315. Copyright in the Public Domain.
316. USGS / Public Domain.
317. USGS / Public Domain.
318. USGS / Public Domain.
319. USGS / Public Domain.
320. USGS / Public Domain.
321. USGS / Public Domain.
322. USGS / Public Domain.
323. Copyright © 2011 by User:Mikenorton / Wikimedia Commons / CC BY-SA 3.0.
324. USGS / Public Domain.
325. USGS / Public Domain.
326. USGS / Public Domain.
327. USGS / Public Domain.
328. USGS / Public Domain.
329. USGS / Public Domain.
330. USGS / Public Domain.
331. Copyright © 2009 by Bill Boaden / CC BY-SA 2.0.
332. USGS / Public Domain.
333. Copyright © by Alex Sanchez. Reprinted with permission.
334. USGS / Public Domain.
335. Ujjwal Kumar / Public Domain.
336. NASA / Public Domain.
337. Copyright © 2004 by Wing-Chi Poon / CC BY-SA 2.5.
338. NASA/GSFC/METI/ERSDAC/JAROS, and U.S./Japan ASTER Science Team / Public Domain
339. U.S. Fish and Wildlife Service / Public Domain.
340. NOAA / Public Domain.
341. NASA / Public Domain.
342. USGS / Public Domain.
343. USGS / Public Domain.
344. USGS / Public Domain.
345. Copyright in the Public Domain.
346. NOAA / Public Domain.
347. Tennessee State Library and Archives / Public Domain.
348. Copyright © 2011 by Tangient LLC / CC BY-SA 3.0.
349. USGS / Public Domain.
350. USGS / Public Domain.
351. USGS / Public Domain.
352. USGS / Public Domain.
353. USGS / Public Domain.
354. USGS / Public Domain.
355. Copyright © 2011 by Tangient LLC / CC BY-SA 3.0.

356. USGS / Public Domain.
357. USGS / Public Domain.
358. USGS / Public Domain.
359. USGS / Public Domain.
360. USGS / Public Domain.
361. USGS / Public Domain.
362. USGS / Public Domain.
363. USGS / Public Domain.
364. USGS / Public Domain.
365. USGS / Public Domain.
366. Copyright © 2010 by CK-12 Foundation / CC BY-SA 3.0.
367. Copyright © 2007 by Hans Hillewaert / CC BY-SA 3.0.
368. USGS / Public Domain.
369. J.F. Poland and R.L. Ireland / USGS / Public Domain.
370. Devin Galloway and Francis S. Riley / USGS / Public Domain.
371. Devin Galloway and Francis S. Riley / USGS / Public Domain.
372. Immanuel Giel / Public Domain.
373. Copyright © 2010 by CK-12 Foundation / CC BY-SA 3.0.
374. Copyright © 2009 by User:Ebyabe / Wikimedia Commons / CC BY-SA 2.5.
375. NASA / Public Domain.
376. User:Robzle / Wikimedia Commons / Public Domain.
377. Copyright © 2006 by Tim Heaton / CC BY-SA 2.5.
378. USGS / Public Domain.
379. USGS / Public Domain.
380. David Bohrer / Public Domain.
381. Copyright © 2003 by User:Solundir / Wikimedia Commons / CC BY-SA 3.0.
382. Copyright © 2010 by Oliver Dixon / CC BY-SA 2.0.
383. Copyright © 2004 by MickofFleet / CC BY-SA 2.0.
384. Copyright © 2007 by Barry Dale / CC BY-SA 2.0.
385. Copyright © 2007 by Richard Webb / CC BY-SA 2.0.
386. Copyright © 2010 by Mike O'Shea / CC BY-SA 2.0.
387. Copyright © 2006 by Brian McMorrow / CC BY-SA 2.5.
388. Copyright © 2011 by RAYANDBEE / Flickr / CC BY 2.0.
389. Jacques Descloitres, MODIS Rapid Response Team, NASA/GSFC / Public Domain
390. NASA / Public Domain.
391. Copyright © 2006 by Walter Siegmund / CC BY-SA 3.0.
392. NASA / Public Domain.
393. Copyright © 2012 by Julian Nitzsche / CC BY-SA 3.0.
394. Copyright © 2009 by Bob Jones / CC BY-SA 2.0.
395. User:User:JGKlein / Wikimedia Commons / Public Domain.
396. Copyright © 2008 by User:Poco a poco / Wikimedia Commons / CC BY-SA 3.0.
397. Copyright © 2009 by User: Eyewall / Wikimedia Commons / CC BY-SA 3.0.
398. User:Bobanny / Wikimedia Commons / Public Domain.
399. Copyright © 2006 by Bruno de Giusti / CC BY-SA 2.5 IT.
400. Angela K. Kepler / Public Domain.
401. Copyright © 2003 by Urs Neumeier / CC BY-SA 2.5.
402. Copyright © 2011 by User:Attia7502 / Wikimedia Commons / CC BY-SA 3.0.
403. NOAA / Public Domain.
404. NASA / Public Domain.
405. Robert Simmon / NASA / Public Domain.
406. NASA / Public Domain.
407. Jeff Schmaltz / NASA / Public Domain.
408. NASA / Public Domain.
409. Jesse Allen and Robert Simmon / NASA / Public Domain.
410. NASA / Public Domain.
411. NASA / Public Domain.
412. NASA / Public Domain.
413. Copyright © 1983 by University of Chicago Press. Reprinted with permission.
414. Copyright © 2008 by Jim Champion / CC BY-SA 2.0.
415. Copyright © 2008 by User:NPVF / Wikimedia Commons / CC BY-SA 3.0.
416. Copyright © 2012 by Tatiana Gerus / CC BY 2.0.
417. Copyright © 2008 by RIJones / CC BY-SA 2.0.

418. Copyright © 2005 by John Allan / CC BY-SA 2.0.
419. Copyright © 2010 by Lewis Clarke / CC BY-SA 2.0.
420. Copyright © 2005 by Tim Bekaert / CC BY-SA 3.0.
421. Copyright © 2009 by Stuart Wilding / CC BY-SA 2.0.
422. Copyright © 2007 by User:Hodur / Wikimedia Commons / CC BY-SA 2.5.
423. Copyright © 2009 by Brocken Inaglory / CC BY-SA 3.0.
424. Stephen West / Public Domain.
425. Copyright © 2007 by User:Antonov14 / Wikimedia Commons / CC BY-SA 3.0.
426. Copyright © 2003 by User:Cookaa / Wikimedia Commons / CC BY SA 3.0.
427. Copyright © 2007 by Nigel Freeman / CC BY-SA 2.0.
428. Copyright © 2010 by Oliver Dixon / CC BY-SA 2.0.
429. Copyright © 2008 by Ian Balcombe / CC BY-SA 2.0.
430. Copyright © 1997 by Trevor Rickard / CC BY-SA 2.0.
431. Copyright © 2008 by Gordon Brown / CC BY-SA 2.0.
432. Copyright © 2007 by Eric Jones / CC BY-SA 2.0.
433. Copyright © 2010 by DS Pugh / CC BY-SA 2.0.
434. User:Smegs07 / Wikimedia Commons / Public Domain.
435. Copyright © 2007 by Alastair Seagroatt / CC BY-SA 2.0.
436. Copyright © 2005 by Nigel Chadwick / CC BY-SA 2.0.
437. Copyright © 2006 by Jim Champion / CC BY-SA 2.0.
438. Copyright © 2008 by Sue Jones / CC BY-SA 2.0.
439. Copyright © 2005 by kevin rothwell / CC BY-SA 2.0.
440. Copyright © 2012 by Nigel Chadwick / CC BY-SA 2.0.
441. Copyright © 1992 by The Southwest Washington Coastal Erosion Study. Reprinted with permission.
442. Copyright © 2008 by Mike Pennington / CC BY-SA 2.0.
443. Copyright © 2008 by Depositphotos / kwest.
444. U.S. Army Corps of Engineers / Public Domain.
445. NASA / Public Domain.
446. User:Ingolfson / Wikimedia Commons / Public Domain.
447. Copyright © 2008 by User:Asteiner / Wikimedia Commons / CC BY-SA 3.0.
448. USGS / Public Domain.
449. Copyright © 2011 by User:Shekk12 / Wikimedia Commons / CC BY-SA 3.0.
450. Copyright © 2010 by User:Bourrichon / Wikimedia Commons / CC BY-SA 3.0.
451. Copyright © 2006 by steverenouk / Flickr / CC BY-SA 2.0.
452. Copyright © 2010 by Depositphotos / Arsgera.
453. Copyright © 2010 by User:Algkalv / Wikimedia Commons / CC BY-SA 3.0.
454. Anthony Bley / U.S. Army Corps of Engineers / Public Domain.
455. Copyright © 2009 by Peter Trimming / CC BY-SA 2.0.
456. Copyright © 2011 by Simon Carey / CC BY-SA 2.0.
457. Copyright © 2002 by Ramon Arrowsmith / CC BY-SA 3.0.
458. Gerald E. Weber and Alan O. Allwardt / USGS / Public Domain.
459. Gerald E. Weber and Alan O. Allwardt / USGS / Public Domain.
460. Landsat / NASA / Public Domain.
461. Copyright © 2010 by User:Kmusser / Wikimedia Commons / CC BY-SA 3.0.
462. Copyright © 2011 by Ian Capper / CC BY-SA 2.0.
463. Copyright © 2010 by Derek Harper / CC BY-SA 2.0.
464. User:Oikos-team / Wikimedia Commons / Public Domain.
465. Copyright © 2008 by Bob Jones / Wikimedia Commons / CC BY-SA 2.0.
466. USGS / Public Domain.
467. Copyright © 2007 by Robert Campbell / CC BY-SA 3.0.
468. User:Feydey / Wikimedia Commons / Public Domain.
469. Copyright © 2011 by Philip Halling / CC BY-SA 2.0.

www.ingramcontent.com/pod-product-compliance
Ingram Content Group UK Ltd.
Pitfield, Milton Keynes, MK11 3LW, UK
UKHW052010041026
15590UKWH00008B/2250

9 781631 890826